ROMANTICISM, ORIGINS, AND THE HISTORY OF HEREDITY

NEW STUDIES IN THE AGE OF GOETHE

New Studies in the Age of Goethe, sponsored by the Goethe Society of North America, aims to publish innovative research that contextualizes the "Age of Goethe," whether within the fields of literature, history (including art history and history of science), philosophy, art, music, or politics. Though the series editors welcome all approaches and perspectives, they are especially interested in interdisciplinary projects, creative approaches to archival or original source materials, theoretically informed scholarship, work that introduces previously undiscovered materials, and projects that re-examine traditional epochal boundaries or open new channels of interpretations.

TITLES IN THE SERIES

Mary Ellen Dupree, *The Mask and the Quill: Actress-Writers in Germany from Enlightenment to Romanticism*
Peter J. Schwartz, *After Jena: Goethe's Elective Affinities and the End of the Old Regime*
Brian Tucker, *Reading Riddles: Rhetorics of Obscurity from Romanticism to Freud*
Benjamin Bennett, *Aesthetics as Secular Millenialism: Its Trail from Baumgarten and Kant to Walt Disney and Hitler*

ROMANTICISM, ORIGINS, AND THE HISTORY OF HEREDITY

Christine Lehleiter

Bucknell University Press

Published by Bucknell University Press
Copublished by The Rowman & Littlefield Publishing Group, Inc.
4501 Forbes Boulevard, Suite 200, Lanham, Maryland 20706
www.rowman.com

Unit A, Whitacre Mews, 26-34 Stannery Street, London SE11 4AB, United Kingdom

Permission to reprint parts of ch 4, which originally appeared as "On Genealogy: Biology, Religion, and Aesthetics in E. T. A. Hoffmann's Elixiere des Teufels (1815/16) and Erasmus Darwin's Zoonomia (1794–96)." *The German Quarterly* 84.1 (Winter 2011): 41–60.

British Library Cataloguing in Publication Information Available

Library of Congress Cataloging-in-Publication Data

Lehleiter, Christine.
Romanticism, origins, and the history of heredity / Christine Lehleiter.
pages cm. — (New studies in the age of Goethe)
Includes bibliographical references and index.
ISBN 978-1-61148-565-3 (cloth : alk. paper) — ISBN 978-1-61148-566-0 (electronic)
1. Romanticism. 2. Heredity, Human. 3. Breeding in literature. 4. Biology in literature. I. Title.

PN603.L44 2014
809'.9145—dc23

2014033371

∞™ The paper used in this publication meets the minimum requirements of American National Standard for Information Sciences—Permanence of Paper for Printed Library Materials, ANSI/NISO Z39.48-1992.

Printed in the United States of America

NSLV

TABLE *of* CONTENTS

A NOTE *on* TRANSLATIONS

Unless otherwise noted, translations are mine. I have tried to stay as close as possible to the syntax and vocabulary of the original German.

ACKNOWLEDGMENTS

My institutional debts are great. A Fellowship from the Center for Eighteenth-Century Studies at Indiana University provided invaluable time in the early stages of this project. A Connaught Start-up Award from the University of Toronto and an Insight Development Grant from the Social Sciences and Humanities Research Council Canada made available crucial funding. The Herzog August Bibliothek Wolfenbüttel generously supported a three-month exploration of its inexhaustible treasures. The excellent collections at the Thomas Fisher Rare Book Library, University of Toronto, made it possible to study the history of science in between classes. When materials were difficult to find the team from Kelly InfoExpress at St. Michael's College was of immense help; I would like to thank in particular Manda Vrkljan.

I am deeply grateful for the incredible enriching time at Indiana University during which this project was started. Most of all I thank Fritz Breithaupt for his intellectual engagement, generosity, and unsurpassed dedication during the PhD years and to this day. Deidre Lynch has been a role model ever since. Bill Rasch and Michel Chaouli have been extraordinary teachers who shaped how I think about the long eighteenth century. At Florida State University, I thank William Cloonan and all my colleagues in the German division for their kind support during my short time there. New colleagues at the University of Toronto, York University, the University of Guelph, and the University of Waterloo have been wonderful conversation partners in the last years. For interest in and support for this project I thank in particular: Tina Young Choi, Angelica Fenner, Willi Goetschel, Alice Kuzniar,

Paola Mayer, John Noyes, Cannon Schmitt, Stefan Soldovieri, Mark Solovey, Anna Shternshis, Marga Vicedo, and John Zilcosky. I am particularly indebted to Markus Stock, a mentor able to combine professionalism and humanity in exemplary ways. I am grateful to my students, especially Teresa Sudenis and Veronika Rummel, and my research assistant Christin Bohnke who worked tirelessly on obtaining permissions for the reproduction of images. For collegiality, friendship, and for stimulating discussion, I thank Christian and Dana Weber, Elliott Schreiber, Andrea Thormählen, and all my colleagues during the student years and beyond.

This book has benefitted greatly from numerous listeners who have shared insightful comments on conference papers. I wish to thank in particular Alice Kuzniar for providing me with the opportunity to present my work at the *Romantic Ecologies* symposium and Alexander Mathäs for organizing and hosting the conference *Defining the Human and the Animal.*

I could not have hoped for a better editor of the New Studies in the Age of Goethe Series than Jane Brown. Her perceptive feedback and depth of knowledge sets a high standard in the profession. The manuscript has also benefitted greatly from the work of anonymous readers who provided decisive comments and suggestions. At Bucknell University Press, Greg Clingham and Pamelia Dailey were of great help. Sam Brawand was an unbelievably efficient copy-editor.

A portion of an earlier version of chapter 4 appeared as "On Genealogy: Biology, Religion, and Aesthetics in E. T. A. Hoffmann's *Elixiere des Teufels* (1815/16) and Erasmus Darwin's *Zoonomia* (1794–96)." *The German Quarterly* 84.1 (Winter 2011): 41–60. I thank the journal for permission to reprint.

My deepest thanks go to my family, I am grateful beyond words. Thank you: Anna Maria, Heinz, Kathrin, and Serban.

La nature fut prise dans ce temps-là comme base, source et type de tout bien et de tout beau possible. La négation du péché originel ne fut pas pour peu de chose dans l'aveuglement générale de cette époque. —*Baudelaire*

[*At that time Nature was taken as ground, source, and type of all possible Good and Beauty. The negation of original sin played no small part in the general blindness of that period.*]

Between Freedom and Determination

IN THESE TWO SENTENCES on the eighteenth century, Baudelaire summarizes and performs what this book discusses in greater detail: the shift from the assumption that nature provides the human being with a *tabula rasa*, free of all burdens of the past, to the nineteenth-century conviction that the individual is biologically determined already at the moment of birth. In the sixty years around 1800, which saw not only the rejection of inherited (legal) privileges but also the emergence of evolutionary theories, the question of the self could become a problem for the first time. The emergence of evolution, the possibility of the transmission of traits by means of heredity, challenged the individual on two levels. First, if parents' habit could transfer to the next generation, the offspring was determined from birth. Second, and connected to the first, if traits could be transmitted, the relationship between individual and species was fundamentally changed. Never before had so many writers contemplated the self and never before had so many authors written texts of an autobiographical nature than in the eighteenth century, but—contrary to Baudelaire's assessment—never before had there also been so much doubt whether the liberation of the self from its natural and determining condition is indeed possible. By inquiring into notions of *heredity* between 1760 and 1820, *Romanticism, Origins, and the History of Heredity* delineates changes in the understanding of the individual's relationship with the ancestors and its effects on the definition of the self.

To write about the formation of the self meant to write about the promise of a growing independence of an individual, but it also meant to write about the conditions, limits, and failures of the process towards ever greater autonomy. While social conditions were certainly an important part of these considerations, natural, material, conditions came increasingly to the attention of writers. Baudelaire's transfer from nature to questions of free will and morality might seem anachronistic at first sight, but it was precisely the changing conception of nature and reproduction that indeed defined the changing assessment of the self and of the possibility of freedom in literary texts around 1800. Both the belief in and the anxiety about the possibility of an autonomous self were, this book will argue, closely connected to contemporary findings in the life sciences and the discipline which was just emerging under the name *biology*.[1]

The abundance of eighteenth-century debates on the precarious status of the self has been noted frequently.[2] Less often discussed are the many doubts that eighteenth-century authors had concerning the possibility of liberation from the material conditions. Even Friedrich Schiller (1759–1805) who is often hailed—or condemned—for his belief in the possibility of freedom and his ignorance of the question of matter is a striking symptom of the development. When he wrote his *Versuch über den Zusammenhang der tierischen Natur des Menschen mit seiner Geistigen* [Essay on the relationship between animal and spiritual nature of the human being] (1780), it needed three attempts to make the text accepted as his medical dissertation. However, as much as his examiners might have been right in accusing the work of being too philosophical for a medical treatise, the *Versuch* is at the same time a work that engages deeply with contemporary studies in medicine and biology in order to ask for the conditions of possibility of freedom. When Schiller returned to the question in his novella *Verbrecher aus verlorener Ehre* [The Criminal of Lost Honor] (1786), he came to the conclusion:

> Stünde einmal, wie für die übrigen Reiche der Natur, auch für das Menschengeschlecht ein Linnäus auf, welcher nach Trieben und Neigungen

> klassifizierte, wie sehr würde man erstaunen, wenn man so manchen, dessen Laster in einer engen bürgerlichen Sphäre und in der schmalen Umzäunung der Gesetze jetzt ersticken muß, mit dem Ungeheuer Borgia in einer Ordnung beisammen fände.[3]

> [If a new Linnaeus were to appear and classify humankind into genus and species according to drives and inclinations, how astonished we would be to find those whose vice must now suffocate in a constricted bourgeois sphere and the narrow confines of the law, together in one and the same species with a monster like Cesare Borgia.][4]

Contrary to many readings of the novella that note the importance of social forces for the protagonist's eventual failure,[5] Schiller acknowledges here a basic biological condition which can be shaped by social forces, but ultimately remains determined by its place in a biological taxonomy. For writers around 1800, the formation of the self and its narrative was not a process which simply started with a blank slate. Long before laws of hereditary transmission were first formulated by Gregor Mendel (1822–1884) and Charles Darwin (1809–1882), authors of literature asked about the material basis of the self and the ramifications of emerging notions of heredity for the conceptualization of selfhood.

This book tells the story of how we started to consider heredity for questions of selfhood. It is about the ways in which selfhood was conceptualized in the context of a new interest in heredity around 1800. While Michael Minden has shown the place and importance of inheritance as the transmission of property in novels from Christoph Martin Wieland's *Agathon* (1764) to Adalbert Stifter's *Nachsommer* (1857), this book looks at transfers between the generations which are defined biologically.[6] It asks: what are the consequences for the formation of the self if the individual is assumed to be born with an inherited burden instead of being born free of any determining forces? In examining debates on heredity at the turn of the nineteenth century, *Romanticism, Origins, and the History of Heredity* aims to complicate the story on heredity and selfhood in German Romanticism on two levels. First, in contrast to established scholarship on the anticipation of evolutionary thought by the German Romantics (in particular Johann Wolfgang Goethe),[7] my aim is to demonstrate that a number of influential Romantic thinkers (including Goethe) were hesitant to embrace evolution (as the transformation of nature by means of trait transmission) as an option. Second, I consider the reasons for which evolutionary thought was rejected and I suggest that precisely because Romantic thinkers rejected the idea of evolution, they could develop a strong notion of selfhood. Without excluding the possibility of other origins of the modern self which have been suggested, such as new theories of the mind and new religious movements such as Pietism, this book proposes that only by revisiting the concept of selfhood in the context of contemporary research on heredity can we understand the reasons behind the emphatic embrace of individuality and selfhood around 1800.

Baudelaire's judgment on the eighteenth century is shaped by nineteenth-century zeal to dismiss the naiveté of the Enlightenment. However, while it recognizes correctly a certain hope formulated in the wake of developments such as Johann Gottfried Leibniz's (1646–1716) rejection of original sin in the context of his discussion of Jansenism, it is also only partly accurate because it ignores the vigor with which the eighteenth century engaged in a discussion of the kind of nature that nineteenth-century authors increasingly appreciated.[8] With unprecedented intensity, scholars in the eighteenth century questioned nature's benevolence and turned to *heredity* as a field of inquiry in order to study its impact on notions such as freedom and morality. In his writings on race, Immanuel Kant (1724–1804) had already wondered whether the transmission of corporeal traits coincided with the transmission of character traits, thereby determining the individual,[9] but where he was unclear or hesitated, Johann Gaspar Spurzheim (1776–1832), in a text considered by E. T. A. Hoffmann (1776–1822) in his legal practice, would eventually express without hesitation "I consider insanity as hereditary."[10]

If questions of freedom and determination were most pressing in the context of morality and liability, inquiries into heredity challenged convictions in a spectrum ranging from those traditionally articulated by religious authorities to those newly proclaimed in the context of natural law debates. Focusing their attention on the transmission of traits, eighteenth-century botanists, animal breeders, and natural historians such as Joseph Gottlieb Koelreuter (1733–1806), Karl Christoph Gottlieb Sturm (1781–1826), and Lorenz Oken (1779–1851) noticed that the transmission of traits had consequences not only for the individual, but also for the species. Realizing that traits were not only determining but could also be modified, they started to wonder whether species transformation was possible. Their inquiry was shaped by both the fascination with the modifiability of nature and the anxiety in light of the potential challenge to the idea of species stability. While the challenge of species stability was understood as a challenge to the idea of divine creation, other aspects of heredity research had the potential to undermine convictions held dearly in natural law philosophy which had often been defined in opposition to religious dogma. With pre-revolutionary pathos, mid eighteenth-century authors such as Johann David Michaelis (1717–1791) could still question (albeit ultimately confirm) the incest taboo as cultural phenomena cemented by religion and church but not justified by nature.[11] However, Johann Wolfgang Goethe, writing less than fifty years later, felt enticed to explore the possibility of hereditary defect as a result of incestuous reproduction and examined what such determination by the parents' reproductive choices or drives would mean for the potential for *Bildung* [formation] of the offspring (see chapter 2). As these examples demonstrate, the fascination with the idea that traits could be transferred from one generation to the next originated not only in the acknowledgment of nature's determining power, but also in its potential for a previously not thinkable openness of creation. In light of the newly realized potential of evolutionary development which came to the fore at the turn of the century, Erasmus Darwin (1731–1802) could proclaim with uninhibited enthusiasm a "world without end!"[12] If Baudelaire located the blind spot of the enlightened century in the denial of nature's dark side, he overlooked the rich variety of notions of nature around 1800. The eighteenth century is, this book argues, the period in which the complexity of nature is discovered by means of heredity.

Romanticism, Origins, and the History of Heredity does not answer the question of whether or not heredity defines our identity and fate. What it sets out to do is to investigate the consequences of the idea that we are greatly shaped or even determined by our biological heritage. More precisely, how is selfhood defined, and redefined, once it is suggested that the human being is determined by (and determines) biology? The question is closely tied to our own place in time and to the discoveries in genetic research in recent decades. The issues raised by this research have motivated the book's conception.

While the last decades of the twentieth century were dominated by the belief that the knowledge of our genetic codes could tell us virtually everything about our origin and future, recent discussions on what is called *epigenetics* have put into question whether the complete mapping of the genetic code can deliver on this promise.[13] The genetic predisposition for Huntington's disease is still a certain prediction of a cruel decline in mental and bodily capacities, but it is much less certain whether the genetic predisposition for breast cancer warrants mastectomy, and even less certain whether Robert Klark Graham's *Nobel Prize Sperm Bank* (the "Repository for Germinal Choice") will ever produce a genius.[14] After the enthusiasm and hope (or desperation) around the Human Genome project, recent years have demonstrated the complexity of the relationship between genetic code and gene expression. Even more, newest epigenetic research suggests the necessity of a conceptual return to the possibility of habit transfer by means of heredity. A space has become possible again between two poles that since the eighteenth century have been defined as nature and nurture.

The book's aim is to demonstrate that the issues around the question of the significance of heredity were first articulated and addressed in the last decades of the eighteenth century, long before the discovery of the gene. Much as authors at the turn of the twenty-first century, authors in the eighteenth century were fundamentally challenged in their belief systems by discoveries in the field that came to be known as *biology*. As twenty-first century contemporaries we share, this book suggests, to an astonishing degree the anxieties around 1800—even though, we might deal very differently with them. These anxieties circle around the question of whether or not the idea of the freedom of will can be maintained in light of a theory which suggests the collapse of body and spirit and whether or not the transformation of nature by the hand of man continues or challenges creation.

By examining eighteenth-century debates on heredity, *Romanticism, Origins, and the History of Heredity* delineates the formation of concepts which inform our current debates and decision-making processes in fields such as medicine, law, and politics. At the same time, it also highlights the degree to which our answers differ from those developed in the eighteenth century. The transformation of species is one such field of inquiry which was opened already in the eighteenth century and raised questions which are still with us today. In 1764, the South-German botanist Joseph Gottlieb Koelreuter announced that he believed to have been successful in transforming one species of tobacco into another by means of repeated artificial pollination. The transformation of species was a blasphemic move since it seemed to challenge the stability of divine creation. Even where it was considered possible, the ideological obstacles were high. For Christian and non-Christian thinkers well into the early modern period and beyond, Aristotle's definition of reproduction as the repetition of a finite set of forms had been of paramount importance.[15] Hence, at times Koelreuter announced enthusiastically that he had been successful with the transformation of species, which he equated with the transformation of lead into gold that the alchemists had tried. At other times, he feared that his transformation challenged species stability and, consequently, insisted that the life forms that he created are infertile and depend entirely on human care and intervention. While Koelreuter acknowledged with fascination that nature is highly modifiable, he also worried about the consequences of his experiments for the understanding of nature and insisted that ultimately nature can neither be changed by the human nor does it evolve into new forms on its own. Referring to the (presumably) natural occurrence of hybrids that occupied the minds of contemporary botanist such as Carl Linnaeus (1707–1778), Koelreuter writes that they are "wohl nichts anders als unzeitige Geburten einer übertriebenen Einbildungskraft" [most likely nothing else than the premature births of an exaggerated imagination].[16]

We might have made great progress in the modification of life forms since Koelreuter, but his fascination and anxiety is in many ways our own. While a majority of citizens in the Western world has embraced the idea that nature has evolved over time and most of these citizens have also accepted genetic engineering either as a fact or as a promise, this acceptance greatly depends on the (perceived) degree to which the genetic modification permanently alters nature as we know her. To provide an example: the production of insulin by means of genetically altered bacteria has met little if any resistance at all. This fact is certainly connected to the clear medical benefits and the alleviation of suffering which the method provides. However, the acceptance is perhaps even more significantly conditioned by the fact that such genetic engineering happens in the lab and that the changes that it introduces in human organisms remain on the microscopic and individual level. The discussion looks very different where changes apply to objects which are visible to the eye, cannot be contained in the lab, and are conceived as permanently changing nature because they can be transmitted by means of reproduction to future generations and thereby from individual to species. The most visible public debates on this issue can probably be found in two areas, namely the worries surrounding the cross-pollination of fields of genetically engineered crops with organically cultivated plants and the debates on gene therapy. While there are vast differences on how genetic engineering of crops is regulated in different countries, the regulations on gene therapy are still relatively universal. Somatic gene therapy, in which genetic modifications are restricted to the individual, experiences growing support.[17] Most jurisdictions, however, prohibit germline gene therapy, in which sperms or eggs are modified, because that would allow the changes to become heritable.[18] The anxiety behind this prohibition is arguably less connected to the potential for human suffering by unknown factors than to the possibility of permanent change in the human germline. Much as eighteenth-century thinkers, we embrace the modifiability of nature but hesitate to introduce permanent change initiated by human intervention. Even more, we are possibly in the process of discovering that the scientific basis of our assumptions is equally shaky as around 1800. The distinction between somatic and germline gene therapy seemed to be guaranteed because it was assumed that soma-to-germline feedback is impossible. However, the so called *Weismann barrier* between soma and germline has been challenged by recent research, raising new worries about "the genetic engineering of the human species rather than just individuals."[19] Lamarckianism seems to have returned.

Our uneasiness, but also fascination, in the context of altering human and other organisms permanently and in such a way that the change can be transmitted to the next generation has much in common with eighteenth-century worries about a lasting transformation of nature by means of human intervention. At the same time, the way in which eighteenth-century authors such as Koelreuter conceptualize their experiments is also very different from our own assessment. While Koelreuter could still insist that human alteration of the creation is only temporary because changes to the individual can be separated from changes to the species, this option has vanished in the Darwinian framework. Based on the conviction that nature's stability is greater than his own power of modification, Koelreuter could experiment without hesitation. Infertility of the hybrid and the necessity of human intervention made it possible to hold on to a notion of nature which existed before humans came and will return to the original state once left alone. Modern researchers and the societies to which they belong have lost this certainty. And it is precisely because of the loss of this certainty that a growing number of people oppose genetic engineering and long for an untouched nature even though they would also admit that there is no nature before history. While the longing might be both nostalgic and anachronistic, it might also be a necessary or unavoidable part of the human condition. This, at least, is what Jean Paul Richter suggests in his essayistic and literary work on questions of heredity and evolution (see chapter 3). What I propose then in this book is that there is much to learn about our modern human condition in light of genetic engineering by looking at the eighteenth century, because the difference between now and then is not so much one of quality than one of degree.

FORMS OF GENERATION: INBREEDING, BASTARDIZATION, AUTOPOIESIS

In search for conceptualizations of selfhood in the context of heredity, *Romanticism, Origins, and the History of Heredity* turns to literary texts around 1800 that discuss the individual's relationship to her genealogy. In doing so, the book revisits core texts of the genre that has been called the *Bildungsroman*, such as Goethe's *Wilhelm Meister* (1795–1796), but it also pays close attention to works that have remained largely unexplored in scholarship on selfhood, like Jean Paul Richter's *Komet* (1820)[20] and E. T. A. Hoffmann's *Elixiere des Teufels* (1815). Although the methodology of this book is not dependent on a model of material influence, the selection of texts discussed here has been shaped by an attempt to present documents written by authors who demonstrably were invested in contemporary fields of biology such as embryology, botany, and morphology.

Romanticism, Origins, and the History of Heredity explores what seem to me the nodal points in eighteenth-century discussions on heredity and selfhood: inbreeding and incest, cross-breeding and bastardization, evolution and autopoiesis. Following the opening chapter, which provides an overview of scientific debates on these issues, each of the succeeding chapters will take up one of them while also always reaching out to the others. The book's particular interest in biological debates guides the selection of materials from a vast number of eighteenth-century literary texts which also discuss topics ranging from genealogy to bastardization. For the incest problem in the context of inbreeding (chapter 2), I have chosen the Mignon episode in *Wilhelm Meister* (1795–1796) over, for example, the incest motive in Goethe's *Geschwister* (1776) because the Mignon episode is more clearly placed in the context of contemporary explorations of heredity. It can be argued, without doubt, that the incestuous quality of relationships is one of the most dominant features in the depiction of the emerging nuclear (bourgeois) family around 1800—Gotthold Ephraim Lessing's *Emilia Galotti* (1772), Heinrich Kleist's *Marquise von O . . .* (1808) or Ludwig Tieck's *Eckbert* (1797), all present fathers and daughters or brothers and sisters in intimate relationships. However, even in the one text in which the relationship might have produced a child (*Marquise von O . . .*), we learn nothing about the fate of this child. It is, however, the hereditary burden of this child which is of interest in the context of this book. The aim in the chapter on incest is less to provide a comprehensive account of the incest motive than to delineate the changes in the conceptualization of selfhood and of *Bildung* in the wake of eighteenth-century experiments with inbreeding in animals, once incest was considered a biological problem.

If scholars such as Hannelore Schlaffer have demonstrated the extent to which mythology informs Goethe's text and how the knowledge of this iconography can enrich a reading of the *Wilhelm Meister*,[21] *Romanticism, Origins, and the History of Heredity* is motivated by a different if complementary fascination. It aims to show the extent to which contemporary debates on heredity inform literary texts around 1800 and what they can contribute to their interpretation. By placing the incest-child Mignon in the context of heredity and inbreeding, I will argue that the figure gains epistemological significance as a hermaphrodite whose body rejects all attempts of reading. The Tower Society's machinery of meaning production only covers the scandal of her monstrous body by declaring her an angel. Where the Hall of the Past at the end of the *Lehrjahre* has been read as a "Kabinett" [cabinet] for art,[22] I want to suggest that there is much to gain for an understanding of the text by reading the Hall in the tradition of the *Kunst- und Wunderkammer*. The iconography on which this book draws is not that of mythology, but the repository of the natural sciences.

Similarly, the discussion of hybridization and the figure of the bastard in Jean Paul Richter's novel *Der Komet* (chapter 3) does not aim to articulate a psychology or sociology of outsiders and outcasts in the context of a rich archive of literary texts focusing on the figure (from Parzival's half-brother Feirefiz to Johannes Niemand in Annette von Droste-Hülshoff's *Judenbuche*); instead, it follows the process of re-definition of selfhood in the light of new relationships between individual and species evidentiated by contemporary research. If the eighteenth century had rehabilitated the individual by conceptualizing it as an organic, self-contained, unity, then the question became what would happen to this individuality once it was dynamically connected to the species as cross-breeding experiments started to suggest. Jean Paul Richter's novel is chosen here because it demonstrates *both* how a vocabulary of heredity and transformation crossed over from botanical into literary texts *and* the ultimate rejection by literature of the hypothesis of species evolution suggested in the life sciences because it would undermine a strong notion of individuality.

Finally, the fourth chapter of this book turns to questions of genealogy and origin in the work of E. T. A. Hoffmann. The novel *Elixiere des Teufels* joins the conversation about the most current findings in the natural sciences by representing the consequences of the assumption that the individual is determined by lineage. Both as lawyer and writer, E. T. A. Hoffmann engages intensively with contemporary convictions in medicine and biology which suggest that mental disease is heritable.

He is interested in the ramifications of this work in particular for the question of personal responsibility in the case of crime. Hoffmann discusses these ramifications to caution us against transferring biological models to the psychological and moral realm too quickly. By placing the *Elixiere des Teufels* in the context of Erasmus Darwin and Johann Gaspar Spurzheim's work on the heredity of insanity, I demonstrate Hoffmann's scathing critique of a science that is proud of its empiricist methodology without reflecting on the imaginative conditions of its observations. For Hoffmann, however, not the fiction is the problem but the fact that fiction is not recognized as such. E. T. A. Hoffmann's *Elixiere des Teufels* present a key moment in the formation of the novel as that sphere of the modern world which functions as a meta-system for the sciences.

If one core model emerges from the discussions of literary texts in this book, it is a strong belief in the power and significance of fiction for the formation of the self—a belief which authors around 1800 can maintain because they reject evolutionary claims. By placing selfhood around 1800 in the context of heredity, the chapters in this book attempt to demonstrate that the kind of emphatic notion of selfhood for which German Romanticism has become famous is not only threatened by considerations of determinism, or evolution, but also a result of it.

Examining the impact of contemporary debates on heredity on formulations of selfhood in and through the novel leads to three sets of observations. First, in discussing heredity, authors revisit the question of determinism which had occupied Enlightenment authors in the wake of French materialism. As Manfred Engel has argued, German Idealism unfolded as a liberation from the "Folgelasten des empirisch-naturwissenschaftlichen Weltbildes"[23] [burdens of the empirical-scientific world view]. While selfhood had been defined in the German tradition in opposition to the empiricism and materialism of Enlightenment, there is a growing uneasiness among German authors around 1800 regarding the possibility of this opposition in the context of debates which suggest that both mental and corporeal features are hereditary. This leads to a heightened interest in figures for which *Bildung* does not fail because of a missing telos or strong social constraints, but because of a hereditary burden. Hence, this first set of observations focuses on heredity as a burden.

Second, in discussing heredity, authors revisit the question of autopoietic production which had been formulated as the (German) solution to Enlightenment empiricism and probability.[24] This new interest in questions of self-organization gains explosive force from the fact that the conceptualization of artistic autonomy had relied heavily on models provided by the natural sciences. While it has been often noted that the term *Bildung* emerged in the context of Blumenbach's idea of an epigenetic *Bildungstrieb* [formative drive] and that for Kant, as Helmut Müller-Sievers has put it in his seminal study on *Self-Generation*, "epigenesis provides the model for the 'self-delivery . . . of our understanding . . . without impregnation by experience,'"[25] what has been missing from scholarship on the novel are studies that ask for the changes in the account of selfhood and autonomy once Blumenbach's *Bildungstrieb* was reconsidered in biology in the context of evolutionary models.

If a primary concern of the novel around 1800 was the formation of the self, how is this self-depicted against the backdrop of emerging evolutionary theories? Can we discern a continued cross-pollination of concepts such as of selfhood and generation or do other models become prevalent? How can autonomy be maintained in light of the new dependency of individual and species? Such questions might suggest that this book wants to locate a pre-Darwinian understanding of individuality in the years following the publication of Erasmus Darwin's *Zoonomia* (1794–1796), in which an evolutionary theory had first been presented. However, in studying texts such as Jean Paul's *Komet* and E. T. A. Hoffmann's *Elixiere*, the chapters that follow attempt to demonstrate how these texts are both indebted to contemporary natural sciences and develop their notions of selfhood in opposition to it. What was new once evolutionary models had been proposed—and what triggered such opposition—was the feeling that the kind of dualism which had shaped eighteenth-century conceptualizations of selfhood was under threat.[26] Hence, this second set of observations is interested in heredity as evolution.

Third, writers revisit the question of the relationship between literature and science. An author like E. T. A. Hoffman, whom I will discuss in chapter 4, focused on the aspect of hereditary theories that suggests the transmission from parents to child as a burden for the child. Rejecting such determinism both for psychological and ethical reasons, Hoffmann grew critical of the sciences which provided the foundation for the claim. Consequently, he inquired into the very conditions behind the formulation of scientific truths. Interestingly, Hoffmann uncovered the imaginative force behind a science which prides itself for its empirical methodology. Evolutionary thought became a principal site of discussion on the relationship between literature and science, because it threatened the distinction which made these two fields possible. In the discussion of evolutionary thought, the question of imagination came once again front and center. In this third set of observations, my primary concern is heredity as imagination.

In all three sets of observations *Bildung* appears as the process through which the character in the novel obtains selfhood. However, this selfhood is not achieved in all the texts discussed. Mignon, for example, is excluded from selfhood. Nikolaus Marggraf in the *Komet* certainly fails the classical ideal of finding a useful place in society, and the life of Medardus, the protagonist in Hoffmann's *Elixiere*, ends with the recognition of an imaginative determinism. All three characters, but in particular Mignon and Nikolaus, also can be read not as characters, but as allegorical accounts of nature and imagination. Even though selfhood fails or is not represented through psychology and identification, nevertheless selfhood remains central. However, *Romanticism, Origins, and the History of Heredity* advances an argument that already surfaces in Morgenstern's definition of the novel and has been revisited by Dennis F. Mahoney, Todd Kontje and others, namely that the *Bildungsroman* is less about the formation of a character than about the formation of the reader.[27] As Mahoney has argued with reference to Hölderlin's *Hyperion* (1797–1799), the novel around 1800 "represents not only the development of a character . . . , but also the envisioned formation of a reader."[28] Following Mahoney's lead, this book does read the protagonist's biography in terms of psychological development, but it also considers formation processes which lay outside the character. I am interested in examining how around 1800 novels became, as Kontje has observed, "less realistic depictions of individuals than meta-fictional reflections on the function of reading and the institution of literature in society."[29] However, while Kontje focuses on the institution of literature in society my interest is more narrow and specific in that it asks for the meta-critical contribution of the novel in a context in which its autonomy—much as that of its protagonists and readers—is perceived as threatened by contemporary findings in the sciences. The texts that form the center of this study present an evolution of conceptualizing selfhood in light of heredity: the Mignon episode questions the possibility of selfhood in the wake of inbreeding research; the *Komet* debates models of autonomy in light of cross-breeding experiments; and the *Elixiere* actively engages with the epistemological consequences of evolutionary thought for the nature of both science and literature.[30]

As scholars in both literary studies and in the history of science have shown, the last third of the eighteenth century witnessed a dramatic intensification of debates on procreation. While scientists were busily leaning over their microscopes and their workbenches covered with mutilated freshwater hydras and chicken embryos in order to establish whether or not the new being was performed and in order to find out to what extent the father or mother was necessary for procreation, authors of literary and philosophical texts wondered how these debates in embryology and biology could help them to solve questions of autonomy, creativity and epistemology. While there has been significant scholarly activity regarding the history of science of heredity and, perhaps even more so, regarding the history of selfhood, much less attention has been paid on the connection of these two fields of research.

This gap in scholarship is certainly greatly conditioned by the disciplinary organization of our research. In this context, the challenge is located on two levels. First, our disciplinary training makes it difficult for us to be attentive for textual clues that point outside established registers of literary studies. And, second, the researcher who is curious to go to these fringes is confronted with the question, and rightly so, how such non-literary discourses are productive for a reading of literary texts. Much has been done, of course, in recent decades to defend such enterprise against the accusation of positivism and to justify a field of literary studies which is interested in the relationship and overlaps of discourses. Scholars such as Friedrich Kittler, Jochen Hörisch, and most recently Andrew Piper have demonstrated the degree to which literary works are connected to the media revolution of their times and, even more, how they are key players in the conceptualization of this revolution. Similarly, and closer to my own project, Müller-Sievers, Stefani Engelstein, and Jocelyn Holland have shown how the knowledge and vocabulary of reproduction and anatomy informs the work of Romantic writers. German Romanticism—here defined in the broader North American sense—has been proven a fertile ground for such interdisciplinary work. This productivity in interdisciplinary work in the Romantic period is, of course, closely tied to the fact the Romantic age was itself an interdisciplinary or pre-disciplinary period.

However, while literary scholars might still have to catch up, there have been a number of publications in recent years that discuss the question of heredity from the perspective of the history of science and of cultural history. For the West European context, extensive work has been done by the Max Planck Institute for the History of Science in Berlin (MPIWG) which has organized several conferences on heredity since 2001.[31] The organizers of these conferences describe their intentions in the following way: "In our project we wish to focus on the scientific and technological procedures, in which the knowledge of heredity was materially anchored and by which it affected other cultural domains."[32] Although the organizers promise to look for the effects of the discussion on heredity in "other cultural domains," most of the papers published in the preprints focus on the development in science and agriculture. In the preprints for the two conferences that cover the time period from the seventeenth to the nineteenth century, there is only one paper on literature (it deals, however, with legal rather than biological issues).[33] The MPIWG conference's omission of literary texts represents a typical situation. In her 2006 publication *Genea-Logik: Generation, Tradition und Evolution zwischen Kultur- und Naturwissenschaften* [Genea-Logic: Generation, Tradition, and Evolution between Cultural Studies and Natural Sciences], which can be considered the most important monograph in recent years on the topic, Sigrid Weigel makes ample reference to literary works, but except in one case[34] these works are taken as mere instances of the scientific developments Weigel focuses on. This omission points to a methodological problem which research projects like this book face when considering the question of the relationship between different disciplines as they emerge around 1800. Although interdisciplinary work is promised, it has been proven difficult to read literary works as productive discussions of scientific findings on heredity and not simply as their illustrations.

The history of the differentiation of disciplines in Western Europe from the late eighteenth century on has been told many times.[35] However, we are still challenged when trying to conceptualize the relationship of these disciplines.[36] Particularly difficult has been the definition of the relationship between what has become known as *science* and *literature*. Part of the challenge lies in the fact that in the eighteenth century, as Jocelyn Holland remarks in *Procreative Poetics*, the "boundaries between different fields of scientific thought were remarkably porous."[37] As twenty-first century readers we are tempted to overlook such porosity and neatly separate lines of inquiry in clusters which have become academic disciplines but which were only just emerging in the late eighteenth century. At the same time, however, we have also often ignored how present the question of the relationship between science and literature was already around 1800. Hence, although the cross-fertilization of the two disciplines at the time of their emergence is now widely accepted, a comprehensive account of available and employed conceptual frameworks is still missing.

The relationship can be thought in many ways. Accepting some simplifications, one can discern five different models that have been used in the past. Applied to the questions discussed in this book, the first possibility is to assume that a strong notion of heredity emerges simultaneously in different disciplines (e.g. law, biology, literature) around 1800. A project that follows such an approach looks for the development that the notion of heredity undergoes in different disciplines in a given time period. The underlying assumption of such an approach is that the development in the disciplines proceeds parallel so that one notion of heredity is reflected simultaneously in all disciplines. An artistic statement about heredity becomes as valid and in fact epistemologically identical with a scientific observation. Following a variation of this model, Robert J. Richards has argued in his 2002 study *The Romantic Conception of Life* that the "fundamental organizing principle in the philosophy of the Romantics . . . was the aesthetic-epistemic tenet of the complementarity of the poetic and scientific conceptions of nature."[38] Similarly, Denise Gigante argues in her 2009 book on *Life: Organic Form and Romanticism* that "once life was viewed vitalistically as power, science and aesthetics confronted the same formal problems." Going beyond Richards's assumption of a complementarity, Gigante considers organicism "an interdisciplinary field responding to the problem of life" in order to argue that Romanticism "made possible the analogy between aesthetic and biological form upon which we still rely."[39]

While scholars such as Richards and Gigante stress simultaneity of responses to one question, the second possibility is to assume that the discussions in the sciences formed the leading discourse at one given point in time which provided a matrix for problems arising in the humanities and literature. To name just one example, new biological notions of hybridity could become models of thinking through problems of genre. Drawing on this option, Müller-Sievers has argued that Johann Friedrich Blumenbach's publications on epigenesis opened for Kant a new way to think through philosophical problems.[40]

The third possibility can be seen as a hyperbole of the second approach, it is the assumption that science does not merely provide a heuristic tool, but a theoretical framework that can explain not only phenomena in its own discipline but the world as such. Since the mid-1990s, there is a growing number of publications in literary studies that have begun to consider literature and culture from a Darwinian perspective. Proponents of this idea argue for a paradigm shift away from "irrational" postmodernism towards a "rational" and empirically proven theory which is argued to be based on Charles Darwin's concepts and that applies them to culture and literature.[41] This last model differs so far from the previous ones that it does not stay in a particular time period, but applies current theories to a historical moment. It is part of a more encompassing trend which strives to approach questions which formerly were discussed primarily in the humanities from a biological perspective. This trend often climaxes in the question of morality and the wish to formulate an ethics within a biological framework.

In a fourth model, partly developed as an answer to the third model and articulated for example by Weigel in her 2006 study *Genea-Logik*, not science but the humanities are promoted as a meta-sphere, with, however, the significant difference that proponents of a Darwinian approach to literature argue on the basis of the truth of their model, while in Weigel's model literature becomes a meta-sphere precisely as a space in which questions about truth claims, scientific or literary, can be discussed.

Finally, in a fifth and last model, fiction is considered as actually anticipating scientific findings. Colin Milburn has stressed recently how the emergence of nanotechnology is closely connected to esoteric and fictional texts, from which nanotechnology's founders drew important inspiration, thereby demonstrating and confirming the function and force of fiction for the scientific endeavor.[42]

It goes without saying that many studies that consider the relationship between science and literature combine—consciously or not—all, or at least some, of these approaches in order to answer the particular question that they have posed. This book is no different. It is based on the hypothesis that biology becomes a leading discourse around 1800 in the sense that it gives thematic impulses for other disciplines. However, in contrast to the attempts to establish biology (or Darwinism) as a set of rules that is able to explain not only biological phenomena but the world as such and therefore can be applied to both natural and cultural products, in this project I follow literary debates that are triggered by such claims. My aim is to show that towards the end of the eighteenth-century, at the very moment when scientific models emerge with unprecedented force, literature increasingly assumes a correlative function, showing the potential limits of the supposedly "rational" scientific claims.

If debates on biology and heredity and their significance for the understanding of selfhood became a space for the negotiation of the relationship between literature and science around 1800, there was at least one author who considered the very activity of "doing" biology a way for obtaining selfhood precisely because it required an involvement in both empirical and imaginative procedures. In the programmatic opening pages of his *Biologie oder die Philosophie der lebenden Natur* (1802–1822)—the work which is often credited for having established the term biology[43]—the German naturalist and botanist Gottfried Reinhold Treviranus (1776–1837) asks rhetorically:

> Und es sollte ein so ganz verdienstloses Werk seyn, das, was *Erfahrung* und *Nachdenken* uns über diesen erhabenen Gegenstand gelehrt haben, und was bisher in den verschiedensten Fächern zerstreut lag, zu einem Ganzen zu vereinigen? Es ist eine längst anerkannte, aber noch nie gehörig angewandte Wahrheit, daß der Mensch nur durch eine gleichförmige Entwicklung aller seiner Geisteskräfte, und nicht durch eine einseitige, wenn auch noch so weit getriebene Cultur zur höchsten Stufe der Humanität gelangt. Der bloße Geometer, der immer nur mit den reitzlosen Bildern des Raums beschäftigt ist, erreicht dies eben so wenig, als der bloße Dichter, der nie das *Zauberland der Phantasien* verläßt. Aber wo ist eine Wissenschaft, die *den Verstand und zugleich die Einbildungskraft* so sehr in Thätigkeit erhält, und daher der Erziehung des Menschen zur Humanität so angemessen ist, als diejenigen, die wir in diesem Werke zu bearbeiten uns vorgesetzt haben?[44]

> [And should it be a work entirely without merit to unite into one whole that which *experience* and *thought* have taught us about this sublime object, and which was until now scattered among different disciplines? It has been recognized for a long time—though not properly applied yet—that the human being can reach the highest level of humanity only through a uniform development of all his mental abilities, and not through a one-sided culture, no matter how highly it is developed. The mere geometer who is always occupied only with the dull images of space reaches this (level) as little as the mere poet who never leaves *the magic country of phantasy*. But where is a discipline which keeps *the mind and at the same time imagination* to such a degree in activity and which is, therefore, appropriate for the education of the human being for humanity, other than the one that we have planned to cover in this work?]

Such celebration of the formative power of the discipline of biology, of its unique way of combining experience and thought, imagination and ratio, must astonish, considering that, as we have argued here, biology becomes defined by its empirical methodology. Treviranus's statement becomes readable, however, if we draw on Max Weber's thesis that modernity is shaped by a differentiation of value spheres. Much as literature invents itself as an autonomous sphere in opposition to a biology which it relegates to empiricism, biology newly defines itself as an independent sphere by differentiating itself from literature which it relegates to fantasy. Each sphere understands itself as uniquely positioned both to make statements about the world and to bring about the modern self-facing this world.

The texts discussed in this book attest to the extent to which literary texts around 1800 were informed by disciplines such as natural history, botany, and embryology, but they also witness an attempt to articulate a sphere in which literature can come into its own. If paying close attention to contemporary debates on heredity can tell us something about the conceptualization of selfhood around 1800, the literary texts under discussion also demonstrate that in a period in which the empirical access to the workings of genetics was still far from reach, fiction became a key player not only in presenting what had been formulated in the sciences but also in negotiating and at times anticipating (later) findings. The larger aim of this book then is twofold: to provide a history selfhood around 1800 against the backdrop of heredity and to contribute to a burgeoning field in German studies and beyond which consider the relationship between literature and science.

EPIGRAPH SOURCE

Charles Baudelaire, *Le Peintre de la vie modern* (1863), quoted in A. E. Pilkington, "'Nature' as Ethical Norm in the Enlightenment," in *Languages of Nature: Critical Essays on Science and Literature*, ed. L. J. Jordanova (New Brunswick, NJ: Rutgers University Press, 1986), 55.

———. *The Painter of Modern Life and other Essays*, trans. and ed. Jonathan Mayne (London: Phaidon Press, 1964), 31.

NOTES

1. The first use of "biology" as the name for a field of study takes place around 1800. Gottfried Reinhold Treviranus, Jean-Baptiste Pierre Antoine de Monet Lamarck, and Karl Friedrich Burdach have all been discussed as introducers of the term in this context. For a concise overview of the debate see Sander Gliboff, *H. G. Bronn, Ernst Haeckel, and the Origins of German Darwinism: A Study in Translation and Transformation* (Cambridge, MA: MIT Press, 2008), 31–32.

2. Cf. Charles Taylor, *Sources of the Self: The Making of the Modern Identity* (Cambridge, MA: Harvard University Press, 1989).

3. Friedrich Schiller, "Verbrecher aus Infamie," in *Werke und Briefe*, ed. Otto Dann, Historische Schriften und Erzählungen II (Frankfurt am.Main: Deutscher Klassiker Verlag, 2002), 7:563. All translations from the original primary and secondary sources are mine unless otherwise indicated.

4. Friedrich Schiller, "The Criminal of Lost Honor," trans. Jeffrey L. High, in *Schiller's Literary Prose Works: New Translations and Critical Essays*, ed. Jeffrey L. High (Rochester: New York, 2008), 39.

5. Otto Dann, "Verbrecher aus Infamie" [Kommentar], in *Werke und Briefe*, by Friedrich Schiller (Frankfurt am Main: Deutscher Klassiker Verlag, 2002), 7:992.

6. Michael Minden, "The Place of Inheritance in the Bildungsroman: *Agathon*, *Wilhelm Meisters Lehrjahre*, and *Der Nachsommer*," in *Reflection and Action: Essays on the Bildungsroman*, ed. James N. Hardin (Columbia: University of South Carolina Press, 1991), 254–92.

7. Robert J. Richards, *The Romantic Conception of Life: Science and Philosophy in the Age of Goethe* (Chicago: University of Chicago Press, 2002).

8. For an account of Leibniz's Anti-Jansenism, see Anthony Krupp, *Reason's Children: Childhood in Early Modern Philosophy* (Lewisburg, PA: Bucknell University Press, 2009).

9. Immanuel Kant, *Kants Gesammelte Schriften*, ed. Herausgegeben von der Königlich Preußischen Akademie der Wissenschaften (Berlin: Georg Reimer, Walter de Gruyter, 1912–1983), 2:438.

10. Johann Gaspar Spurzheim, *Observations on the Deranged Manifestations of the Mind; or, Insanity* (Boston: March, Capen & Lyon, 1833; repr. Gainesville, FL: Scholars Facsimiles and Reprints, 1970), 96.

11. Johann David Michaelis, *Abhandlungen von den Ehegesetzen Mosis welche die Heyrathen in die nahe Freundschaft untersagen*, Diss., n.p. (1755; repr. Zweite und vermehrte Auflage. Göttingen: Verlegt Abram Vandenhoecks seel. Witwe, 1768).

12. Erasmus Darwin, *Zoonomia; or, The Laws of Organic Life*, (London: Printed for J. Johnson, in St. Paul's Church-Yard, 1794–1796), 2:505.

13. For two recent accounts on this research written for a general audience see Nessa Carey, *The Epigenetics Revolution: How Modern Biology is Rewriting our Understanding of Genetics, Disease, and Inheritance* (London: Icon Books, 2011); and Richard C. Francis, *Epigenetics: The Ultimate Mystery of Inheritance* (New York: W. W. Norton, 2011).

14. In the late 1970s, the millionaire Robert Klark Graham established a sperm bank initially targeting Nobel Prize winners as sperm donors. 217 children were born with help of this sperm bank. The highly controversial project ceased to exist after Graham's death in the late 1990s. Paul Olding, "The Genius Sperm Bank," *BBC News*, 15 June 2006, http://news.bbc.co.uk/2/hi/5078800.stm (17 March 2014).

15. Aristotle, *Generation of Animals*, trans. A. L. Peck (Cambridge, MA: Heinemann and Harvard University Press, 2000), 7 and 401.

16. Joseph Gottlieb Koelreuter, *D.J.G.K.'s Vorläufige Nachricht von einigen das Geschlecht der Pflanzen betreffenden Versuch und Beobachtungen* (Leipzig: Gleditsch, 1761), 36.

17. See James Gallagher, "Gene therapy: Glybera approved by European Commission," November 2, 2012, http://www.bbc.co.uk/news/health-20179561 (accessed June 24, 2013); Gina Kolata, "After Setbacks, Small Successes for Gene Therapy." *New York Times*, November 5, 2009, A19, also at http://www.nytimes.com/2009/11/06/health/06gene.html?_r=0 (accessed January 19, 2014); and Sue Pearson, Hepeng Jia, and Keiko Kandachi, "China approves first gene therapy," *Nature Biotechnology* 22, no. 1 (2004), 3–4.

18. For regulations in Germany see A. Lindemann, F. M. Rosenthal, S. Hase, P. Markmeyer, and R. Mertelsmann. "Guidelines for the Design and Implementation of Clinical Studies in Somatic Cell Therapy. German Working Group for Gene Therapy," *Journal for Molecular Medicine* (*Berlin, German*) 73, no. 4 (April 1995): 207–11; and *Germline Therapy. Human Genetic Modification: Embryo Protection Law* [Germany], n.d., http://www.dnapolicy.org/policy.international.php?action=detail&laws_id=38 (accessed June 24, 2013). For

the position of the U.S. American National Institutes of Health see *Germline Gene Therapy*, March 2006, http://www.genome.gov/10004764 (accessed June 24, 2013).

19. *Weissmann-Barrier*, http://en.wikipedia.org/wiki/Weismann_barrier (accessed May 15, 2013).

20. See, however, Dennis F. Mahoney who provides an introduction to the text in his *Der Roman der Goethezeit, 1774–1829* (Stuttgart: Metzler, 1988), 150–53.

21. Hannelore Schlaffer, *Wilhelm Meister: Das Ende der Kunst und die Wiederkehr des Mythos* (Stuttgart: Metzler, 1980).

22. Ibid., 67.

23. Manfred Engel, *Der Roman der Goethezeit. Anfänge in Klassik und Frühromantik: Transzendentale Geschichten* (Stuttgart: J. B. Metzler, 1993), 1:24.

24. While the Enlightenment insists on probabilism and, therefore, has to exclude imagination, it is precisely imagination which stands, in Jane Brown's words, "as the climax of the subject matter of poetry" in Goethe and Schiller's poetology (Brown 161).

25. Helmut Müller-Sievers, *Self-Generation: Biology, Philosophy and Literature around 1800* (Stanford: Stanford University Press, 1997), 49.

26. As Engel has pointed out, even for Goethe—who has often been read as a monist—dualism remains at the core of his understanding of selfhood (*Der Roman der Goethezeit*, 75–77; 317).

27. On the *Bildung* of the reader in Morgenstern see Martini 256.

28. Dennis F. Mahoney, "The Apprenticeship of the Reader: The Bildungsroman of the 'Age of Goethe,'" in *Reflection and Action: Essays on the Bildungsroman*, ed. James N. Hardin (Columbia: University of South Carolina Press, 1991), 103.

29. Todd Kontje quoted in ibid., 105.

30. The question of whether or not the novel can be read as the depiction of an ideal *Bildungsweg* [path of formation] represented in the protagonist or whether it forms the reader who synthesizes such a path from the multiplicity of figures in the novel was already debated shortly after the publication of *Wilhelm Meister*. For a discussion of this question by means of Körner's and Humboldt's reaction to the *Meister*, see Engel, *Der Roman der Goethezeit*, 236–47.

31. The results are published in preprints available at the institute's website: Max Planck Institute for the History of Science (hereafter MPIWG), *A Cultural History of Heredity I: Seventeenth and Eighteenth Centuries*, preprint 222 (Berlin: Max-Planck-Inst. für Wiss.-Geschichte, 2002), available at http://www.mpiwg-berlin.mpg.de/Preprints/P222.PDF (accessed January 17, 2014). See also the selection of essays in Hans-Jörg Rheinberger, Peter McLaughlin, and Staffan Müller-Wille, "Introduction," in *A Cultural History of Heredity 1: 17th and 18th Centuries*, Max Planck Institute for the History of Science, preprint 222 (Berlin, Max-Planck-Inst. für Wiss.-Geschichte 2002), 3–5.

32. Rheinberger, McLaughlin, and Müller-Wille, *Cultural History*, 3.

33. Ulrike Vedder, "Literature and the Law of Succession," in *A Cultural History of Heredity II: 18th and 19th Centuries*. Max Planck Institute for the History of Science, preprint 247 (Berlin, Max-Planck-Inst. für Wiss.-Geschichte, 2003), 175–85.

34. Sigrid Weigel, "Der Text der *Narrenburg* [Stifter]—die Gesetze der Erbschaft zwischen Naturgesetz und Schrift," In *Genea-Logik: Generation, Tradition und Evolution zwischen Kultur- und Naturwissenschaften* (München: W. Fink, 2006), 148–53.

35. See for example Wolf Lepenies, *Das Ende der Naturgeschichte: Wandel kultureller Selbstverständlichkeiten in den Wissenschaften des 18. und 19. Jahrhunderts* (München: C. Hanser, 1976); and John V. Pickstone, *Ways of Knowing: A New History of Science, Technology and Medicine* (Manchester: Manchester University Press, 2000).

36. For a recent consideration of C.P. Snow see Kagan 2009.

37. Jocelyn Holland, *German Romanticism and Science: The Procreative Poetics of Goethe, Novalis, and Ritter* (New York and London: Routledge, 2009), 4.

38. Robert J. Richards, *The Romantic Conception of Life: Science and Philosophy in the Age of Goethe* (Chicago: University of Chicago Press, 2002), 329. For a similar approach see Astrida Orle Tantillo, *The Will to Create: Goethe's Philosophy of Nature* (Pittsburgh, PA: University of Pittsburgh Press, 2002). I would also consider Clifford Geertz' notion of "Thick Description" a variation of this approach. On methodological questions in the context of writing cultural history see his "Thick Description: Toward an Interpretive Theory of Culture," in *The Interpretations of Cultures: Selected Essays* (New York: Basic Books, 1973), 3–30.

39. Denise Gigante, *Life: Organic Form and Romanticism* (London and New Haven, CT: Yale University Press, 2009), 3.

40. Müller-Sievers, *Self-Generation*. See also Michel Chaouli's work on the function of chemical tropes in Romantic concepts of poetry in *The Laboratory of Poetry: Chemistry and Poetics in the Work of Friedrich Schlegel* (Baltimore: Johns Hopkins University Press, 2002).

41. Cf. Joseph Carroll, "Pluralism, Poststructuralism, and Evolutionary Theory," *Academic Questions* 9, no. 3 (Summer 1996): 43–57; and see also his "Evolution and Literary Theory," *Human Nature* 6, no. 2 (June 1995): 119–34.

42. Colin Milburn, *Nanovision: Engineering the Future* (Durham, NC: Duke University Press, 2008).

43. Marjorie Grene and David J. Depew note that the term had been used as early as the mid-eighteenth century by the philosopher Michael Christoph Hanov (1695–1773) in his work *Philosophia Naturalis* (1766), but that it was Treviranus who made it popular in the current meaning. See Grene and Depew, *The Philosophy of Biology: An Episodic History* (New York and Cambridge: Cambridge University Press, 2004), 122–23.

44. Gottfried Reinhold Treviranus, *Biologie, oder Philosophie der lebenden Natur für Naturforscher und Aerzte* (Göttingen: Johann Friedrich Röwer, 1802), 1:6, italics mine.

CHAPTER ONE

Man sieht z.B. an den Aurikeln, daß, wenn man sie aus der Wurzel zieht, man sie alle nur von einer und derselben Farbe bekommt; wenn man dagegen aber ihre Samen aussät: so bekommt man sie von ganz andern und den verschiedensten Farben. Die Natur hat also doch die Keime in sie gelegt, und es kommt nur auf das gehörige Säen und Verpflanzen an, um diese in ihnen zu entwickeln. So auch bei dem Menschen! —*Kant*, Über Pädagogik

[*It is observable, for instance, that the auricula only bears flowers of one and the same color when cultivated from the root. On the other hand, if it is grown from seed one gets flowers of totally different and most varied colors. Thus nature has after all placed the germs in these plants, and it is merely a matter of proper sowing and planting that these germs develop in the plants. The same holds true with human beings.*]

The Discovery of Heredity

CONCEPTUALIZING HEREDITY AROUND 1800

IN WHAT IS NOW a well-known idea, Reinhart Koselleck has called the time period between 1770 and 1830 "Sattelzeit," thereby stressing the significance of these decades for the development of modernity.[1] He has characterized the period as a time in which a society marked by impermeable class hierarchies built on inherited privileges is changing towards a mobile bourgeois society that is differentiated according to functions based on achievement.[2] This belief in achievement and perfectibility had become possible on the basis of the Enlightenment assumption that each individual is born free and neither privileged by lineage nor determined by an inherited burden.[3] As a consequence, the reference to ancestors and kinship lost significance for the understanding of society and of the individual.[4]

The philosophical and legal considerations which prepared and shaped the French Revolution had been decisive for this abandonment of the past. Quoting contemporary legal proclamations addressed against the heritage principle of the *ancient régime*, Pierre Nora's "Generations" provides a comprehensive picture of the extent to which the idea of ancestry was rejected by the representatives of the French Revolution. Nora describes a shift from a diachronic (father–son power transfer) to a synchronic (shared power in the community of brothers) understanding of society and political institutions. A new concept of generation emerged: a group of peers, independent of their parents, driven by the ideas of equality, progress, and youth. For this group of brothers, kinship was not established by blood but by reference to the state, which took over the parent function in the name of adoption.[5] Similarly, Marc Shell has shown that the equality of legitimate and illegitimate children concerning the line of succession, as stated in the *Code Civile*, can be understood as an attempt of the new French state to constitute the nation as an adoptive family in which all human beings are brothers.[6]

The rejection of inherited privilege and the proclamation of equality which took place in the French Revolution originated from the understanding that nature has equipped all human beings with equal qualities. It was natural law philosophy that prepared the legal framework for the French Revolution by stating that differences between humans are results of socio-historical circumstances rather than expressions of divine will or specific biological conditions. The conviction that nature has created all humans as equals became a widely adopted thesis which found its way outside philosophical and legal circles and into popular encyclopedia. A typical formulation of this understanding of nature can be found in Georg Simon Klügel's *Enzylopädie* from 1784. Under the heading "Natural Law," we find the sub-heading "Angebohrne Rechte der Menschheit" [Innate Rights of Humanity]:

> [I]n dem absoluten Stande der Natur sind alle Menschen einander gleich. Alle Menschen haben *dieselbe geistige und körperliche Natur*, dieselbe Art zu empfinden, dieselben Bedürfnisse, dieselbe Bestimmung zu einer immer wachsenden Vollkommenheit; also haben alle gleiche Rechte zu den Mitteln, ihre Glückseligkeit zu erhalten und vermehren. Bey aller Ungleichheit, welche die bürgerliche Gesellschaft nothwendig gemacht hat, muß der Vornehmste doch in dem Geringsten *die Aehnlichkeit der menschlichen Natur* erkennen und ehren.[7]

> [(I)n the absolute state of nature, all human beings are equal. All humans have *the same spiritual and corporeal nature*, the same way to feel, the same needs, the same determination for an ever growing perfection; hence, all

(humans) have the same rights regarding the means (necessary) to obtain and increase their happiness. Despite all inequality made necessary by civil society, even the noblest person must recognize and honor *the similarity of human nature* in the lowest person.]

There can be no doubt that from the seventeenth century on we observe a decisive abandonment of privileges and inheritance as the justification of political power and as the basis of individual self-understanding, which is grounded on the conviction that nature has equipped all human beings equally. However, this philosophical-political victory of "égalité," equality, hides a second much less known story that does not only involve genealogical lineage, but actually accentuates it beyond the previous centuries. Modernity and the emergence of modern selfhood cannot simply be explained by the rejection of genealogy in the wake of natural law and of the French Revolution.

Informed by questions first developed in anthropology and sociology, David Warren Sabean and Simon Teuscher in the introduction to a recent publication on *Kinship in Europe* highlight that a "great deal of comparative discussion about kinship has been focused on the level of explicit rules in codifications of law and custom."[8] Although they agree that the "analysis of legal doctrines, judicial decisions, and innovations in legal instruments certainly remain a crucial task of analyzing kin organization," they also stress that recent historical studies show that "law can be a very flexible instrument for quite different ways of doing things and that practice cannot be deduced from legal norms."[9] Hence, instead of focusing on philosophical and legal conceptualizations of kinship, the authors in the volume edited by Sabean and Teuscher analyze socio-historical data. By doing so, they come to the surprising conclusion that from the mid eighteenth century on, marriages became not less but "more endogamous, both in terms of class and milieu and among consanguineal kin."[10] On the basis of this fundamental observation, the authors argue against the dominant thesis that the decline of kinship is one of the defining developments in Western history and against the thesis that the long-term development can be best described in terms of "successive contractions toward the modern nuclear family."[11] Instead, Sabean and Teuscher describe two major transitions in the understanding of kinship from the Middle Ages to the nineteenth century. According to their proposition, the shift at the end of the Middle Ages "can be associated with an increasing stress on vertically organized relationships." The second shift described by Sabean and Teuscher takes place in the mid-eighteenth century and is marked by "a stronger stress on horizontally ordered interactions."[12] By defining the second shift via a greater stress on horizontality and despite their focus on kinship instead of adoption, Sabean and Teuscher are in agreement with scholars such as Nora who observes the emergence of the idea of brotherhood and "generation" in the decades before and after the French Revolution.

Although this book remains greatly indebted to Koselleck and his suggestion that the interval between roughly 1770 and 1830 and its discovery of historical time was crucial for the emergence of what we call modernity, it also aims to show that what has been overlooked so far is the extent to which genealogy makes a comeback in modern self-understanding. While the authors in Sabean and Teuscher's volume focus on socio-historical data and Nora on legal and philosophical discussions, this book proposes that it is necessary for a full understanding of modern genealogy to closely examine the observations on kinship and heredity made in the natural sciences around 1800.

THE EMERGENCE OF "HEREDITY"

Speaking of heredity in the second half of the eighteenth century means to speak of a concept which is not defined yet as transfer of genetic material from generation to generation. Such genetic transfer was far from being conjectured and the discovery of the organic unit in whose strands the genetic material is located, the chromosome, was still some seventy years in the future.[13] However, contrary to popular understanding that the discovery of laws of heredity starts only with Charles Darwin (1809–1882) and Gregor Mendel (1822–1884), the transfer of traits to the developing individual and next generation became a major concern and site of conceptualization already in the eighteenth century.

As Roger J. Wood and Vítězslav Orel observe in their study *Genetic Prehistory in Selective Breeding* (2001), the term "heredity" was first introduced as a biological concept in the early sixteenth century in the context of human monstrosities and diseases which were conceived to be transmitted from parent to child.[14] To deduce the emergence of a biological understanding of heredity from occurrences of the word in the context of human disease has become a standard part of the narrative on heredity.[15] The current scholarly understanding is that, once the concept of heredity had emerged in the context of diseases, observations in animal and plant breeding became a major contributor to the formulation of laws according to which heredity works. However, it is important to note that most likely not doctors, but animal breeders were the first professionals whose livelihood was dependent on knowledge derived from careful and systematic observations of the phenomenon of transmission of traits. In German publications, the word stem "erb" was used in the breeding context already in the seventeenth century. In 1670, Georg Simon Winter employs the word "Erbmängel" [hereditary defect] in his publication *Gründliche und kurtz durchgehende Beschreibung, wie eine Stutherey anzurichten* [Careful and comprehensive description how to establish a stud farm] when discussing features which the breeder should avoid in horses used for breeding purposes:

> Die Landsart betreffend, wovon ein großer Herr oder Cavallier sein Gestüth anzurichten willens, ist nach eines iedwedern belieben zu erwehlen, sintemahln es in allen Nationen gute unnd böse oder schlimme Pferde zu geben pflegt, und ligt eben nichts daran, wann man sagt: Diß ist ein Türkisch, Persianisch, Arabisch, Englisch, Spanisch oder Italianisch Pferd, daß es drumb gerecht un gut seyn müsse, sondern man muß das Gewächs, seine Güte, Stärke, wie auch Fehler, und *in specie* die Erbmängel, zuvor hero betrachten, alsdann kan man eine Nation zur Pferdezucht wehlen welche man will.[16]

> [Regarding the nationality (of the horses) from which a lord or gentleman wants to establish his stud farm, he can choose as he likes, since in all nations there are good and wicked or bad horses, and it is irrelevant when one says: This is a Turkish, Persian, Arab, English, Spanish or Italian horse and thereby assumes that it has to be just and good. Instead, one has to look first at the growth, its quality, strength, as well as defects, and similarly hereditary defects. Afterwards one can choose the nation which one wants for horse breeding as one pleases.]

What is so surprising and innovative about Winter's statement is not only his early reference to hereditary defects and their biological understanding, but also his rejection of the significance of geographical regions for breeding. Anticipating breeding strategies which became widespread only in the last third of the eighteenth century, Winter rejects the idea that a certain region—a certain climate one would call it later—is significant for the breeding success. Rather, he stresses the importance of paying attention to bodily features and the avoidance of "Erbmängel." The eighteenth chapter of his manual is devoted entirely to the description of the "hereditary defects" which should be avoided in both stud and mare chosen for breeding.[17]

Winter closes his meditation on hereditary defects with what he calls an "old proverb," which is in fact a statement from Horace's epistles: "*Naturam expellas furcâ tamen usq recurret*"[18] [You may drive nature out with a pitch fork, but she will keep coming back], thereby cementing his rejection of the idea that contingent conditions could mend a given natural precondition.[19]

While the term heredity appeared already in the late sixteenth century, a broader discussion of the notion started only in the last third of the eighteenth century. It is remarkable that this heightened attention to heredity and hereditary defects came at a time when the political repercussions of the natural law philosophy—with its claim of natural equality—were strongest. A sign of the new and lively interest in biological inheritance are a number of essay competitions on hereditary disease. In 1748, the *Académie de Dijon* launched an essay competition which asked for "les causes des maladies héréditaires"[20] [the causes of hereditary diseases] and just around the outbreak of the French Revolution, in the years 1788 and 1790, two other competitions were launched by the Parisian Royal Society of Medicine on the question of whether heritable diseases existed and, if yes, whether they were curable.[21] López-Beltrán suggests that the intensified attention to hereditary diseases was motivated by more than theoretical ambition. Rather, he argues, it originated from the wish to demonstrate social relevance of the medical profession. While hereditary transmission of disease had been associated with aristocratic families and certain sicknesses were closely associated with privilege (e.g. gout which was considered hereditary)[22] or with particular aristocratic families (e.g. the haemophilia of European royalty), such transmission of disease was now turned into a question of public hygiene.[23] The new democratization of society made it possible and politically necessary to think hereditary disease outside the aristocratic class. I would add to López-Beltrán's astute observation which stresses the biopolitical turn that the new attention to hereditary disease could only emerge against the backdrop of the promise of natural equality. Once this promise had been stated, the question about its limits could be asked.

The extent to which we still struggle with distinctions first sought in the late eighteenth century becomes clear when we consider the differentiation between inherited and contracted diseases which was highlighted in the debates of the time. In the justification for awarding the first prize in the second essay competition of the Royal Society to Joseph Claude Rougemont (1756–1818), at that time a professor of medicine in Bonn, the jury mentions explicitly that "Sehr gut werden darin die erblichen Krankheiten von jenen unterschieden, welche das Kind im Mutterleibe oder im

Augenblick der Geburt sich zuziehen kann"[24] [In this (essay), hereditary diseases are very well distinguished from those, which the child can contract in the womb or in the moment of birth.] Rougemont's distinction between those forces that are transferred from parent to child and those forces that shape the developing being during pregnancy is one that scholars in the eighteenth century just started to outline without being yet able to clearly differentiate—despite claiming to do so. It should be noted that the certainty with which we have distinguished inherited and contracted diseases in the high time of the discovery of the genetic code has vanished in recent years and given way to a more nuanced understanding of the relationship between genetic code and gene expression, thereby aligning us closer again with the uncertainty in the conceptualization of heredity around 1800.[25]

The challenges which were faced by eighteenth-century inquiries into heredity can be summarized in seven questions:

1. How can the observations of trait transmission—in the human mostly observed in the case of defect—be accommodated within a framework that assumes nature as a stable entity?
2. To what extent is the development of body parts, and by extension of new beings, mere growth of a pre-established program and to what extent is it governed by a self-organizing force shaping amorphous matter?
3. How can the similarities between parents and children or, even more puzzling, between grandparents and grandchildren be explained?
4. How can we conceptualize the origin of the human race and the great number of variations, which are observed to be consistent and apparently transmitted, at least as long as a population group lives in a stable climate?
5. Do results in animal breeding depend on climate or seed and can results obtained with animals be transferred to human reproduction?
6. If new forms can be obtained by ways of breeding, how can nature and artificiality—God's creation and human intervention—be distinguished, if at all?
7. What is the role of gender in hereditary transmission: if traits can be transferred who carries the traits to the next generation, the father or the mother?

In line with these questions and the uncertainty regarding generational transmission in the age of its discovery, the term "heredity" is used in this book to cover a range of phenomena which are connected to the transmission of characteristics from parent to child within the eighteenth-century understanding of reproduction. These phenomena include the similarities between parents and children (including hereditary diseases), the similarities within families, but also the apparent mismatch between parents and their children and the occurrence of monstrosities.[26]

FITTING HEREDITY INTO EXISTING MODELS: PREFORMATION AND EPIGENESIS

One of the biggest challenges for eighteenth-century scientists was to accommodate observations of trait transmission within a framework that assumed that nature is a stable entity. For Christian thinkers, the stability of nature was central and guaranteed by the act of creation. However, the belief in such stability was older. Since antiquity, reproduction had been understood as a process that provided the preservation and continuation of natural forms. In *Generation of Animals*, Aristotle supported his claim that similarity, not difference, is the aim of generation by arguing that nature strives for finiteness:

> If . . . the products were dissimilar from their parents, and yet able to copulate, we should then get arising from yet another different manner of creature, and out of their progeny yet another, and so it would go on ad infinitum. Nature, however, avoids what is infinite, because the infinite lacks completion and finality, whereas this is what Nature always seeks.[27]

Since Aristotle understood nature as a finite entity, he concluded that every derivation from the original form had to be considered degeneration. He expresses this idea in an often quoted statement:

> [A]nyone who does not take after his parents is really in a way a monstrosity, since in these cases Nature has in a way strayed from the generic type. The first beginning of this derivation is when a female is formed instead of a male, though this is required by Nature. [28]

In Aristotle, the similarity between parent and child was not understood to be personal but generic, a similarity that was shared by all human beings. If a son was similar to his father, it was not because he inherited decisive traits from his father, but rather because he, like his father, was similar to the original form of a human being. Therefore, every deviation from the original form was understood as a degeneration from humanity, and as such, it was considered a monstrosity (τό τέρας), which could have no lasting influence on nature's overall design.[29] Aristotle assumed that in the case of a monstrosity, the original form of the father had not been able to form the female matter to its fullest extent.[30] Aristotle saw this distortion of the male form in close connection to women's sensual and imaginative powers.[31] The mother was seen as the medium that endangered the exact representation of the father's image. Hence, in Aristotelian thinking, the creative ability to shape offspring belonged to the woman and was assigned a negative value because it disturbed the male set of forms. The transmission of traits was difficult to explain in an Aristotelian framework except as a momentary aberration due to the mother's influence which would have no lasting influence on a particular species and, therefore, on nature as such.

Aristotelian accounts of reproduction remained influential well into the eighteenth century and new theoretical frameworks, which offered innovative and in some respects vastly different approaches to thinking about reproduction, did little to enhance the conceptualization of the transmission of traits. A significant part of eighteenth-century discussions on reproduction took place within the larger debate of whether preformationists or epigenesists were better able to explain embryonal development. Although the work of seventeenth- and eighteenth-century naturalists certainly cannot be neatly separated in the two camps—as it seems sometimes because their depiction is so dominant in scholarship on the period—it is nevertheless helpful to recapitulate the two approaches, because they illustrate both the observations that eighteenth-century scientists were able to make and the enormous challenge they faced in explaining those observations.

Preformationists believed that the germs of all living beings were preformed since creation. Germs merely had to be stimulated by the right environment to start expanding their already existing forms.[32] The idea that beings are already preformed was triggered by observations made possible by the microscope. Naturalists started to believe that the spermatozoon which they were able to see was a minuscule body which would be stimulated into growth in the womb's environment. One version of this approach imagined the germs encapsulated in one another like Russian dolls which unfold, or evolve, with each new generation; therefore, sometimes also the somewhat confusing term "evolution" in contemporary treatises.[33] Since it was assumed that all beings were preformed, the transmission of traits was unthinkable in this approach. Even degenerative changes were difficult to explain, because—considering preformation—their mere existence posed the question of whether God had created defective creatures.

Only with the German physiologist Caspar Friedrich Wolff (1733–1794) it became possible to acknowledge that microscopic observations failed to provide evidence for preformed beings. In his 1759 *Theoria Generationis*,[34] Wolff explained the emergence of the new organism not as a gradual unfolding of preformed germs, but as a formation out of amorphous material (see fig. 1.1). Wolff called the energy behind this development the *vis essentialis*[35] [the formative force].

However, Wolff's approach remained unsatisfactory for the German naturalist Johann Friedrich Blumenbach (1752–1840) because Wolff's *vis essentialis* mainly regulated the development of nutrition and as such always depended on nutritive material. In his 1781 publication *Über den Bildungstrieb und das Zeugungsgeschäfte* [On the formative drive and its relation to the (business) of procreation], Blumenbach, instead, described the *formative drive* as an organic force that shapes the organism's reproduction and definite form and that is able to preserve it for a lifetime independent from nutritive matter.[36] Blumenbach's formative drive provided an explanation for the transition from unorganized matter to an organic cooperation that was not dependent on nutrition and thereby clearly distinct from growth.

In a study which opened a whole field of inquiry into the relationship between science and literature, Helmut Müller-Sievers has pointed out that the precondition for this new, epigenetic, understanding of nature as productive force was provided by the discovery of time as historical time, in which events can have a beginning and an end. In the wake of Michel Foucault, Müller-Sievers describes the shift from preformationism to epigenesis as a shift from space to time: from the taxonomy of species and the idea of encapsulated beings to the temporal origin of the generative process.[37] Despite the acknowledgment of time within nature and the implied potential for change, however, the transition of traits remained unanswered in the framework of epigenesis.

Wolff and Blumenbach are widely recognized as epigenesists who believed in the new formation of every being and helped to prepare a strictly biological understanding of embryology. However, they had remarkably different careers. While Wolff was still denied access to German academia by preformationists like Albrecht von Haller (1708–1777), Blumenbach, a former Haller student himself, was successful in making epigenesis *the* leading model in the life sciences and beyond (in the *Critique of Judgment*, Kant, for example, refers several times to epigenesis and Blumenbach's *formative drive*).[38] Albrecht von Haller's resistance to epigenesis came af-

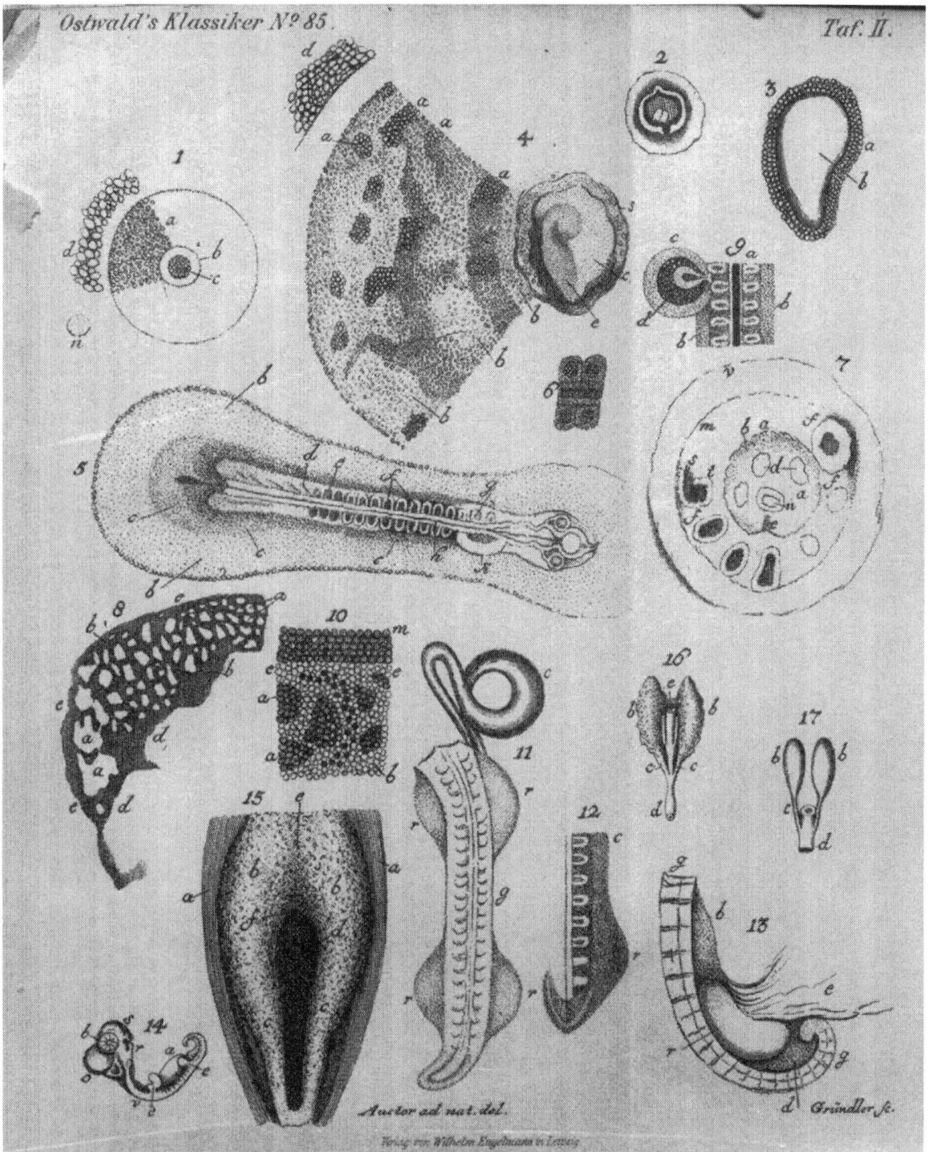

fig. 1.1. Development of the embryo. Courtesy of Gerstein Science Information Centre, University of Toronto. Caspar Friedrich Wolff. Theoria Generationis *(1759). Edited and translated by Paul Samassa. 2 vols. Leipzig: W. Engelmann, 1896.*

ter an initial favoring of the approach and had much to do with new and puzzling experimental research. One of the most influential experiments that shaped the discussion on the possibility of preformed germs for several decades was Abraham Trembley's (1710–1784) work on the "polyp" (freshwater hydra) and his observations in the 1740s that the animal was able to regenerate amputated parts.[39]

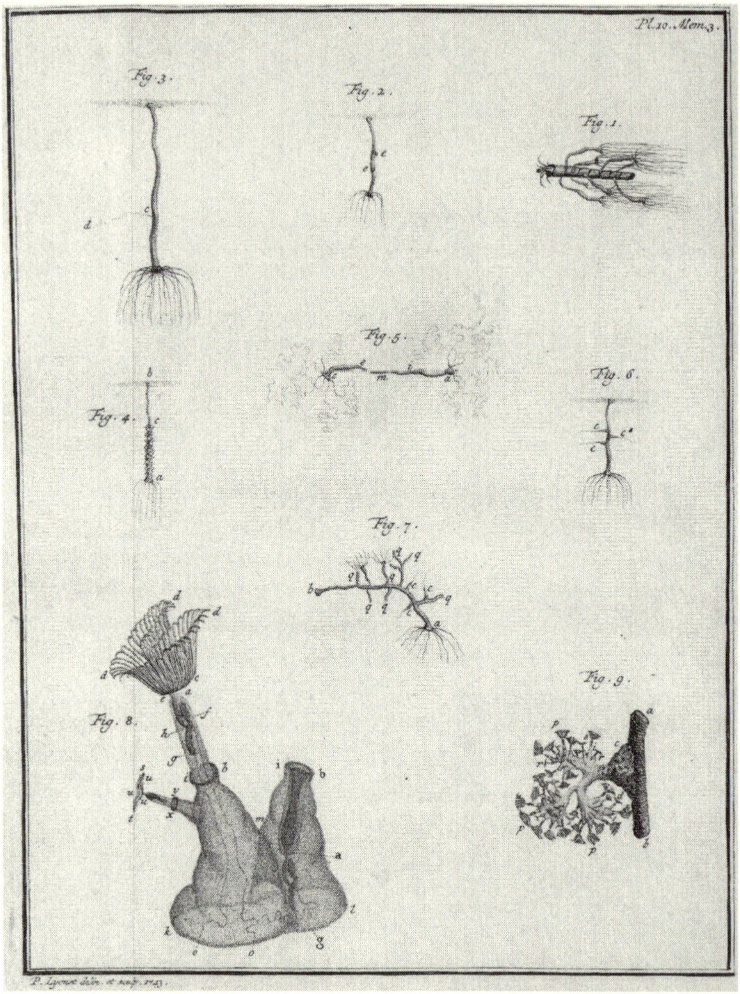

fig. 1.2. Freshwater Hydra. Herzog August Bibliothek Wolfenbüttel: Nh4° 19. Abraham Trembley. Mémoires pour servir à l'histoire d'un genre de polypes d'eau douce *(Mémoires concerning the polyps). Leiden: Jean & Herman Verbeek, 1744.*

The regenerative force challenged many scientists because it seemed impossible to explain such spontaneous and at the same time precise force of regeneration without assuming preformed parts which only grew once called into action by outside stimuli (e.g., the destruction of a body part). Even Blumenbach conceded that he could only observe the effects (*Wirkungen*) of the *Bildungstrieb*, but could not name its cause.[40]

Although the dominant historiography narrates the relationship between preformationism and epigenesis as a sequence, it is probably more accurate to state that by the end of the eighteenth century, the decisive question for scientists remained to what extent the development of body parts, and by extension of new beings, was mere growth of a preestablished program and to what extent it was governed by a self-organizing force dealing with amorphous matter. This question and related pressing questions such as the similarity between parent and child or even grandparent and grandchild had to remain unanswered within the frameworks of epigenesis and preformationism. While preformationists assumed all beings already shaped at the moment of creation and, therefore, could not allow for the transmission of traits from parent to child, the epigenesists assumed new formation from amorphous matter, which did not allow for the transfer of qualities either. Hence, both approaches had to rely heavily on contingent factors like climate or food to explain changes in an individual's form as temporary modifications.[41]

Only towards the end of the eighteenth century did a third model appear (preformationism and epigenesis being the other two) that promised to account for generational transfer. One could call this third model the *genealogical model* because it discussed the relationship between parent and child, individual and species along a time line which made it possible to consider change from generation to generation—without, however, yet giving up teleology, a step which only Charles Darwin took in what we call today the *evolutionary model.* By allowing inter-generational transfer and habit induced change, this third genealogical model added tensions to the relationship of individual and species, while using both as valid concepts.

THE HISTORY OF AN IDEA: "ANLAGE"

Discussions on whether or not the embryo is preformed had to rely on evidence obtained by means of scientific instruments (e.g., the microscope), which was—as a result of the mediated access to the object of inquiry—easier to "coerce" into established religious and philosophical frameworks.

New studies on race and breeding provided evidence which was more obvious and thus had greater potential to challenge pre-existing frameworks, thereby, triggering the formulation of the genealogical model. I will discuss contemporary race and breeding studies in greater detail below, but here I want first to focus on the term "Anlage" (investment, trait), which became a central concept in both breeding and race studies and for the question of the educability of the individual.

The term, eventually used to describe the potential for future development, was originally taken from a primarily financial context (in the sense of investment), but—used metaphorically—had early on the pre-biological meaning of "disposition." The transition of the financial meaning to a biological one can be traced in the leading lexicons and encyclopedias from the late eighteenth century onwards. Its occurrence demonstrates a rising awareness for hereditary transmission in popular works. While there is no mention of "Anlage" in the sense of an innate quality in the *Zedler* of 1732,[42] sixty years later, in the 1793 publication of *Adelung*, we find a preparation of such meaning. According to *Adelung*, "Anlage" is first the activity of "anlegen" [invest] and the object which is "angelegt" [planned, outlined]. However, under points (10) and (11) of the entry, an understanding of "Anlage" as the origin of something which needs to be developed is listed and the example provided refers explicitly to human qualities and talents: "10) figürlich, den Anfang, den Grund einer Sache. Ihr Mißtrauen ist die erste Anlage zum Menschenfeinde. Noch mehr aber, 11) natürliche Geschicklichkeit oder Fähigkeit zu etwas. Sie haben eine nicht gemeine Anlage zur Dichtkunst" [43] [10) figurative, the beginning, the basis of something. Her distrust is the first beginning to (being a) misanthropist. Even more, however, 11) natural skill or talent for something. You have an uncommon talent for poetry]. Although it is not at all clear whether "natürlich" is here understood in biological terms, such an interpretation is certainly prepared.[44]

A similar development towards greater biologization can be observed in encyclopedic works around 1800. In the *Deutsche Enclyclopädie* from 1778, "Anlage" is already introduced as "dispositio":

> Es giebt manche Handlungen, die uns zwar möglich sind, also das Maas unserer Kräfte nicht übersteigen, gleichwohl aber zur wirklichen Ausführung nicht gebracht werden können, es seye dann, daß wir uns vorher gewisse Voraussetzungen zugeleget. . . . Weil man die Möglichkeit etwas zu thun das Vermögen nennet; so ist die bloße Anlage ein entferntes

> Vermögen welches durch die Acquisition der Voraussetzungen in das eigentliche oder nächste Vermögen (facultas proxima) und dieses endlich durch vielfältige Uebung in die Fertigkeit (habitus) verwandelt wird. . . . Ein geringes Nachdenken zeiget, daß der Mensch zu den allermeisten Handlungen, wovon nur sehr wenige ausgenommen sind, nichts als die bloße Anlage von der Natur hat.[45]
>
> [There are some deeds, which we are able to do—thus they do not exceed the measure of our powers—which, however, cannot be brought to real execution, unless we have acquired certain preconditions beforehand. . . . Since one calls the potential to do something capacity, the mere trait (Anlage) is a distanced capacity which is transformed into real or next capacity (facultas proxima) by the acquisition of the preconditions and this finally (is transformed) into skill (habitus) through diverse exercise. . . . A brief reflection shows that for most deeds—excluding only a few—nature provides the human being with nothing else than the trait (Anlage).]

Here, "Anlage" is discussed as potential. This allows an examination of each individual, but also group or even race, whether or not it lives up to its potential, thereby implying the distinction between the potential (disposition) and the actual. However, much as in the *Adelung*, despite the use of "Natur," it is not clear to what extent such disposition is interpreted in terms of what one would later call biology. Nor is it clear to what extent this "Anlage" is specific to individuals. "Anlage" seems to be described as a potential which is common to all human beings and which finds individual expression depending on its use and practice. In a later paragraph, the entry mentions "Anlage" in terms of talent and suggests for this interpretation of the term the modification "vorzügliche Anlage" [particular trait]. Hence, the entry distinguishes between "Anlage" as general ability in all human beings and "Anlage" as a particular quality.

Only forty years later, however, the possibility of a biological interpretation of the term emerges. In the *Allgemeine Encyclopädie der Wissenschaften und Künste* from 1819, the entry on "Anlage" states explicitly that both psychological *and* organic qualities of individuals can be described with the term: "in anthropologischer Hinsicht gebraucht, sowol in Beziehung auf die organische als psychische Natur des Menschen, in sofern die Natur die Grundlage zu etwas in ihm gemacht hat, den Anfang eines Auszuführenden"[46] [used in anthropological respect, both regarding the organic and the psychic nature of the human being, to the extent that nature has provided the basis for something in the human being, the beginning of something which needs to be executed]. While contemporary breeders use "Anlage" already in the context of transfer of traits (see below), "Anlage" is here still considered only as disposition with which each person is equipped independently. This is even more astonishing considering the fact that the anthropological interpretation of the term is introduced in the medical context and, therefore, the proximity to hereditary diseases and other traits is given; nevertheless hereditary dispositions are not mentioned. Similarly, "Anlagen" are seen as something given, not acquired: "Da die Anlagen des Menschen nicht von ihm erst erworben werden, sondern Geschenke der Natur sind und schon zum Grunde liegen müssen, ehe er überhaupt Etwas erwerben kann, so müssen dieselben zu den Gaben, namentlich zu den Naturgaben gerechnet werden"[47] [Since the traits (Anlagen) of the human being are not acquired by him, but gifts from nature and as such have to form a basis already before anything can be acquired, they have, therefore, to be considered gifts, namely gifts from nature].

Although the *Allgemeine Encyclopädie* stays within a framework in which trait is interpreted as potential, it is noteworthy that the definition already limits the individual's options in ways which run counter to earlier promises within the framework of natural law philosophy. Concerning pedagogical consequences, the author of the entry draws the following conclusion:

> Für den Erzieher ist es Pflicht, die vorzüglichsten Anlagen seines Zöglings zu erforschen, deshalb auf die angegebenen Kennzeichen zu achten, und, um sie hervor zu locken, denselben zu mannigfaltigen Arten von Thätigkeit zu veranlassen. Denn es ist wichtig, daß der Mensch, so viel als möglich, dem hauptsächlich sich widme, wozu er die meiste Anlage von der Natur empfangen hat.[48]
>
> [It is the duty of the educator to explore the pupil's most remarkable traits, hence, to pay attention to the respective signs (of these traits), and, in order to trigger them, to encourage him to varied kinds of activities. Because it is important, that the human being devotes himself as much as possible to that to which nature has given him the most potential (Anlage).]

The entry makes visible the changes in the understanding of the task of education and the formation process in the context of a more and more determining understanding of "Anlage." While Klügel's *Enzyclopädie* (1784) explains under the heading "Innate Rights" that "all human beings have the same spiritual and corporeal nature,"[49] in the *Allgemeine Encyclopädie* from 1819 education and formation become extrapolation and training of potentials that determine the individual child's options of development from birth.

In looking at the preformationism-epigenesis debate, it remained difficult to account for heredity because neither framework allowed for transmission of specific qualities (traits) from one individual to the next. In fact, the emergence of the very idea of a "trait" was difficult within these frameworks because they focused on the embryological level of development and, therefore, on a stage in development in which individual traits could not be observed or chemically proven at that point in time. With the transfer of the term "Anlage" from a primarily financial meaning to a biological interpretation it became possible to think the development as a potential that can, but does not have to be actualized. While preformationism also allowed for the development according to environmental stimuli, the idea of "Anlage" suggested that there are "potentials" which are specific to individuals and that these specific traits define a person already at the moment of birth. The relative openness of epigenesis and preformationism which was concerned with finding developmental rules in general was replaced by an interest in how individuals (or specific groups) were differently equipped by nature.

"ANLAGE" IN THE BIOLOGICAL CONTEXT: KANT'S WRITINGS ON RACE

Although the biological interpretation of the term "Anlage" was slow to enter lexicons and encyclopedias, the term already had a distinct biological meaning in Kant's writings on race from the 1770s and 1780s. In fact, it was Kant (1724–1804) who defined the term in ways that became influential for the discourse in natural history. Thirty years after the publication of Kant's texts on race, Friedrich August Carus (1770–1807) still credited the philosopher for having urged and furthered terminological clarity in the discussion.[50] Kant's conceptualization of the term is important, because it illustrates that the ideas of reproduction and heredity around 1800 do not easily fit into the preformationism-epigenesis frameworks and how seemingly outdated frameworks, like preformationism, were successfully revived and redefined to explain reproduction and heredity.

Kant gave his most comprehensive account of race in two essays written in 1775 and 1785. The urgency of the topic might be deduced from the fact that his "Von den verschiedenen Rassen der Menschen (1755)" [Of the different races of humankind] is the only publication during his so-called "silent decade" preceding the *Kritik der Reinen Vernunft*.[51] Ten years later, he wrote "Bestimmung des Begriffs einer Menschenrasse" [Definition of the concept of a human race], which was first published in the *Berliner Monatsschrift* in November 1785. In both essays, Kant discussed the question of race in the larger framework of the monogenesis/polygenesis debate—the question of whether we have to assume one or several beginnings of the human race. The latter, which assumed several local beginnings, had been claimed by Carl Linnaeus (1707–1778), Soemmering (1755–1830) and Georg Forster (1754–1794), and was later defended by August Wilhelm Schlegel (1767–1845); the former approach required a notion of nature that was—within certain limit—able to change over time, an idea that was supported by Kant and Blumenbach.[52]

Kant opened his essay "Von den verschiedenen Rassen der Menschen (1755)" with a reference to Georges Louis Leclerc Buffon (1707–1788). He took over Buffon's definition of "Gattung" [species] as an entity defined by the potential of its members to reproduce fertile offspring with one another.[53] Buffon had defined species as follows:

> [T]here is . . . advantage of knowing the species of animals, and distinguishing them one from another, which is by regarding those as one and the same species, who, by means of copulation, produce and perpetuate beings like themselves; and as a different species, those from a connection between whom nothing is produced, or whose product are unlike their parents. Thus a fox will be a different species from a dog, if nothing results from a copulation of a male and a female of these two animals, and when even there should result a bipartite, or a kind of mule, which cannot generate, that will be sufficient to establish the fox and dog of two different species.[54]

Buffon's definition was in competition with Linnaeus's taxonomic tables which were organized by reproductive organs and visual features (by phenotype, we would say today). Buffon was willing to neglect the criteria of visual similarity and considered only the ability to reproduce, hence the criteria of kinship (within limits, one could call this a definition by genotype). Buffon accomplished two goals: first, with this definition he allowed for the fact that organisms might look different although they belong to the same species *and*, secondly, he provided for a confirmation of the conviction that the number of species is finite and does not change over time. Since Buffon did not insist on the constancy of what we would call today phenotype, he opened a space for speculations on possible changes of this phenotype. Even if Buffon held on to the conviction that the number of species remains constant, he allowed for the possibility that nature has developed over time and, therefore, has history.[55] On the basis of Buffon's definition, Kant concluded that all human races must belong to one human species since they can produce fertile offspring with each other:

> Nach diesem Begriffe gehören alle Menschen auf der weiten Erde zu einer und derselben Naturgattung, weil sie durchgängig mit einander fruchtbare Kinder zeugen, so große Verschiedenheiten auch sonst in ihrer Gestalt mögen angetroffen werden.[56]

> [According to this concept, all human beings on the wide earth belong to one and the same natural species because they consistently beget fertile children with one another, no matter what great differences may otherwise be encountered in their shape.][57]

By further developing Buffon's claim, Kant came to the conclusion that the variety of human races as we know them today must have developed over time from one original couple. But this conclusion confronted him—much as his epigenesis and preformationism contemporaries—with the problem of how exactly to explain the variety of the human races.[58]

In general, Kant followed the line of argumentation that his contemporaries pursued, namely that the differences among human races can be explained with climatic and other local circumstances.[59] Using this climatic framework, it seems at first difficult to develop any idea of biological heredity. However, Kant used an original approach to describe the relationship between climate and organism, by introducing the notion of "Auswickelung" [unfolding, "outfolding"]. The idea has connections to the preformationist point of view, which assumed that seeds for all beings of the earth are present since creation and that they just unfold at different points in time, after having been stimulated into action by certain environmental conditions. Once triggered, the germ develops into the full organism that had been created at the beginning of time. Although the new being (and its descendants) might look different than previous beings, there was no reason for these theoreticians to assume any connection to hereditary transmission. Instead, the assumption was that only now did the environmental conditions for this specific being occur. Kant's idea of "Auswickelung," however, was somewhat different. In contrast to most preformationists, Kant assumed that an organism is composed of *a number of different germs* which react to different outside stimuli.[60] As a consequence, one organism (i.e., one species) can have different appearances although the set of germs with which it is equipped is always the same. In a way, Kant's concept of "Auswickelung" transferred what the preformationists saw on the level of the individual organism to the level of the "germs" inside of an organism. Kant's notion of "Auswickelung" and the preformationist's notion of development are the same in the sense that they describe the development of something reacting to a specific outside stimuli. However, Kant's approach allows for variations within a species because different germs are realized within each race of one species.

To illustrate his claim, Kant introduced two concepts in connection with "Auswickelung": "Keim" [germ] and "Anlage." Kant defined "Keim" as that "Auswickelung" that has to do with specific parts of the body. As an example, he described birds that have "Keime" for an extra layer of feathers. Depending on the climate in which they are living they develop these "Keime" or they do not. He defined "Anlage" as that "Auswickelung" that has to do with the proportion of different parts of an organism. As an example he used a wheat germ, which has an "Anlage" for different kinds of thickness of its skin. Depending on the climate, its skin will be thicker or thinner. Kant drew the conclusion that the variety of forms in nature is too often defined as new species. Observed carefully, one realizes that significant differences within one species can be explained with adaptation to the specific environment, namely the development of "Keime" and "Anlage" to different degrees.[61]

There is one further feature which distinguished Kant's idea of the interplay between environment and body, which he considered in his discussion of race, from that of the preformationists. While Kant assumed that "Anlage" and "Keime" react to contingent factors and as such are potentially reversible, he also acknowledged that there is a point where certain features are maintained even when the initial stimulating factor ceases to exist. In this case, Kant defined a race as "völlig eingeartet" [completely specified, fully developed]. If an individual of such a fully developed race is transferred to another environment, skin color and other features will not change any longer and the quality is transmitted to the next generation. With the establishment of a theory which combines environmental influences and hereditary features without giving up the idea of reversibility of these features and without giving up the stability of species (see the reproduction criteria), Kant had formulated a pattern for conceptualizing heredity which was influential well into the nineteenth century. In *Ideen zur Geschichte der Menschheit*, Carus used the same conceptual framework when he defined race: "Die Form ist also nicht angeboren, wenigstens nicht dem ersten Menschen, vielmehr nur durch allmälige Vererbung entstandene Erbeigenthümlichkeit der sich gleich bleibenden Merkmale der Jungen"[62] [Hence, the form is not innate. At least not with the first human being; rather, it is an inherited quality, obtained by means of progressive inheritance, of features which stay the same in the offspring].

Kant offered with his notion of "Anlage" and "Keim" innovative ways to think about the mechanics of heredity at a time when units like nucleus and chromosome were not yet known. While contemporary reference works used "Anlage" generically without a clear location of the potential, Kant defined the term biologically. By locating the potential for development in multiple units smaller than the organism, Kant was able to guarantee both malleability and species stability. While the units remain species specific and, therefore, ultimately do not allow an evolutionary development, the idea of small units which might be expressed or dormant shares interesting features with the current model of gene expression. Although Kant's model is based on potential reversibility of features, his idea of "völlig eingeartet" allows that environment-bound qualities can be inherited by the next generation.

ANIMAL BREEDING: FROM CLIMATE TO SEED

At the same time as Kant's theoretical inquiries into human origin, a new arena of experimental genealogical research opened up: animal breeding. Kant developed his thoughts on "Anlage" in the context of his writings on human race, but the same crucial question, whether or not breeding results depended on climate or seed, was equally important for animal husbandry. The question was not only of scientific interest, but it had also important economic implications. Assuming that environmental factors are decisive for the modification of livestock, seventeenth-century breeding treatises had paid most attention to how to house and feed the animal, not how to select the best animals for breeding by means of reproduction. Typical chapter headings in early modern manuals like Winter's *Gründliche und kurtz durchgehende Beschreibung, wie eine Stutherey anzurichten* (1670) are: "Wie der Ort soll beschaffen seyn, da die Stutherey soll gehalten, und der Hof hin gebauet werden" [How the place should be, in which the stud farm is supposed to be established and where the farm is supposed to be built], "Von dem Trächtigen StuthenStall, und was darinn zu *observirn*" [Of stables for pregnant mares and what to *observe* therein], "Vom rauhen Futter, als Hew und Strew" [Of rough fodder, like hay and straw], etc.[63] Even Joachim Christian Bergen's *Anleitung für die Landwirthe zur Verbesserung der Viehzucht* (1781), published more than a hundred years later, focused still entirely on feeding and by doing so illustrates how quickly this paradigm of breeding shifted towards a new one defined in the context of heredity in the last third of the eighteenth century.[64]

What the polyp had been for the question of epigenesis vs. preformationism was the Merino sheep concerning breeding and hereditary transmission of traits. While it was known that Spanish Merino sheep produced one of the finest wools in Europe, attempts to raise Merino in other European countries had mostly failed, despite great efforts to provide them with proper food and housing. In fact, the exclusive focus on these conditions probably caused the failure of these early breeding attempts. Only once a new breeding paradigm was followed—which the observations of Merino breeds helped to shape—could the Merino also thrive in other countries.

The Merino sheep had been introduced into Spain by the Moors in the twelve century and its wool was soon considered of superior quality. There were strict regulations imposed by Spain on the export of Merino until the eighteenth century, when sheep breeding became a focus of the agricultural industry in many European states. As one German author summarized in 1807: "Veredelung der Wolle durch spanische Schafe gehört seit dem Schlusse des achtzehnten Jahrhunderts zu den Lieblingsgesichtspunkten unsrer neubelebten landwirtschaftlichen Industrie" [refinement of wool by means of Spanish sheep has been amongst the strongest interests of our revived agricultural industry since the end of the eighteenth century].[65] When the first Merino sheep were exported, it was asumed that the sheep would degenerate in another climate, and, therefore the export of Merino sheep would be no danger to Spain's success as a wool exporting nation.[66] By the 1780s extensive research on sheep breeding quickly changed this opinion. The international interest in the topic was great. The 1782 treatise on sheep breeding by Louis-Jean-Marie Daubenton (1716–1800)—Buffon's collaborateur who had introduced the Merino in France—was translated, adapted, and published already two years later for German readers in the form of August Christian Wichmann's Katechismus der *Schaafzucht* (1784) (see fig. 1.3).

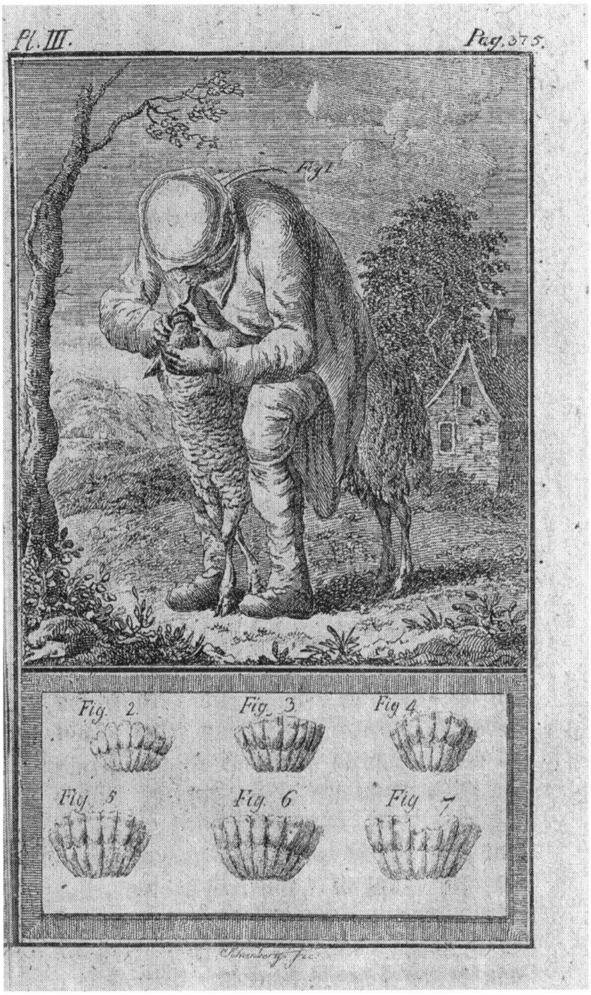

fig. 1.3. Sheep teeth. Herzog August Bibliothek Wolfenbüttel: O 522.8° Helmst. Louis Jean Marie Daubenton. Katechismus der Schaafzucht . . . bearbeitet und herausgegeben von M. Christian August Wichmann *(Catechism of sheep breeding . . . revised and edited by M. C. A. Wichmann). Leipzig, Dessau: Breitkopf, 1784.*

How deeply intertwined observations in sheep breeding were with scholarly deliberations on the monogenesis/polygenesis question, and how results obtained in sheep breeding were connected to philosophical and political questions, can be seen when we consider the fact that Kant—in a text published around the time of Wichmann's treatise on sheepbreeding—used the example of the Spanish shepherd and sheep to illustrate his support of the climate theory in the debate on human races:

> Der Mohr (Mauretanier), der, in seinem Vaterlande von Luft und Sonne braun gebrannt, sich von dem Deutschen oder Schweden durch die Hautfarbe so sehr unterscheidet, und der französische oder englische Kreole in Westindien, welcher, wie von einer Krankheit kaum wieder genesen, bleich und erschöpft aussieht, können um deswillen eben so wenig zu verschiedenen Klassen der Menschengattung gezählt werden, als der spanische Bauer von la Mancha, der schwarz, wie ein Schulmeister, gekleidet einher geht, weil die Schafe seiner Provinz durchgehends schwarze Wolle haben. Denn, wenn der Mohr in Zimmern, und der Kreole in Europa aufgewachsen ist, so sind beide von den Bewohnern unsers Weltteils nicht zu unterscheiden.[67]

> [The Moor (Mauritanian), who is burned brown by air and sun in his native country and who is so much distinguished from the German or Swede by the skin color, and the French or English Creole in the West Indies, who looks pallied and exhausted, as though scarcely recovered from an illness, can be as little counted among different classes of human species because of this, as can the Spanish peasant form La Mancha, who goes around dressed in black like a schoolmaster because the sheep in the province all have black wool. For if the Moor has grown up in rooms and the Creole in Europe, they are both indistinguishable from the inhabitants of our part of the world.][68]

Comparing skin color with clothes and their common dependency on regional circumstances, Kant argued that there is only one human race.[69] Kant's comment on (Spanish) sheep, the peasant's clothes, and (the Moor's) skin rests on the assumption that the appearance of both will change under different climatic conditions. Interestingly, however, the sheep to which he referred were about to demonstrate the opposite. More and more sheep breeders saw that they were able to maintain the features of the Merino sheep in places like Saxony or even as far away as Sweden (both regions became centers for sheep breeding). The fact that many breeds kept their features despite the geographical displacement led to doubts concerning the climate theory. As Wood and Orel summarize "[t]he evidence pointed to a link between parents and offspring independent of climatic or other natural influences, a powerful connection between generations. This was the essence of a breed (race), its 'blood,' which would resist a change of climate or soil."[70]

The first questioning of the climate theory came not from scholars of Natural History, but from practitioners in the field and authors of breeding manuals. Already in 1790, only five years after Kant's publication on race and its promotion of a modified climate theory and nine years after Bergen's *Anleitung für die Landwirthe zur Verbesserung der Viehzucht*, Johann Heinrich Fincke opened his treatise, *Schreiben über ungekünstelte und sichere Verfeinerung aller groben Wolle aus eigener Erfahrung und Thathandlungen aufgestellt*, with the rejection of the climate theory as superstition ("alte Sage"):

> Es ist eine bekannte und alte Sage der meisten Schäfer, auch vieler Hauswirthe und Pächter, die keine wirkliche und richtige Erfahrungen vor sich gehabt haben: daß die Feinheit oder Qualität der Wolle, von der Weide und von dem Himmelsstriche abhänge; daß es nicht möglich sey, daß eine Gegend, die jetzto haarige, spröde Wolle hervorbringt, eben so feine Wolle auf ihrer Weide und Futterung erziehen könne, als eine viele hundert Meilen entfernte Landeschaft; ja nicht einmal so gute Wolle, als eine etwa zehn bis zwanzig Meilen entlegene Gegend, die im Besitz von guter Wolle ist. . . . Geschieht an einem Orte eine würkliche Veränderung, nämlich Verschlimmerung der Wolle in der Nachkommenschaft einer feinen Heerde Schaafe; so rührt solches von schlechter oder unvorsichtiger Auswal der Zuchtböcke her, nicht von der Weide oder Weltgegend.[71]

> [It is a well-known and old legend of most shepherds, and also of many landlords and tenant farmers, who don't have real and correct experiences: that the fineness or quality of wool is dependent on meadow and region; that it is not possible that a region which produces now hairy brittle wool on its meadows and food, can produce fine wool as well as a landscape which is many hundred miles away; nay, not even so good wool as a region only ten or twenty miles away which is in the possession of good wool. . . . If in a place a real change happens, namely worsening of the wool in the descendant of a fine herd of sheep; then this originates from the bad or careless selection of breeding rams, not from the meadow or region.]

The question of whether climate or seed are the decisive factor remained hotly debated in the following decades—and, considering new epigenetics research, to a certain extent until today—but in 1833, Röttger von Veltheim (1781–1848) at least came to the apodictically articulated conclusion: "Also der Samen, und nicht das Klima constituirt die Raçen" [Hence, the germ, and not the climate constitutes the races].[72] From now on, the tendency was to see climate as a modifying factor, but not a constitutive one.

FROM CROSSING TO INBREEDING (DEGENERATION)

The best example for the rapid and revolutionary developments within animal husbandry and its effects on the conceptualization of heredity is probably Robert Bakewell (1725–1795) of Dishley in Leicestershire, one of the most active and internationally recognized breeders of the eighteenth century. As for many other breeders of the time, the main focus of Bakewell's breeding efforts was the improvement of sheep stock.[73] Noting the importance of parentage for the quality of the offspring, Bakewell started to use inbreeding as his main approach in the attempt to perfect his animals. By doing so, Bakewell challenged old beliefs in animal breeding and probably even older cultural norms, most importantly the incest taboo. Possibly under the influence of such taboos, older breeding manuals stressed the importance of crossing in order to obtain good breeding results. In 1724, less than fifty years before Bakewell's most active period, Johanne Ernesto Francisco Kraus still stated:

> Daß, von Anbeginn des erschaffenen Welt-Crayses, und von GOtt darüber gesetzten menschlichen Geschlechtes, nachdem die menschliche *Socie*tät sich vermehret oder erweiteret hat, . . . viel Fleiß und Nachdenckens, wie eine Landes-Art mit der anderen vermenget, und dadurch etwas beser-*formir*tes, als sonsten Land-gewöhnlich, hervor gebracht werden möge . . .[74]
>
> [That, from the beginning of the created world-circle and (from the beginning) of the human race, which was put by God to dominate it, there was also a lot of diligence and thinking—after the human society has multiplied or scattered . . . (about the question of how) a race of a specific country could be bred with another and how something better could be produced than usually common in this region . . .]

Such statements are not only found in breeding manuals, but are also expressed in what can be considered the most advanced scientific discussions of the time. Buffon, for example, expressed "the necessity of always crossing, or mixing the breed of horses to prevent their degenerating."[75] The basis of Buffon's statement was the idea that all forms in nature are variations of prototypes (an idea which would become influential for morphology in the German tradition):

> There is in nature a general prototype of each species, from which each individual is modeled . . . the model of what is handsome and excellent is dispersed through all parts of the earth, and that in each climate there is a portion thereof, which perpetually degenerates, unless united with another portion taken from a distant country; . . . Without this being attended to, corn, flowers, and animals, will degenerate, or rather take so strong a tincture of the climate as to deform and bastardize the species; the form remains, but disfigured in all the lines which are not essential thereto; by mixing on the contrary, the kinds, and, above all, by crossing their breed with foreign species, their forms seem to become more perfect.[76]

Since the perfect expression of this prototype is distributed throughout all variations of this prototype, crossing alone can guarantee that the prototype is expressed most ideally. Although this approach includes the possibility of perfection, the underlying assumption is that the process of perfection develops towards a form which is theoretically conceivable at any moment. Development in a specific direction of the spectrum of possibilities can be considered as degeneration because it limits the number of expressed possibilities. The ideal and most perfect form is an average of *all* possible forms. For Buffon, as it is clearly expressed here, inbreeding needs to be avoided in order to obtain the most perfect form.[77] Cross-breeding was a way of perfection on the way towards regaining that original, most perfect and most *encompassing* form. This is an important difference to selective breeding which started to be practiced in animal husbandry at the same time and which aimed to highlight *specific* features.

Ignoring available theoretical frameworks and cultural norms, Bakewell departed from the call for crossing. As Wood and Orel observe, among breeders like Bakewell, "the drive to succeed encouraged indifference to certain pressures of convention, social, religious or even intellectual."[78] Bakewell's desired outcome was not the perfected prototype, but specific qualities which were economically desirable. Using inbreeding as the preferred approach, Bakewell was able to significantly alter the stock and to bring out specific desired features, in particular larger carcass, combined with thin legs and small heads.

Although Bakewell himself published little, his work was known well beyond the borders of England. His breeding successes were publicized and published by his disciple George Culley (1735–1813) and his farm was frequently visited by national and international visitors. Hans Moritz von Brühl (1736–1809), ambassador of the Elector of Saxony to the Court of Great Britain (and a close friend of the poet Christian Gellert), for example, inquired of Bakewell in 1787 about the best methods of breeding of livestock. Brühl's interest is not astonishing, considering that Saxony became a center in the eighteenth century for sheep breeding, which eventually brought forward the highly successful Saxon Merino, still bred today in Australia.[79] News of Bakewell's achievements reached Germany fairly quickly also via a number of other publications. Arthur Young's *The Farmer's Tour through East of England* (1771), which reported on the "modern system of breeding," was translated into German already in 1775 and the second edition of George Culley's *Observations on Livestock* (1794), was translated into German in 1804. Besides a number of reports on Bakewell's methods which had also appeared in the *Annals of Agriculture* and which were translated into German, the most important transfer of Bakewell's knowledge for German audiences were probably the publications of Albrecht Thaer (1752–1828), agrarian and physician to King George III, who published extensively on English agriculture and its potential benefits for German farmers and whose work was influential for Goethe's discussion of agriculture in the *Wanderjahre*.[80] In his *Einleitung zur Kenntnis der englischen Landwirthschaft* (1798), Thaer reported on Bakewell's method of inbreeding to achieve desired traits in sheep and cattle:

> *Bakewel [sic] zu Dishley in Leicestershire*, ein reicher Pachter, war der erste, der auf folgende Grundsätze in der Viehzucht verfiel: Er behauptete, alles Kreutzen tauge nichts; man müsse by allen Thieren jede Race nur in sich selbst veredlen, die vollkommensten Individuen einer Race zur Zuzucht auswählen und die Vorzüge eines Stücks mit denen eines andern, aber von derselben Race, zusammen zu setzen suchen. . . . Diese Grundsätze haben sich unter dem Nahmen [sic] des Backewellschen [sic] Systems, oder *System of breeding in and in*, berühmt gemacht.[81]

> [*Bakewel (sic) from Dishley in Leicestershire*, a rich tenant farmer, was the first who recognized the following principles of breeding: he claimed that

> all crossing is useless; (he claimed instead that) one has to improve each animal race only within itself, (that one has) to select the most perfect individuals of a race for breeding and to combine the advantages of one piece with the ones of another, but from the same race . . . These principles have become famous under the name Backewell's (sic) system, or *system of breeding in and in.*]

Thaer's report demonstrates not only that Bakewell's methods were conceived as innovative and closely connected to his person, but that they challenged old taboos:

> Ich habe schon mehrmals des Vorurtheils gegen Verwandtschaft erwähnt, wovon manche eine besondere Degeneration und Schwächung besorgten. Es ist nun aber durch die Erfahrung der Engländer und auch mancher Deutschen hinlänglich widerlegt. Wenn die Fehler in einer Familie sind, so pflanzen sie sich freilich durch solche Paarungen um so mehr fort, und werden größer. Aber eben so die Vollkommenheiten, und es giebt kein sichreres Mittel, als einen Stamm, worin sie forterben, in sich selbst fortzupflanzen.[82]
>
> [I have already mentioned several times the prejudice regarding kinship, of which some fear a particular degeneration or weakening. However, it is now sufficiently refuted by the experiences of the English and some Germans. If defects are in a family, then of course they are even more transmitted by means of mating, and become greater. But by the same measure perfections (become greater), and there is no safer means (to achieve perfection) as inbreeding in a stem, in which they (the perfections) have become hereditary.]

Anticipating criticism, Thaer rejected the incest taboo in animal breeding[83] by adding a remark which rationalized possible defects as the result of an accumulation of degenerative traits.[84]

Thus, Bakewell's new breeding methods had decisive impact on the formulation of rules of heredity and simultaneously challenged the religious and cultural taboos against incest. The acceptance of Bakewell's breeding methods required the separation of religious and biological questions—a step which many were not yet willing to take. (For a more detailed discussion on the question of inbreeding and the distinction between biology and ethics as well as Goethe's position in the debate see chapter 2.)

STABLE NATURE NEVERTHELESS

Despite the German fascination with Bakewell's breeding methods and the plasticity of livestock which he demonstrated—including the possibilities for human intervention (and achievement)—there remained confusion about the degree to which Bakewell's work was an artificial modification of nature, and as such reversible, and to what extent the product of his breeding efforts was nature herself. The anxiety that breeders could permanently alter what many considered God's creation was great. Anticipating twenty-first-century debates on genetic engineering, the explosive question for eighteenth-century authors was whether or not an "original" nature could be separated—theoretical and practical—from one which had been shaped by human intervention. One of the German authors commenting on Bakewell's methods and one to whom Goethe referred explicitly when considering questions of heredity was Karl Christian Gottlieb Sturm (1781–1826). Sturm was an animal breeder from Hohenleuben in Thuringia. He led the agricultural institute in Tiefurt (Weimar), which had been established by Duke Carl August following Sturm's initiative, taught since 1807 in Jena, and from 1819 onwards was a professor for "Cameralwissenschaften" [state finance and public administration] in Bonn.[85] On Bakewell's breeding experiments, Sturm commented:

> Merkwürdig ist in dieser Hinsicht die in England bekannte Rindviehraçe von Backwell [sic]. . . . Eigentlich ist dieß aber nur eine Kreuzung, keine eigentliche Raçe, oder vielmehr ein durch Kunst mit Inzucht gebildeter Schlag. Dieses Beispiel von Backwell [sic] zeigt in der That, wie sehr der Landwirth die Natur zu seinen Absichten benutzen kann.[86]
>
> [Strange in this respect is Backwell's (sic) cattle race. . . . But actually this is only a cross-breed, not a real race, or rather it is a variety which is produced artificially (literally "by art") via inbreeding. This example from Backwell demonstrates indeed, how the farmer can use nature to pursue his own goals.]

In the wake of breeding results such as those provided by Bakewell, we can observe in Germany an attempt to separate "true" nature from a nature which is considered to be artificially ("by art") modified, in particular by means of inbreeding. This separation coincided with the conviction that these two notions of nature can and should be distinguished.

How difficult it is to assess eighteenth-century descriptions and definitions of nature becomes clear when we examine Sturm's consideration of nature's productivity. He was convinced that there is an infinite number of examples that show how "die immerfortwährende Natur noch gegenwärtig beschäftiget ist, neue Varietäten und Raçen unter den Pflanzen und Thieren zu erzeugen"[87] [eternal nature is today still active to create new varieties and races in plants and animals]. Read in isolation, the statement could suggest that Sturm supported an evolutionary concept of nature in which an infinite and unpredictable number of forms are possible. He also seems to assume that nature is the agent of this development in a way that is not compatible any longer with the deterministic definition of nature provided by the Enlightenment. However, when read within the context of the entire treatise, it becomes evident that Sturm considered nature's creativity within clearly defined limits. He was convinced that nature varies infinitely without, however, changing the number of species. Sturm's aversion and possibly fear in light of an abundance of new forms obtained by breeding experiments is perhaps best visible in his introduction of terms such as "künstlich" [artificial], "Verkünstelung" [to make something artificial], and "künstliche Umbildungen" [artificial transformation], which he attributed to nature that has been changed by selective breeding.[88] Similarly, he defined Bakewell's breeding results as "durch Kunst mit Inzucht" [by means of inbreeding artificially] obtained and denied Bakewell the ability to create a new race.[89]

The preference for nature over artifice, which had shaped many eighteenth-century discussions, had been based on the conviction that the distinction between nature and artifice could not be collapsed; nature was defined as that which is *not* artifice or fashion. This distinction still informs Sturm's 1825 attempt to come to terms with nature's new forms. To both address his uncertainty and account for new forms, Sturm established a new distinction within the concept of nature. He distinguished a stable nature from that nature which has been manipulated by human intervention and which he described with vocabulary taken from the realm of art and artifice.[90] (For further discussion of the question of a nature "durch Kunst" and its problematization within the context of plant breeding see chapter 3.)

While breeders already put their knowledge about the transformation of one race into the other into practice, theoreticians were more skeptical, or, more bound to traditional frameworks concerning the significance and permanence of such changes. For the most part, they denied that there is change via breeding efforts beyond mere variation of an underlying permanent form. This is perhaps best expressed and illustrated by Sturm's *Gesetz des Zahnwechsels* [Law of Dentition] and Goethe's reaction to it. Sturm assumed that when two varieties of one race were crossbred, the features transmitted from the father would supersede the features from the mother within a few generations. Sturm formulated the following "Grundsatz" [theorem]:

> Die Umänderung des Typus oder auch die Hervorbringung eines neuen geht nach dem Gesetze des Zahnwechsels in der Regel vor sich, fängt vom Kopfe an und erstreckt sich von da aus immer weiter nach hinten. . . . Jeder Zahnwechsel bezeichnet eine neue Periode des Lebens und das Thier ist erst vollendet, wenn die Zähne vollkommen entwickelt sind. So viele Jahre aber erforderlich sind, um die Zähne vollkommen ausgebildet zu nennen, so viele Generationen sind nothwendig um eine Raçe in eine andere durch den Vater umzuändern, oder eine neue constante hervorzubringen. Dieß Gesetz darf ich als meine eigene Entdeckung ansehen, und es trift bei allen Thierarten sehr genau zu.[91]

> [The transformation of a type or the creation of a new one usually takes place according to the Law of Dentition; it starts at the head and extends from there further and further backwards. . . . Each dentition marks a new period of life and the animal is complete when the teeth are completely developed. The number of years necessary to fully develop the teeth equals the number of generations necessary to transform a race into another via the father, or to create a new and stable one. I can call this law my own discovery and it can be applied very precisely to all animals.]

Sturm proudly presents us with a law that he considers a discovery of his own. He claimed that the number of generations necessary to transform one race into another equals the number of years that it takes for the formation of teeth in this race. The transformation of one race into another in the number of years corresponding to the development of teeth might seem esoteric today, however, even Goethe took notice of Sturm's law and commented on it.[92] When Goethe, in 1824, considered the question of race and heredity, he referred to Sturm:

> Doktor Sturm in den Beiträgen der deutschen Landwirtschaft, spricht mit Überzeugung das Naturgesetz aus: 'daß sich eine Race durch die andere verändern lasse und dann konstant bestehe, nach dem Gesetz

> des Zahnwechsels.' Wir halten dies für ein höchst bedeutendes Enunziat das nach vielen Seiten des Organismus hinweist, freilich muß die Umwandlung eine Grenze haben und nur die Vollkommenheit des Geschöpf kann sie bestimmen.[93]

> [Doctor Sturm, in the "Beiträgen der deutschen Landwirtschaft" (Contribution to the German Agriculture), expresses with conviction the following natural law: "that one race can be transformed into another and then stays stable, according to the Law of Dentition." We consider this to be a highly significant statement, which points to many sides of the organism; of course, the transformation must have a limit and only the perfection of the creature can determine it.]

Whether or not Goethe can be considered as having anticipated evolutionary thinking has been a question of debate for several decades. A number of scholars have argued recently that Goethe developed the idea of a limitless permeability of nature (Robert J. Richards) and even supported the idea of a will-driven nature (Astrida Orle Tantillo).[94] Goethe's comment on Sturm's law, however, complicates the issue. It suggests that Goethe indeed was fascinated by the powers of transformation and was willing to see them working on a grand scale, but also that he—at least in his last decade—hesitated to embrace an evolutionary concept of nature. For Goethe, the transformation of a race has its limits. Even more, Goethe stated this limit apodictically: the transformation *must* have a limit. Goethe assessed the limit as defined by the "Vollkommenheit" [perfection, full development], which a specific creature can achieve. This "Vollkommenheit" is a vague criterion, since it is hard to identify before it has been achieved, but it indicates that Goethe thought of nature as teleological. Hence, even though Goethe developed a notion of limitless transformability, he did not think about a mechanism of selection or adaptation, nor about an intentional movement; instead he imagined the necessity of an external force to limit these tendencies. In this sense, I think it would be misplaced to call Goethe an evolutionary thinker. Nevertheless, as I will indicate below, Goethe shared another idea with Darwin and proto-evolutionary thinkers: the economy of life force. In fact, it is Charles Darwin who—although only partly agreeing with Goethe—traces the "law of compensation or balancement" to Goethe and Geoffrey.[95]

As Goethe's reaction to the suggestion of species transformation shows, eighteenth-century thinkers in Germany remain safely within the idea of a stable nature. Following Sturm, Goethe ultimately held on to types and their variation and excluded—much like Buffon—the possibility of species evolution. While the degree to which German authors around 1800 accepted modifications of natural form is astonishing, their insistence on a limited and unchangeable number of species or types is equally surprising. When they thought in terms of progressive development, they considered nature's development as the realization of an already existing basic plan.[96]

The insistence on species stability is based on three (at times competing) principles promoted in various degrees by German thinkers. The first principle is taken over by Kant from Buffon and, as discussed above, demands that all members of one species can produce fertile offspring among each other. This *principle of the production of fertile offspring* allows a wide range of modifications without having to give up the stability of the total number of species, because it excludes the possibility of species transition which would suggest the potential for species evolution. Although there were rising doubts regarding this principle in the context of hybridization experiments, the mule, being unable to produce fertile offspring among each other, was considered the case study for the validity of the principle.[97]

The second principle assumes that there are a number of animal "types" which are stable, but can be modified according to *a model of equilibrium or economical distribution.* The assumption is that the higher development of one organ within an organism goes hand in hand with the lesser development of another organ in the same organism. According to this principle of equilibrium, the total number of modifiable organs remains the same within one type. Despite great modifiability, however, there is no transfer from one type to another. This principle is nicely illustrated in Karl Friedrich Kielmeyer's (1765–1844) *Verhältnisse* from 1793. Here, Kielmeyer explained:

> [Es] kann nun bei nochmaligem Ueberblik der Organisation nicht entgehen, daß gerade da, wo ein Sinnorgan verloren geht, und also die Mannigkfaltigkeit der Empfindung vermindert wird, nur ein freierer Raum für eines der übrigen gewonnen wird, und da wo eines weniger entwikelt dasteht, das andere desto ausgearbeiteter erscheine.[98]

> [When surveying organizations (of life forms) once again, one cannot fail to notice that precisely there where one sense organ is lost—and, therefore, the variety of sense impressions is diminished—a freer space

> for one of the other (sense organs) is won, and there where one is less developed, the other seems more elaborate.]

By making the development of one organ dependent on the lesser development of another, Kielmeyer introduces a life force that is bound to a space but, within this space, plays freely. In his *Einleitung in die Vergleichende Anatomie* two years later, Goethe formulated the same idea in terms of economics:

> Hier sind die Schranken der tierischen Natur, in welchen sich die bildende Kraft auf die wunderbarste und beinahe auf die willkürlichste Weise zu bewegen scheint, ohne daß sie im mindestens fähig wäre den Kreis zu durchbrechen oder ihn zu überspringen. Der Bildungstrieb ist hier in einem zwar beschränkten, aber doch wohleingerichteten Reiche zum Beherrscher gesetzt. Die Rubriken seines Etats, in welche sein Aufwand zu verteilen ist, sind ihm vorgeschrieben, was er auf jedes wenden will, steht ihm, bis auf einen gewissen Grad, frei. Will er der einen mehr zuwenden, so ist er nicht ganz gehindert, allein er ist genötigt an einer andern sogleich etwas fehlen zu lassen; und so kann die Natur sich niemals verschulden, oder wohl gar bankrutt werden.[99]

> [These are the bounds of animal nature; within these bounds the formative force seems to act in the most wonderful, almost capricious way, but is never able to break out of the circle or leap over it. The formative impulse is given hegemony over a limited but well-supplied kingdom. Governing principles have been laid down for the realm where this impulse will distribute its riches, but to a certain extent it is free to give to each what it will. If it wants to let one have more, it may do so, but not without taking from another. Thus nature can never fall into debt, much less go bankrupt.][100]

While acknowledging freedom to a certain extent, Goethe stresses that both the "budget" and the "rubrics" in which it is invested are prescribed. In a later passage of the text, he called the principle according to which nature distributes her powers in the described limits the "Idee eines haushälterischen Gebens und Nehmens" [idea of a economical give and take].[101] The economy to which he referred is not the one of speculation, but the one of the oikos in which a central force governs within limits.[102]

The third principle is a derivation of the first. This *genealogical principle* defines the relationship between different modifications within one species as one of descent. In other words, it is assumed that all members of one species are descendants from one origin, one species couple or one progenitor. The common origin connects variations with each other and by doing so distinguishes them from descendants of other origins. Unity is created not by common features, but by common origin.

ONE FATHER ORIGIN: QUESTIONS OF GENDER

The question which arises here is how this origin is to be imagined. While Kant had deduced the variation of human races from one couple, we can observe around 1800 an attempt to derive species modifications not only from one couple, but from one individual.[103] This individual is defined as male. In his writings on breeding, Sturm provided an example for this tendency: "Offenbar ist die Meinung die richtige: daß alle Varietäten und Raçen ursprünglich nur einem Stammvater angehören"[104] [apparently the view is correct: that all varieties and races originally belong to one progenitor (Stammvater) alone]. As illustrated in Sturm's breeding manual from 1825, the male ancestor is considered origin and medium of transfer of features along the genealogical line. This focus on one individual and one male is at the same time a continuation of traditional conceptualizations of genealogy as it is a break from early eighteenth-century assessments of reproduction.

To express the genealogical relationship, writers often referred to the stem–branch metaphor. A comparison with the tree—similar to Johann Gottfried Herder's use of the tree metaphor in his writings on the genealogy of history—is one way to account for this vocabulary.[105] An additional possibility opens when we consider contemporary research on the polyp (freshwater hydra). While the early eighteenth century was fascinated in general by the regenerative abilities of the hydra, it is especially interesting to see the relationship between generations described as stem and branch. The terminology was certainly triggered by the visualization of laboratory experiments with the freshwater hydra (see fig. 1.2). The freshwater hydra's amazing ability to regenerate had a female connotation at first. The polyp was seen as a "stem-mother" of many children who extended like branches from the mother's body and became independent by separating their bud-like bodies from her stem. Over the course of the eighteenth century, the interpretation of the freshwater hydra slowly changed, with the result that at the end of the eighteenth century, the freshwater hydra was considered a male organism. Experiments with the freshwater hydra contributed decisively to a new conceptualization of reproduction and heredity and thereby shaped the role attributed to gender in the hereditary process.

In his *Mémoires pour servir à l'histoire d'un genre de polypes d'eau douce* (1744), Abraham Trembley (1710–1784) had depicted the polyp as a female being, despite the fact that in French the animal is grammatically masculine:

> L'orsqu'un jeune Polype commence à pousser, on ne voit d'abord qu'une petite excrescence, qui ordinairement se termine en pointe. . . . La couleur de cette excrescence, de ce petit bouton, est d'ordinaire plus foncée que celle du corps de la *mere* [sic]. . . . Ce jeune Polype ne conserve pas long-tems [sic] la figure d'un cylindre, son bout postéreur, par lequel il tient à sa *mere* [sic], s'étrécit peu à peu, il s'étrangle, & enfin il ne paroit la toucher que par un point. Le jeune Polype, qui, dans ses commencements, était beaucoup plus large a son bout postérieur, n'est nulle part si mince, après qu'il est formé.
>
> . . .
>
> Ces Observations m'ont appris, qu'un jeune Polype peut, quatre ou cinq jours après qu'il a commencé à sortir de sa *mere* [sic], avoir lui-même un jeune qui commence à pousser, c'est-à-dire, qu'on peut voir tous ces quatre ou cinq jours une nouvelle génération de Polypes. . . . J'ai supposé dans ce que je viens de dire, que tous les Polypes produisoient des petits, qu'ils étoient tous *meres*; & c'est ce que j'aurai bientôt occasion de prouver.[106]
>
> [When a young polyp starts to grow, one does not see but a small growth, which usually ends in a tip. . . . The color of this growth, of this small button, is usually darker than the body of the *mother*. . . . The young polyp does not keep for long the shape of a cylinder, its lower end, through which it is connected to its *mother*, it slowly shrivels and strangles itself, and eventually it touches only at a point. The young polyp who in its beginning was much larger at its lower end, is nowhere smaller after it is formed.
>
> . . .
>
> These observations taught me that a young polyp can, four or five days after it began to separate from its *mother*, have itself a young one which starts to grow, that is, that we can see each four or five days a new generation of polyps. . . . I have assumed in what I have said that

> all polyps produce small ones which all become *mothers*; and that is what I will have the opportunity to prove soon.]

In the course of more than twenty pages, Trembley described the relationship between mother-polyp and child-polyp. The question of how the child can be masculine ("un jeune Polype") although the parent-polyp is feminine and the birth process is thought of as a process of branching is not answered. It seems as if the birthing being cannot only be thought of as mother, while the born being can only be conceptualized as son.

Trembley's experiments were copied throughout Europe. One of his followers in Germany was Jacob Christian Schäffer (1718–1790). In his *Die Armpolypen in den süßen Wassern um Regensburg* [The Hydra in the Sweet Waters around Regensburg] (1754), he claims to describe his experiments and observations, but in many parts the treatise reads like a translation of Trembley's work.[107] What is astonishing, however, is the stress on the young animal's female sex:

> Wann *eine junge Polype* anfänget auf der Oberfläche des Leibes *ihrer Mutter* sichtbar zu werden, so hat *sie* zuerst das Ansehen eines kleinen fast runden Hügelgens, oder Knöpfgens. . . . Ist *die junge Polype* so weit herangewachsen, so bleibet *sie* einige Zeit ohne große Veränderung an der *Mutter*. *Sie* fänget mit ihren Armen Würmer, wie die *Mutter*, und ziehet sie wie jene in sich. Ist es *ihr* aber nicht mehr angenehm mit *ihrer Mutter* in Verbindung zu stehen, oder ist auch die *Mutter* der Jungen überdrüssig; so sondern sie sich von einander ab.[108]

> [When a young polyp starts to become visible on the surface of her mother, then she has first the appearance of a small, almost round hill, or of a little button. . . . When the young polyp has grown, she stays for a while without great change connected to the mother. She catches worms with her arms, as the mother, and pulls them inside like the mother. However, when she does not like to stay in contact with her mother any longer, or when the mother is tired of the young ones; then they separate from each other.]

In Schäffer's description, both parent and child are marked as female. The animal's development is anthropomorphized by the emotional investment which shapes the mother-daughter relationship. The polyp's separation from the mother (and vice versa) is psychologically motivated and reads like a coming of age story. Only thirty years later Blumenbach reconsidered the experiments with an astonishing shift in gender. In *Ueber den Bildungstrieb*, Blumenbach wrote:

> [W]enn denn die Tiere [Polypen] lebendige Junge austreiben wollen, so schwillt blos eine Stelle dieses ihres aus so einfachen Stoffe gebauten Körpers ein wenig an, und aus dieser ungeformten, aber durchsichtigen kleinen Geschwulst wird gleichsam unter unsern Augen zuerst der cylindrische Leib des jungen Polypen und dann auch *seine* Arme ausgebildet.[109]
>
> [When the animals (polyps) want to produce alive young ones, a place of their body, which is built from simple matter, just swells a bit and from this unformed, but transparent small growth develops—basically in front of our eyes—first the cylindrical body of the young polyp and then also his arms.]

Blumenbach seems to avoid the question of sexuality by choosing the plural, but at the one point where he switches to the singular, he describes the polyp as a male being. The reinterpretation of the polyp and its reproductive powers regarding its sexuality is perhaps most visible in Erasmus Darwin's influential *Zoonomia, or, The Laws of Organic Life* (1794–1796). Ten years after Blumenbach, Erasmus Darwin (1731–1802)—grandfather of Charles—wrote:

> Those who have attended to the habits of the polypus, which is found in the stagnant water of our ditches in July, affirm, that the young ones branch out from the side of the parent like the buds of trees, and after a time separate themselves from them. This is so analogous to the manner in which the buds of trees appear to be produced, that these polypi may be considered as *all male* animals, producing embryons, *which require no mother* to supply them with a nidus, or with nutriment, and oxygenation.[110]

In Erasmus Darwin's work, the shift in the assessment of whether the polyp is male or female is clearly decided in favor of the latter. In the new paradigm in which (and with which) Darwin articulated his theory of generation, the reproductive powers of the polyp organism are assigned to the father. Darwin assigned traditional "female" qualities, like reproduction and nourishing, to the male and this reassignment is so pronounced that he even rejected any necessity for female contribution.

Darwin's description of the polyp opens his much longer elaboration on "generation." In the new theory of generation that Darwin developed in the *Zoonomia*, the shift from a mother to a father origin coincides with a reinterpretation of the relationship between the generations. Having established the male as a reproductive force, Erasmus Darwin tries to better understand the parent-child relation within the generational process:

> Owing to the imperfection of language the offspring is termed a new animal, but is in truth a branch or elongation of the parent; since part of the embryon-animal is, or was, a part of the parent; and therefore in strict language it cannot be said to be entirely new at the time of its production; and therefore it may retain some of the habits of the parent-system.[111]

Drawing on the polyp analogy (and in opposition to earlier and influential claims by Buffon who had insisted on the importance of both female and male contribution in generation), Darwin concluded that the parent-child relationship is so intricate that it is even difficult to separate the two as individuals. Although Darwin was aware that reproduction in mammals works differently than polyp branching, the branching model provided him with a way to explain the similarity between parent and child and the assumed transfer of habit. While scholars so far had major difficulties to account for similarities between parent and child without having to rely on the power of imagination (discussed already in Aristotle) or climatic conditions (cf. Kant), Darwin offered a new approach by suggesting that parent and child form one organic unity. Since the reproducing organism is thought as male, the transmission of traits is also assumed to happen via the male trait carrier.[112] Darwin's work was quickly translated into German (1795–1797) and made almost instantly a strong impression not least because it promised a new approach to the definition of the generational relationship and a model to think male creativity (see also chapter 4 on this question).

GENEALOGICAL TRANSFER

The shift from a climate oriented theory of reproduction and modification to one which is based on the genealogical line thus coincided with the conviction that the genealogical development depends on the male contribution. In the context of breeding efforts, this renewed focus on the male found its expressions in theorems which suggested that species transformation takes place via the father. Already in 1790 when many breeders still followed the climate approach (and before the publication of Erasmus Darwin's theoretical considerations), Fincke recommended the improvement of sheep stock by means of the ram: "Weit sicherer [als durch Mutterschafe] geschieht die Verbesserung der Wolle, und mit geringen Kosten, lediglich durch angeschaffte Zuchtböcke"[113] [It is much more certain, and for little money, to improve wool by means of acquired breeding rams (than by mother sheep)]. With reference to Joseph Gottlieb Koelreuter (1733–1806), one of the most industrious German plant breeders in eighteenth-century Germany (see chapter 3), Lorenz Oken (1779–1851) stated in 1805:

> Endlich hat Kölreuter die Umwandlung einer Art in die andere durch den männlichen Samen für die Pflanzenwelt erwiesen . . .; ja die neueste Zeit ging noch weiter, sie hat Kölreuters Versuche in die Thierwelt verpflanzt, und wenig fehlt, wenn sie das Problem noch nicht vollständig gelöst hat. . . . Ist es nicht genug, wenn uns Hellenius in den Abhandlungen der schwedischen Academie etc. 1801 berichtet, dass der Schafbock mit einem Reh sich fortpflanze, und die Abkömmlinge bei jeder Generation dem Schafbocke ähnlicher werden?[114]
>
> [Finally, Kölreuter has proven the transformation of one plant species into another via the male seed . . .; nay, modern times have even pushed further, they have transplanted Kölreuter's experiments into the animal world, and only little is missing, if they have not yet solved this problem completely. . . . Is it not enough, when Hellenius in the treatises of the Swedish Academy etc. reports in 1801 that the ram propagates with a deer and that the offspring in each generation assimilates more and more to the ram?]

The effects of new assumptions in reproduction and heredity on the question of gender are complex. On one side, we can observe a shift from mother to father as the decisive reproductive entity. This shift happened despite the fact that the model according to which hereditary transfer was conceptualized in the beginning remains the one which had been dominant in conceptualizing a-sexual reproduction (the mother polyp simply becomes a father polyp without a change in the understanding of how they reproduce). On the other side, it is astonishing how little impact the novelty of the approaches to animal breeding and heredity had on the assessment of the significance of the male contribution compared to older, Aristotelian, assessments of gender. Much as in Aristotle's *Generation of Animals* in texts by Oken, Sturm, Erasmus Darwin and others who published on heredity and breeding, it is the male contribution which guarantees genealogical transmission.

The influence of Aristotelian conceptualizations of reproduction and sex goes beyond the mere continuation of the assessment of the importance of the male contribution; it also shapes the definition of this contribution. Discussing the question of race transformation in animal husbandry in 1825, Sturm presented the following observation:

> daß der Kopf in der Regel im Jungen nach dem Vater und die Hintertheile nach der Mutter gebildet werden. Der Vater, als der Erhalter der Raçe, daher auch der Erzeuger des wesentlichen Theils seyn müsse. Es ist dieß eine Behauptung, die unsers Wissens ganz neu ist. [115]
>
> [(T)hat the head is usually formed following the father and the hindquarters following the mother. The father, as preserver of the race, hence, has to be the provider of the essential parts. This is a claim which is, according to our knowledge, entirely new.]
>
> . . .
>
> Ferner, daß es bei der Verwandlung der einen Raçe in die andere vorzüglich auf den Vater ankomme, und durch diesen allein mit der Zeit die Raçe der Mutter in die seinige umgeändert werden kann.
>
> [Furthermore, that the transformation of one race into the other depends especially on the father, and that it is through him alone that the race of the mother can be transformed into his in the course of time.][116]

Like the introductory gesture to the "Law of Dentition," Sturm advertises the rule on the impact and quality of the male contribution as a new one. And, indeed, it is new considering the precision with which it assigns the formation of specific body parts to the influence of sex and the conviction with which it allows for the transformation of one race into another. However, regarding the overall assessment of sex, the strength of the continuation of Aristotelian approaches is astonishing. The male contribution in Sturm's account is affiliated with the head and this affiliation is strongly reminiscent of the Aristotelian definition of the male principle as form (in contrast to female matter). This becomes particularly clear when we consider Sturm's affiliation of the female contribution with the animal's hindquarters. Although male and female contributions are assigned in

Sturm to specific body parts, the distinction between head and body seems to be only a slight variation of the Aristotelian distinction between male form and female matter.[117] Much as in Aristotle, it is the male contribution that shapes the female contribution by imposing its form. And much as in Aristotle, it is also the male contribution that guarantees the continuation of the race.[118] Since the head is, according to Sturm, the main site of racial differentiation,[119] the mother's contribution is indirectly defined as less significant in terms of heredity. By applying this gendered criterion, Sturm virtually excluded female animals from the order of races—and ultimately also species. While the father's contribution, the head, can be brought in a racial hierarchy, the text suggests that the mother's contribution is less important for the establishment of such a hierarchy.[120]

To conclude, origin and genealogical transfer in texts on generation and heredity by authors like Sturm, Oken, and Erasmus Darwin is attributed to the father.[121] In their theories, the origin is considered masculine—it is a "Stammvater" who governs the species[122]—and the generational transfer is considered as taking place along the male line.[123] Although these authors stress the fact that they discover and formulate new laws according to which heredity works and despite an increasing insistence on empirical methods, their theories are significantly shaped by traditional accounts of reproduction. (We will show in chapter 4 how E. T. A. Hoffmann dismantled the claim of scientific empiricism by highlighting its imaginative precondition.)

EVOLUTIONARY THINKING

The transformation of one race into the other is introduced by Sturm in the framework of permanence: the progenitor, and the line that originates from him, is able to maintain his form despite the mother belonging to a different race. However, having thus described the way in which one race can be transformed into another, the luring anxiety (or enthusiasm) for scholars around 1800 was whether such human selection and intervention in copulation would have the potential to permanently change races, species, and eventually nature. For all practical purposes, breeders were convinced that they were able to alter a stock significantly and relatively quickly by means of breeding. In 1804, in his *Praktische Anleitung zur Führung der Wirthschafts-Geschäfte* [Practical Guide for the Management of the Farm], Gustav Gericke (1755–1817) expresses the conviction that such transformations can take place within a few generations:

> Wenn man unter eine Heerde der klufthaarigen Schaafe Spanische Widder bringt, und von diesen allein die Mutter-Schaafe bespringen läßt, so erhält man ganz fein- und kraushaarige Lämmer. . . . Diese nennt man Spanische Schaafe erster Generation. Läßt man nun die Mutter-Schaafe dieser Generation, im 3ten Jahre ihres Alters, wieder durch Spanische Böcke belegen, so werden die Lämmer, die davon fallen, wieder grobwolligter. . . . Die . . . Schaafe heißen nun Schaafe zweiter Generation. Läßt man nun aber von dieser Generation die Mutter-Schaafe wieder von Spanischen Widdern belegen, dann schlagen . . . die Lämmer ganz nach den Vätern und bekommen nun eine ganz feine, . . . fettige und elastische Wolle. Diese heißen nun Schaafe dritter Generation, oder völlig veredelte Spanische Schaafe, und von dieser dritten Generation an arten sie nun nicht mehr aus, sondern werden immer schooner und schooner, je mehr sie sich mit den Spanischen Böcken begehen, und von dieser Generation an kann man erst selbst Böcke zuziehen. [124]

> [If one introduces Spanish rams into a herd of cleft haired (klufthaarige) sheep, and if one covers the mother sheep only with those (rams), then one obtains very fine- and curly-haired lambs . . . One calls these Spanish sheep of the first generation. If one covers the mother-sheep of this generation, in their third year, again with Spanish rams, then the lambs which are produced are again coarse woolen. . . . The . . . sheep are now called sheep of the second generation. If one covers now again the mother sheep of this generation with Spanish rams, then the lambs become . . . like the fathers and get now a very fine, . . . fat, and elastic wool. These are now called sheep of the third generation, or completely refined Spanish sheep, and from this third generation onwards, they don't degenerate any longer; instead, the more they are covered by Spanish rams, the more beautiful they become, and from this generation onwards one can breed rams for breeding.]

Gericke's *Anleitung* attests not only to the strength of the belief in the power of breeding around 1800, it also demonstrates the breeders' careful observation concerning the appearance and disappearance of phenotypical signs in different generations. Breeders worked with this knowledge in order to improve their livestock, but such phenomena were difficult to account for in existing theoretical frameworks without being able to refer to the genotype. Even within Erasmus Darwin's model, which first offered a viable model to account for the transfer of qualities by blurring the line between parent and offspring, such disappearance and reappearance of features was not explainable. This would be true for many decades.[125]

As we have seen when discussing Erasmus Darwin's comments on the polyp, one of the new convictions in theories of reproduction was not only that traits could be transferred to the next generation and manipulated by different crossing methods, but also that the parents' habit would be transferred to the offspring. While each individual was shaped by the parent generation, she was also considered to develop own habits and thereby shape the next generation. Towards the end of the eighteenth century, heredity research, although it rediscovered the dependence of the individual, also promised the potential for change on the level of humankind.

Drawing on Blumenbach's work, Erasmus Darwin developed what we might call a first theory of evolution. In his above quoted *Zoonomia*, Darwin explained his theory of a gradual development of the organisms from a simple "living filament" to complex organisms. Instead of assuming that all life on earth was created at one moment in time, Darwin imagined at the beginning of the world a simple organic unit, a "living filament," which developed over many generations in today's and future forms of life.[126] Full of enthusiasm about his idea, Darwin exclaimed:

> [W]ould it be too bold to imagine, that in the great length of time, since the earth began to exist, . . . that all warm-blooded animals have arisen from one living filament, which THE GREAT FIRST CAUSE endued with animality, with the power of acquiring new parts, attended with new propensities . . .; and thus possessing the faculty of continuing to improve by its own inherent activity, and of delivering down those improvements by generation to its posterity, world without end![127]

Assuming a teleological framework, Darwin was certain that nature develops to an ever-greater complexity and perfection. The higher development of species and the variety of species are explained by heredity of habits, environmental stimuli, and bastardization.[128] A few late eighteenth- and early nineteenth-century biologists like Darwin or also Lamarck were convinced that individuals are shaped by the parents' biological inheritance, but also that they can change this inheritance and transfer new habits to the next generation, thereby eventually forming new species. In fact, it was the last possibility—the formation of new species—which was the most decisive and provocative aspect of the Erasmus Darwin's theory and which went beyond what most German authors would accept at the time. In general, however, we can observe that the oscillation between an interpretation of heredity and genealogical transfer as determining (hereditary disease) and an interpretation which contained a liberating potential (new forms) became a prevalent feature in the assessment of the intergenerational relationship around 1800.

BREEDING THE HUMAN

The question of the impact of human intervention in nature became most pressing when the observations made in animal breeding were considered in relation to human development.[129] Whether or not hereditary patterns observed in animal populations under the pressure of selective breeding could be valid for human beings was a particularly explosive question when it came to faculties that were traditionally considered of "spiritual" origin—precisely the ones considered to elevate the human being above the animal state. The conviction of the exceptional status of man had been deeply rooted in the Christian tradition and was intensified by the Cartesian reference to what we could call intellectual capacities. However, the extensive eighteenth-century interest in primates resulted in a questioning of the human-animal divide and the whole model of a hierarchically structured chain of being (see also chapter 3).[130] Upholding the divide, however, was of importance not only to cement human superiority but also especially to preserve free will.[131] If not only bodily features but also intellectual capacities could be transmitted by means of biological trait, the independence of moral faculties and judgments could be questioned.

The particular challenge to the idea of free will and the promise of an open future was the hypothesis—suggested by scholars such as Erasmus Darwin and Lamarck, and growing in appeal around 1800—that the habit of the parent generation could be transferred to the next generation, thereby shaping it. Although habit transfer had been assumed before, for example in form of socialization, the most recent conviction within the bourgeois framework of selfhood had been that the individual can and should free herself from such inherited values on the way to full adulthood (this is arguably what the *Bildungsroman* is about). The difficulty habit transfer in humans posed even for scholars who had already accepted it in animal populations can be seen in Sturm's 1825 comments on the issue:

> Es ist vom Menschen wie vom Thiere eine anerkannte Erfahrung, daß sich alle diejenigen Theile, die am meisten geübt werden, bloß durch die Gewohnheit ausbilden. Der Fechter erhällt im Arm die größte Stärke,

> der Säckträger im Rücken, und wenn die Beschäftigung beider fortgepflanzt würde, so würden wir in einer gewissen Reihe von Generationen (deren Zahl ich aber nicht zu bestimmen wage) lauter junge, auf der rechten Seite überbaute Fechtmeister, oder im Rücken starke d.i. gebohrne Säckträger haben. So wie beim Menchen finden wir dieß auch bei Thieren, wo diese Eigenschaften allerdings forterben, z.B. die Muskelkraft des Pferdes von Eltern abstammend, die immer in schwerem Zuge gebraucht wurden u.s.w.[132]
>
> [For both human being and animal it is well known that all those parts which are exercised the most are developed by mere habit. The fencer obtains the greatest strength in the arm, the sack carrier in the back, and if the activity of both were transmitted, then we had in the course of a certain number of generations (of which the number, however, I don't dare to determine) nothing but young fencers who are strongly built on the right side, or people who are strong in the back, i.e., born sack carriers. As in the human being, we find this also in animals. However, these qualities are hereditary in animals, for example the muscle strength of the horse from parents which were always used in heavy pulling etc.]

Sturm formulates the possibility of habit transfer only in the subjunctive. Even though he focused on bodily transmission, he eventually rejects habit transfer even for bodily features for the human being. The reason for this rejection becomes visible when Sturm turns to the question of the transmission of intellectual abilities. Sturms coined for those abilities the term "geistige[n] Anlagen,"[133] thereby combining a term used in contemporary treatises for biological features, "Anlagen" [traits], with a term coming from a religious, spiritual background, "geistig" [spiritual, intellectual]. He discussed the question of whether such traits are heritable:

> Bei dem Menschen ist der Uebergang der geistigen Anlagen höchst selten, denn selten zeugen ausgezeichnete Männer dergleichen Söhne. Es scheint indeß hier zwischen den verschiedenen Geistesfähigkeiten ein nothwendiger Unterschied gemacht werden zu müssen; das eigentliche Genie, ich möchte es das Göttliche im Menschen nennen, ist ein bloßer Anhauch Gottes, fleischlicher Mitteilung unfähig, daher wird es auch höchst selten vorkommen, daß ein Genie wieder eines anzeugt, vielmehr finden sich solche von gewöhnlichen, oft geistig ganz unbedeutenden, Eltern.[134]

> [In humans, the transmission of spiritual traits is very rare, because it is very rare that exceptional men sire such sons. It seems that we have to make here a necessary distinction between different spiritual abilities; the real genius, which I want to call the divine in the human being, is a mere aura of God, unable to be transmitted via the flesh, therefore, it will be very rare that a genius sires another, it is rather so that we find that such (geniuses) are sired by usual, often intellectually very ordinary, parents.]

Sturm's rejection of habit transfer in the human being is motivated by the wish to safeguard an area which cannot be touched by the biologically interpreted transfer of traits. He hopes to preserve a non-material, divine area for the human being. Similar to his strategy regarding nature—namely the introduction of two notions of nature, one true and one artificial—Sturm rescues here an area in the human being which is not touched by heredity by introducing two notions of intellectual abilities, of which only one can become a transmittable trait.[135]

How difficult it became to keep intellectual qualities independent from bodily features and to rescue a space for human capacities which are not shaped by hereditary transfer is perhaps nowhere clearer than in Kant's anthropological writings. The split between the world of phenomena and of noumena had formed the basis of Kant's ethics and his conceptualization of the human being as a moral being in the *Kritiken*. However, in his own writings on race this split is constantly in danger of being undermined. Extrapolating on the idea of "Anlage" in the context of the effects of tropical climates on corporeal features, Kant stated:

> Die größte feuchte Hitze des warmen Klimas muß hingegen an einem Volke, das darin alt genug geworden, um seinem Boden völlig anzuarten, Wirkungen zeigen. . . . Der Wuchs der schwammigen Teile des Körpers mußte in einem heißen und feuchten Klima zunehmen; daher eine dicke Stülpnase und Wurstlippen. . . . kurz, es entspringt der Neger, der seinem Klima wohl angemessen, nämlich stark, fleischig, gelenk, aber unter der reichlichen Versorgung seines Mutterlandes faul, weichlich und tändelnd ist.[136]

> [In a people which has grown sufficiently old in the greatest humid heat of the warm climate to have adapted completely to its soil, there must be effects . . . The growth of the spongy parts of the body had to increase

> in a hot and humid climate; hence the thick turned-up nose and the thick lips. . . . in short, this results in the Negro, who is well suited to his climate, namely strong, fleshy, supple, but who, given the abundance provision of his mother land, is lazy, soft and trifling.][137]

These highly problematic statements demonstrate that Kant was not free from the old discourse of a hierarchy of races. If Kant's insistence on the monogenesis hypothesis can be understood as an expression of an enlightened universalism, it becomes also clear that this universalism has its limits and that the equality that it promises (all human beings belong to the same species) is more potential than reality. As we have seen in the discussion on Kant's notion of "Anlage," once the final stage of full development of an "Anlage" is reached, bodily and mental features cannot be reversed (in the short and medium term) and they become independent from climatic conditions.[138] Hence, even if an inhabitant of a hot climate is moved to Europe, this person will remain determined to be "lazy, weak, and dallying." The urgency of this consideration was connected to two aspects: first, a close relationship between racial and mental capacities could suggest that there is a hierarchy of races regarding their ability to act as free, moral, beings. Second, since race is understood as an expression of contingent forces which, however, ultimately become heritable and permanent, the newborn individual would be determined already at the moment of birth, and this also in her capacity as a moral being.

The problem became perhaps most pronounced around the notion of "perfection" (Vollkommenheit). Kant, in *Physische Geographie*, stated that "Die Menschheit ist in ihrer größten Vollkommenheit in der Rasse der Weißen. Die gelben Indianer haben schon ein geringeres Talent. Die Neger sind weit tiefer, und am tiefsten steht ein Teil der amerikanischen Völkerschaften" [Humanity has its highest degree of perfection in the white race. The yellow Indians have a somewhat lesser talent. The Negroes are much lower, and lowest of all is part of the American races.][139] Such hierarchy could remain relatively insignificant for Kant's definition of the human being as a free and moral being, if a clear split between bodily and intellectual abilities could be maintained so that all races would have the same access to full humanity regardless of their different corporeal features. However, precisely this split is in danger, since Kant seems to establish a correlation between bodily features and the degree of humanity which can be achieved given those features ("highest degree of perfection in the white race"). Much, of course, depends on how "Vollkommenheit" [perfection] and "Menschheit" [humanity] are to be understood. The two terms are closely related and defined, among other places, in Kant's second chapter in *Die Religion innerhalb der Grenzen der bloßen Vernunft* (Religion within the Limits of Reason Alone): "Das, was allein eine Welt zum Gegenstande des göttlichen Ratschlusses und zum Zwecke der Schöpfung machen kann, ist die Menschheit (das vernünftige Weltwesen überhaupt) in ihrer moralischen, ganzen Vollkommenheit. . . ." [That which alone can render a world an object of the will of God and a purpose of Creation, is humanity (the rational being in general) in its moral, entire perfection. . . .][140] Humanity is here not defined in the sense in which it is often used today, namely as a collective term describing the entirety of all human beings on earth. Rather, humanity defines here the ideality of human being, or the homo noumenon in contrast to the homo phaenomenon, as Kant calls it in *Metaphysik der Sitten* [Metaphysics of Morals].[141] Perfection, then, is not only referring to corporeal aspects or talents associated with these features, but to the state of a race's "moral . . . perfection." Despite all the efforts to keep moral and corporeal aspects separated that are found in Kant's critical writings, Kant's statement on the superiority of the white race suggests that the bodily determination of the human being is in danger to coincide not only with a certain mentality, but also with moral abilities and thereby with the degree of humanity developed in a certain race. Kant's writings on race illustrate the emerging conceptualization of heredity around 1800, but they also allude to the social and political implications which would become part and parcel of future discussions on heredity.[142]

One of the features of those races in Kant's writings in which humanity is not developed to its highest degree is that these races are not able to discipline their laziness. In the best case, their laziness is disciplined by an outside force. Considering Kant's argument and the colonial discourse of his time, it is not surprising that the disciplining power cannot be represented by an individual of the same race. As we have seen in the discussion of "Anlage," this individual would be determined by the environment, and ultimately inherited features, in the same way as her fellows. Hence, an outside force of another "higher developed" race is called upon:

> Man sieht es auch an den wilden Nationen, daß, wenn sie gleich den Europäern längere Zeit hindurch Dienste tun, sie sich doch nie an ihre Lebensart gewöhnen. Bei ihnen ist dieses aber nicht ein edler Hang zur Freiheit, wie Rousseau und Andere meinen, sondern eine gewisse Rohigkeit, indem das Tier hier gewissermaßen die Menschheit noch nicht in

sich entwickelt hat. Daher muß der Mensch frühe gewöhnt werden, sich den Vorschriften der Vernunft zu unterwerfen.[143]

[It is also observable in savage nations that, though they may be in the service of Europeans for a long time, they can never grow accustomed to the European way of life. But with them this is not a noble propensity towards freedom, as Rousseau and others believe; rather it is a certain raw state in that the animal in this case has so to speak not yet developed the humanity inside itself. Therefore, the human being must be accustomed early to subject himself to the precepts of reason.][144]

Kant—explicitly rejecting Rousseau—stated that the laziness of the "savage" races is controlled by Europeans, but that this control often fails because these indigenous populations have apparently learned this discipline not early enough in their lives. Several issues have to be highlighted: first, the governing of the "savage" nations is justified by the fact that—as a result of environmental pressure—in these population groups humanity is developed to a lesser degree. Second, these groups are compared to animals. Third, reason is defined as the disciplining power (of animality). Fourth and finally, Europeans, apparently having developed reason already to a higher degree, perform the disciplining function. Despite these divisive categories, however, it is also important to note that the quote is shaped by an idea of development ("entwickelt") which suggests that these categories are still malleable to the extent that a development to the highest degree of perfection is possible for all beings. If Kant described the human being on one side as having (in some cases) limited access to the realm of freedom, because her bodily determination, and on the other side as a being defined by having access to the realm of freedom, then, he was not formulating a paradox, but he was describing the human being in two different states. The gap between these states could be bridged because they were thought as stages in a development. The promised universalism is guaranteed as a universal potential for development towards the state of perfection.

If Kant assumed that individual races and humanity in general are on a path of development, one crucial question in our context of heredity is by what means this development is unfolding. In the "Lectures on Pedagogy," Kant outlined:

Disziplin oder *Zucht* ändert die Tierheit in die Menschheit um. Ein Tier ist schon alles durch seinen Instinkt; eine fremde Vernunft hat bereits

> Alles für dasselbe besorgt. Der Mensch aber braucht eigene Vernunft. Er hat keinen Instinkt und muß sich selbst den Plan seines Verhaltens machen. Weil er aber nicht sogleich im Stande ist, dieses zu tun, sondern roh auf die Welt kommt: so müssen es Andere für ihn tun. Die *Menschengattung* soll die ganze *Naturanlage* der Menschheit durch ihre eigne Bemühung nach und nach von selbst herausbringen. Eine Generation erzieht die andere.[145]
>
> [Discipline or breeding changes animal nature into human nature. An animal is already all that it can be because of its instinct; a foreign reason has already taken care of everything for it. But the human being needs his own reason. He has no instinct and must work out the plan of his conduct himself. However, since the human being is not immediately in a position to do this, because he is in a raw state when he comes into the world, others must do it for him. The *human species* is supposed to bring out, little by little, humanity's entire *natural predisposition* by means of its own effort. One generation educates the next.][146]

Kant envisioned the development as a trajectory from animality to humanity, which is defined by the independent use of reason. Strangely oscillating between individual life and life of humanity, Kant acknowledged that the human being (individual and species) is not born in the state of humanity, but has to develop towards it. Although the focus in this quotation is on the development of reason and the *lack* of instinct in the human being, it is remarkable to what extent Kant drew on vocabulary which was taken from the context of natural history and animal breeding in order to describe the development towards greater humanity. As we have seen, terms like *Zucht* [breeding, cultivation], *Gattung* [species], and *Naturanlagen* [natural traits] appear frequently in contemporary texts on natural history, in breeding manuals, and in Kant's own writings on race. Similar to the development of "Anlage" and its stimulation by environmental forces which Kant described in the writings on race, here he seems to define natural trait as a potential that gradually unfolds when stimulated. Using the compound "Naturanlage" (instead of the simple "Anlage"), Kant stressed the biological framework in which the concept can be understood. The new focus on what could be understood as biological factors coincided with the suggestion that the developmental process does not start newly with each individual, but is an accumulative process which takes place over the course of many generations. Much as the breeder shifted his attention to the successive improvement via generational transfer, so did Kant shift his attention to the development of human kind on the basis of genealogy ("Die *Menschengattung* soll die ganze *Naturanlage* der Menschheit durch ihre eigne Bemühung nach und nach von selbst herausbringen" [The human species is supposed to bring out, little by little, humanity's entire natural predisposition by means of its own effort]).

As we have seen in Kant's writings on race, the development of traits within the climatic approach depends on a stimulating environment. For the human being and human qualities, such stimulating environment is provided also by education ("One generation educates the next"). In fact education is a necessity for reaching the highest level of humanity. Kant's statement is clear and strong: "Der Mensch ist das einzige Geschöpf, das erzogen werden muß" [The human being is the only creature that must be educated].[147] And, similarly, "Der Mensch kann nur Mensch werden durch Erziehung. Er ist nichts, als was die Erziehung aus ihm macht" [The human being can only become human through education. He is nothing except what education makes out of him].[148] However, assuming education as the stimulating environment necessary on the path to full development of humanity, we are confronted with a tautological structure. If we assume—as Kant's equation between the "savage" and the child suggested—that the human being, and by extrapolation humanity, started out in a less perfect state, *and* that a development to a higher state can only take place via education from generation to generation, then, the question is how can there be a guided development to greater perfection at all. Kant quickly realized this problem and hoped for a kind of *deus ex machina* to solve it:

> Es ist zu bemerken, daß der Mensch nur durch Menschen erzogen wird, durch Menschen, die ebenfalls erzogen sind. . . . Wenn einmal ein Wesen höherer Art sich unserer Erziehung annähme, so würde man doch sehen, was aus dem Menschen werden könne. Da die Erziehung aber teils den Menschen einiges lehrt, teils einiges auch nur bei ihm entwickelt: so kann man nicht wissen, wie weit bei ihm die Naturanlagen gehen.[149]

> [It must be noted that the human being is educated only by human beings, human beings who likewise have been educated. . . . If some day a being of a higher kind were to look after our education, then one would see what the human being could become. But since education partly teaches the human being something and partly merely develops something within him, one can never know how far his natural predispositions reach.][150]

"Naturanlage" is a potential for the development of highest humanity of which, however, we don't know the expression. Only if a higher being were to represent the stimulating environment, could we know the full potential of the human being. Not hoping for a divine educator, Kant nevertheless projected the possibility that a plan of education could be outlined by humans themselves:

> Wir können an dem Plane einer zweckmäßigen Erziehung arbeiten und eine Anweisung zu ihr der Nachkommenschaft überliefern, die sie nach und nach realisieren kann. Man sieht z.B. an den Aurikeln, daß, wenn man sie aus der Wurzel zieht, man sie alle nur von einer und derselben Farbe bekommt; wenn man dagegen aber ihren Samen aussät: so bekommt man sie von ganz andern und den verschiedensten Farben. Die Natur hat also doch die Keime in sie gelegt, und es kommt nur auf das gehörige Säen und Verpflanzen an, um diese in ihnen zu entwickeln. So auch bei dem Menschen! . . . Die ersten Eltern geben den Kindern schon ein Beispiel, die Kinder ahmen es nach, und so entwickeln sich einige Naturanlagen. . . . Soviel ist aber gewiß, daß nicht einzelne Menschen bei aller Bildung ihrer Zöglinge es dahin bringen können, daß dieselben ihre Bestimmung erreichen. Nicht einzelne Menschen, sondern die Menschengattung soll dahin gelangen.[151]
>
> [What we can do is work out the plan of an education more suited to the human being's purpose and hand down instructions to that effect to posterity, which can realize the plan little by little. It is observable, for instance, that the auricular only bears flowers of one and the same color when cultivated from a root. On the other hand, if it is grown from seed one gets flowers of totally different and most varied colors. Thus nature has after all placed the germs in these plants, and it is merely a matter of proper sowing and planting that these germs develop in plants. The same holds true with human beings. . . . The first parents already give the children an example which the latter imitate, and that way some natural predispositions are developed. . . . But this much is certain, that individual human beings, no matter what degree of formation they are able to bring to their pupils, cannot make it happen that they reach their vocation. Not individual human beings, but rather the human species, shall get there.][152]

The vocabulary of seed and cultivation helps Kant to think through the problem of the development of humanity. It stresses Kant's belief in the power of the stimulating environment, education, and demonstrates once again that Kant hoped for a step by step perfection in which the achievements—much as skin or blossom color—accumulate from generation to generation until they are virtually stable in a stage of higher perfection. However, as Kant replaced the divine plan with a human one, the tautological structure of his outline of human development remains. It remains unclear how human beings can purposefully improve without this change being triggered by an outside force. When considering the development of humanity, human beings in Kant's model are both the triggering environment—the authors of the plan—and the seed which unfolds as a reaction to this trigger. The complementary relationship of the two entities, seed and environment, collapses in the one entity of the human being.

The transfer of a preconceived plan for education and its realization over many generations which Kant demanded must also astonish any reader who knows Kant's essay "Was ist Aufklärung?" (1784).[153] In this essay—very much in tune with declarations informing the American and French and Revolutions by thinkers like John Locke and Thomas Paine—Kant insisted that no generation can impose its value judgments on the next.[154] The tension between the two statements (transfer of an educational plan to the next generation and no right to dictate to the next generation) demonstrates how much Kant—at least in the *Physische Geographie*—still assumed a teleological trajectory for the development of humanity.

Kant's text demonstrates the ambivalence in the assessment of the relationship between parent and child, past and future which became typical for texts by authors who wrote around 1800. The interpretation of the relationship between parents and children and the transfer from one generation to the next oscillated between determination by the parent generation *and* the potential to liberate oneself from this heritage. This ambivalence in the assessment of the relationship between parent and child became a defining feature of the concept of heredity around 1800. We can find it in Kant's understanding of the generational development of humanity, in Erasmus Darwin's conceptualization of habit transfer in *Zoonomia* (1794–1796), and in Carus's *Ideen zur Geschichte der Menschheit* (1809). Carus seems to follow Kant's analogy when explaining the function of education for the development of human potential: "Alle Bildungsanstalten für den Menschen sind Weckungs-, Belebungs- und Befruchtungsmittel derjenigen Keime, die Anlagen heissen"[155] [All educational measures for the human being are reviving, stimulating, and fertilizing media of those germs, which are called traits]. At the same time, however,

Carus did not want to make the development of "Anlage" entirely dependend on the stimulating environment. He insisted that such environmental forces, among which he counted both climatic circumstances and genealogy,[156] can be shaped by the human individual. Carus concluded: "Doch eben dieser Einfluß der Aussenwelt ist keineswegs zwingend, niederdrückend oder gar vernichtend. Vielmehr ist er sehr bedingt, nur veranlassend und Kraftweckend . . ."[157] [However, this influence of the environment is not at all coercive, oppressive or even destructive. Rather, it is very limited, only stimulating and awakening forces . . .].

BETWEEN DETERMINATION AND FUTURE CHANGE

While the powers which led to the French Revolution presented a model of radical change that came along with a total rejection of the past and inherited privileges, from the mid-eighteenth century on we can observe the development of a notion of past and genealogy which had determining powers, but also carried the possibility of future change. By the beginning of the nineteenth century this tension between past and future had already become part of the very definition of the German general term "Erbe" [inheritance, heritage, heredity]. The apparent paradox which defined this new notion of inheritance was noted quite consciously. The dictionary of the Grimm brothers, arguably the most important German dictionary at the time, can serve as an example of this development. Under the heading "Erbe," we find the term introduced first as a piece of land, and in a second step also as all other property that one can inherit.[158] However, of particular interest is the discussion of the word heir ("der Erbe"). The Grimms suggested, first, that the word "Erbe" might be derived from the Slavic "rab" or "rob" but, in a second step, rejected this possibility because "rab" would mean servant or serf, but never heir: "rab *bezeichnet einen knecht oder hörigen, niemals einen erben, umgedreht unser* erbe *keinen knecht*"[159] [rab means servant or bondsman, never heir, conversely, our heir never means servant]. Hence, for the Grimm brothers, the notion of bondsman and heir are mutually exclusive. The person that inherits something cannot at the same time be dependent. To inherit means to be free.[160] As a consequence, the Grimms derived "Erbe" [heir] not from "rab" but instead from "arfi," which has the meaning of "son." Furthermore, the Grimms discussed how these two interpretations—of the heir as a free man and as a dependent son—might have contradictory tendencies because often the meanings of servant and child merge, for example when the power of the father is compared with that of the master:

> altn. [Altnordisch] arfi, das neben heres zugleich den sinn von filius hat, wie wir sagen er hinterläszt keine erben = kinder; die benennungen des sohnes in vatergewalt und des dienenden in herrngewalt treten aber oft ineinander über, man erwäge . . . subditus neben τέκνον neben παῖς und puer, welche neben kind auch servus ausdrücken. . . .[161]
>
> [oldn. (Old Norse) arfi, which has, besides heres, also the meaning of filius, as we say he does not leave behind heirs = children; but the naming of the son in the power of the father and the servant in the power of the master also often overlap, consider . . . subditus next to τέκνον next to παῖς and puer, which mean besides child also servus. . . .]

To describe the meaning of heir, the Grimms highlighted a combination of independence, "not a servant," and dependence, "a son." Inheriting, then, meant at the same time to become adult *and* to confirm one's status as a son.[162] It is this combination of dependence and freedom, origin and change, which emerged in the course of the long eighteenth century in a number of discourses and which became decisive for the formulation of the self around 1800.

LEGAL IMPLICATIONS

The tension between a new independence of the individual and her simultaneous reference to the parent, which we can observe both in biological discourses and lexical considerations, found also its parallel in German law. This new formulation of the relations within the family is undertaken with reference to its biological foundation. While the French *Code Civile* (1804) was arguably formulated on the basis of the assumption that the human being is born as a *tabula rasa* and was marked by the attempt to build the nation as a community of adopted brothers, in the reformed Prussian law of 1794, the *Allgemeine Landrecht*, blood relationships obtained a new significance—foreshadowing, one could argue, a focus on "Gemeinschaft" [community] instead of "Gesellschaft" [society].

When Prussia started to reform its laws at the end of the eighteenth century, the reviewers of the proposal demanded a more precise definition of the term "family." One of the commentators had criticized the vague definition: ". . . das Wort Familie ist von sehr vielfacher Bedeutung. Es kann der Hausvater und seine Hausgenossen, oder Verwandte, die einerlei Namen führen, oder Verwandte von mütterlicher Seite, oder gar verschwiegerte Verwandte bedeuten"[163] [(T)he term family has many different meanings. It can mean the housefather and his fellow tenants or relatives which have the same name or relatives from the mother's side or even in-laws]. Svarez, arguably the principal author of the *Allgemeine Landrecht*, agreed with this objection. However, instead of defining more precisely what such a family household would look like, the term family became newly defined by blood relationships, with stress on the patrilineal line. The third edition of the new Prussian law from 1796 declares:

> Durch die Abkunft von gemeinschaftlichen Stammältern [sic] werden Familienverhältnisse begründet. [164]
>
> [Family relations are established by descent from common parents (ancestors).]
>
> . . . .
>
> Personen, welche gemeinschaftliche Stammältern haben, heißen Blutsverwandte.[165]
>
> [Individuals who have common parents (ancestors) are called blood relatives.]

By marrying, the wife became part of the husband's family. However, she took part in the husband's family only for her person. Thus, the wife's family had no right to profit from her husband's inheritance.[166] While the father's authority was re-established via the reference to lineage, the wife's position was defined via her biological function and role as mother. One of the article's in the Prussian law for example reads: "It is the duty of mothers to breastfeed their children."[167]

Referring to the ideals of the French Revolution and to sociological data, commentators have often interpreted the transfer from a stratified society in which a patriarch reigns over a household to a differentiated society of nuclear families as a rejection of heritage and a weakening of fatherhood and masculinity in light of a new autonomy of the child that attains the right of an individual.[168] However, as legal and biological treatises seem to suggest, this new autonomy of the child[169] did not stand in contradiction to the growing importance of kinship and patrilineal lineage. What we observe is rather a more precise definition of the roles within the family in such a way that the autonomous child (individual) becomes possible not despite the father figure, but precisely against the backdrop of a father figure that establishes the connection to ancestry and of a mother figure that provides nurturing qualities. In short, the emergence of the individual coincided with a growing significance of the father figure as the origin of heritage. Increasingly the individual turned to the progenitor in search of identity. Modernity, we can say, is not the rejection of heritage, but its reevaluation.

As a consequence of the growing awareness for lineage on the biological, legal, and political level, the state made new attempts to regulate this lineage. One of the publications which moved in this direction as early as 1779 is the *System einer vollständigen medicinischen Polizey* by the physician and hygienist Johann Peter Frank (1745–1821). In his publication, Frank described the tasks and goals of the state vis-à-vis the family:

> Das Augenmerk der medicinischen Polizey ist, die Natur und ihre Kräfte, so viel als möglich ist, geltend zu machen, und so zu benutzen: daß man von jedem gegebenen Paar von Menschen beiderley Geschlechts, unter der Aufsicht guter Gesetze, die besten, gesundesten und dauerhaftesten Früchte erhalte.[170]
>
> [The focus of the medical police is to bring nature and its powers as much as possible into their rights and to use them in order to get, under the supervision of good laws, from each given heterosexual couple, the best, healthiest, and most durable fruits.]

The political attempts to regulate generations are part of a paradigm shift that Foucault and Giorgio Agamben have described as a shift from a notion of life that belongs outside the political to one in which life is infiltrated by the political and which takes control over the regenerative forces in the family.[171] The state's interest in the purity of the biological heritage of his citizens entered more and more in the center of such *biopolitics*.[172]

Biological questions became decisive not only for the state's attempt to control reproduction, but also for the family's attempt to determine legitimate heirs. For the greater part of the eighteenth century, and this is still true for the first editions of the *Allgemeine Landrecht*, most lawyers were willing to accept all children born in a marriage as children and heirs of the married couple, and ignored questions of blood relationships for the sake of social peace. Towards the end of the eighteenth century, however, law makers were more and more insisting that questions of blood relationship had to be solved. In 1800, the lawyer Johann Christian Majer (1741–1821) stated that:

> [N]ichts natürlicher, als daß der größere Grad von Gemeinschaft des Bluts, und das geringere Maaß von Vermischung mit fremdem Blute—dem geringeren Grade der Gemeinschaft des Blut und dem größeren Maße von Vermischung mit fremdem Blute—in der Succession vorgehe. . . .[173]
>
> [Nothing (is) more natural than the fact in the line of succession that the greater degree of blood community and the lesser degree of mixture with foreign blood supersedes the lesser degree of blood community and the greater degree of mixture with foreign blood. . . .]

Such attempts to establish kinship relations were supported by the medical community. In his *Ueber die physische Erfordernisse der Erbfähigkeit der Kinder* (1779), the doctor and later professor of medicine Wilhelm Gottfried Ploucquet (1744–1814) insisted that only those children of whom it is proven by means of corporeal signs, such as facial similarity, that they are indeed the children of their parents, should be allowed to inherit:

> Die Natur der Sache bringt es mit sich, daß die sogenannte Leibes Erben denn erst diesen Namen verdienen, und also die Erbschaft der Güter fähig seyn können, wenn die Eltern sich im Kinde gleichsam aufs neue herfürgebracht sehen.[174]
>
> [It lies in the nature of things that the so called corporeal heirs are only then entitled to carry this name, and to inherit the estate, if the parents see themselves reproduced in the children.]

Since there was no test to prove blood relationships yet, supporting sciences such as physiognomy were called upon. What had been a structure of privileges that was expressed in visual signs of clothing and profession became internalized as a trait. However, precisely because it could not be proven, it had to be claimed. Generational inheritance was, after all, not yet a fact, but a hypothesis. Literary texts were concerned to point out the hypothetical character of this supposed biological inheritance.

EPIGRAPH SOURCE

Immanuel Kant, "Über Pädagogik," in *Kants Gesammelte Schriften*, ed. Herausgegeben von der Königlich Preußischen Akademie der Wissenschaften (Berlin: Georg Reimer, Walter de Gruyter, 1912–1983), 9:445.

———. "Lectures on Pedagogy (1803)," in *The Cambridge Edition of the Works of Immanuel Kant* (Cambridge: Cambridge University Press, 2011), 440.

NOTES

1. Reinhart Koselleck, "Einleitung," in *Geschichtliche Grundbegriffe. Historisches Lexikon zur politisch-sozialen Sprache in Deutschland*, ed. Otto Brunner, Werner Conze, and Reinhart Koselleck (Stuttgart: Klett, 1972), 1:xv. The temporal focus of *Geschichtliche Grundbegriffe* (*GG*) is approximately the period between 1750 and 1850. In the introduction to *GG*, Koselleck mentioned 1770 as a decisive year for the beginning of modernity. However, he later downplayed the significance of a precisely defined periodization. See Melvin Richter and Michaela W. Richter, "Introduction, Translation of Reinhart Koselleck's 'Krise' in 'Geschichtliche Grundbegriffe,'" *Journal of the History of Ideas* 67, no. 2 (April 2006): 347 and 356.

2. For the concept of differentiation see also Niklas Luhmann, *The Differentiation of Society* (New York: Columbia University Press, 1982). For a detailed discussion of Luhmann's notion of modernity see William Rasch, *Niklas Luhmann's Modernity: The Paradoxes of Differentiation* (Stanford, CA: Stanford University Press, 2000).

3. In the Enlightenment, we can observe a fundamental reconsideration of the traditional idea of providence and original sin. First impulses for a critique of the idea came from French speaking areas with thinkers like Fontanelle and Bayle. Answering Bayle's suggestions, Leibniz' position of an optimistic deism becomes formative for the German Enlightenment. Leibniz' idea of a world that is, in principle, the best of all possible excludes the notion of absolute evil. Only the earthquake of Lisbon (1755) triggers new discussions on the question of theodicy. See Rudolf Behrens, *Umstrittene Theodizee, erzählte Kontingenz: Die Krise teleologischer Weltdeutung und der französische Roman (1670–1770)* (Tübingen: Niemeyer, 1994); and Elmar J. Kremer, "Leibniz and the 'Disciples of Saint Augustine' on the Fate of Infants Who Die Unbaptized," in *The Problem of Evil in Early Modern Philosophy*, ed. Elmar J. Kremer and Michael J. Latzler (Toronto: University of Toronto Press, 2001), 119–37. For an overview of the changes in the notion of "nature" in respect to ethics throughout eighteenth-century European literature see A. E. Pilkington, "'Nature' as Ethical Norm in the Enlightenment," in *Languages of Nature: Critical Essays on Science and Literature*, ed. L. J. Jordanova (New Brunswick, NJ: Rutgers University Press, 1986), 51–85.

4. See Reinhart Koselleck, *Vergangene Zukunft. Zur Semantik geschichtlicher Zeit* (Frankfurt am Main: Suhrkamp, 1979).

5. Pierre Nora, "Generation," in *Realms of Memory: Rethinking the French Past*, ed. Pierre Nora and Lawrence D. Kritzman, trans. Aurthur Goldhammer (New York: Columbia University Press, 1996–1998), 1:499–531. In *The Family Romance of the French Revolution* (Berkeley: University of California Press, 1992), Lynn Hunt has argued that the model of society propelled in the wake of the French Revolution was that of the community of brothers which was maintained by reference to the *dead* father.

6. See Marc Shell, *Children of the Earth: Literature, Politics, and Nationhood* (Oxford and New York: Oxford University Press, 1993).

7. Georg Simon Klügel, *Enzyklopädie oder zusammenhängender Vortrag der geminnützigsten Kenntnisse. Dritter Theil, Erste Abtheilung, welche das Naturrecht, die praktische Mechanik, die bürgerliche Baukunst, die Kriegsbaukunst und die Schiffsbaukunst enthält* (Berlin and Stetting: Friedrich Nicolai, 1784), 6, italics mine.

8. David Warren Sabean and Simon Teuscher, "Introduction," in *Kinship in Europe: Approaches to Long-Term Developments (1300–1900)*, ed. David Warren Sabean, Simon Teuscher, and Jon Mathieu (New York, Oxford: Berghahn Books, 2007), 2.

9. Ibid.

10. Ibid., 3. See also Jon Mathieu, "Kin Marriages: Trends and Interpretations from the Swiss Example," in *Kinship in Europe: Approaches to Long-Term Developments (1300–1900)*, ed. David Warren Sabean, Simon Teuscher, and Jon Mathieu (New York, Oxford: Berghahn Books, 2007), 211–30.

11. Sabean and Teuscher, "Introduction," 1.

12. Ibid, 3.

13. Cf. A. Lima-de-Faria, *One Hundred Years of Chromosome Research and What Remains to be Learned* (Dordrecht and Boston: Kluwer Academic Publishers, 2003).

14. Roger J. Wood and Vítězslav Orel, *Genetic Prehistory in Selective Breeding: A prelude to Mendel* (New York and Oxford: Oxford University Press, 2001), 38; 61–76.

15. Carlos López-Beltrán, "The Medical Origins of Heredity," in *Heredity Produced: At the Crossroads of Biology, Politics, and Culture, 1500–1870*, ed. Staffan Müller-Wille and Hans-Jörg Rheinberger (Cambridge, MA: MIT Press, 2007), 105–32.

16. Georg Simon Winter, *Gründliche und kurtz durchgehende Beschreibung, wie eine Stutherey anzurichten, dieselbe in gutem Esse zu erhalten, und wie man sich deroselben mit gutem Nutzen gebrauchen könne* (Onoltzbach: In Verlegung deß Authoris, 1670), 42.

17. Ibid., 48.

18. Horace quoted in ibid., 50.

19. The original Epistle 1:10, 24 can be found in Horace, *Epistles, Book 1*, ed. Oswald Ashton Wentworth Dilke (London: Methuen & Co. Ltd., 1954), 53.

20. Roger Tisserand, *Au temps de l'Encyclopédie. L'Académie de Dijon de 1740 à 1793.* (Paris: Boivin & Cie, 1939), 582.

21. See Carlos López-Beltrán, "Heredity Old and New. French Physicians and l'Hérédité Naturelle in Early 19th Century," in *A Cultural History of Heredity II: 18th and 19th Centuries*, Max Planck Institute for the History of Science, preprint 247 (Berlin, Max-Planck-Inst. für Wiss.-Geschichte, 2003), 11. See also his "Forging Heredity: From Metaphor to Cause, a Reification Story," *Studies in the History and Philosophy of Science* 25, no. 2 (April 1994): 211–35. Rougemont's essay was translated into German in 1794, a time when the topic became most virulent in Germany. Cf. Joseph Claude Rougemont, *Abhandlungen über die erblichen Krankheiten. Eine gekrönte Preißschrift*, trans. Friedrich Georg Wegeler (Frankfurt am Main: Fleischer, 1794).

22. See Philip K. Wilson, "Erasmus Darwin and the 'Noble' Disease (Gout): Conceptualizing Heredity and Disease in Enlightenment England," in *Heredity Produced: At the Crossroads of Biology, Politics, and Culture, 1500–1870*, ed. Staffan Müller-Wille and Hans-Jörg Rheinberger (Cambridge, MA: MIT Press, 2007), 133–54.

23. López-Beltrán, "Medical Origins," 111.

24. Rougemont, *Abhandlungen*, iii. See also ibid.

25. This recent approach to biological heredity which considers the multifaceted context of gene expression is called "epigenetics." For three 2011 publications on the topic see: Nessa Carey, *The Epigenetics Revolution: How Modern Biology is Rewriting our Understanding of Genetics, Disease, and Inheritance* (London: Icon Books, 2011); Richard C. Francis, *Epigenetics: The Ultimate Mystery of Inheritance*.(New York: W. W. Norton, 2011); and Benedikt

Hallgrímsson and Brian K. Hall, eds., *Epigenetics: Linking Genotype and Phenotype in Development and Evolution* (Berkeley: University of California Press, 2011).

26. To formulate these sets of inquiries, I have drawn on suggestions made by Justin E. Smith. Justin E. H. Smith, ed. "Imagination and the Problem of Heredity in Mechanistic Embryology," in *The Problem of Animal Generation in Early Modern Philosophy* (Cambridge: Cambridge University Press, 2006), 80–99.

27. Aristotle, *Generation of Animals*, trans. A. L. Peck (Cambridge, MA: Heinemann and Harvard University Press, 2000), 7.

28. Ibid., 401.

29. The tendency to associate hereditary patterns with degeneration remains strong even in the nineteenth century when nature's creativity starts to be acknowledged as a positive feature. Friedrich August Carus for example still mentions the word "Vererbung" in the context of hereditary disease and degeneration: "Die Natur . . . strebt zum Ideal; der Mensch aber erscheint immer mehr als der höchste Typus auf Erden; . . . Es sind unnatürliche Erscheinungen, Körper- oder Seelenkrankheiten, wenn manche Menschen für kürzere oder längere Zeit, durch Naturtriebe, Leidenschaft oder Wahnsinn zu den Thieren zurüksinken,—wohin auch die sprachlosen wilden Kinder, und stumpfsinnigen Menschen durch Vererbung von Schwächen des Stammes, die Kakerlaken in Ostindien, die Albinos in Afrika und die Kretinen in dem Süden von Europa gehören" [Nature strives for the ideal; but the human being appears more and more as the highest type on earth; . . . These are unnatural appearances, corporeal or mental sicknesses, when for shorter or longer periods of time some human beings fall back into the state of animals through natural drives, passions or madness—among which also belong the speechless wild children, and human beings retarded because of hereditary weakness, the Albinos of East India, the Albinos of Africa and the Cretins of Southern Europe], in (Carus, *Nachgelassene Werke. Sechster Theil. Ideen zur Geschichte der Menschheit* [Leipzig: Iohann Ambrosius Barth und Paul Gotthelf Kummer, 1809], 94).

30. Albrecht Korschorke, "Inseminationen, Empfängnislehre, Rhetorik und christliche Verkündigung," in *Kunst—Zeugung—Geburt: Theorien und Metaphern ästhetischer Produktion in der Neuzeit*, ed. Christian Begemann and David E. Wellbery (Freiburg im Breisgau: Rombach, 2002), 95.

31. For a detailed discussion of female imagination see Marie Hélène Huet, *Monstrous Imagination* (Cambridge, MA: Harvard University Press, 1993).

32. For an overview see Ilse Jahn, *Grundzüge der Biologiegeschichte* (Jena: G. Fischer, 1990).

33. Cf. James L. Larson, *Interpreting Nature: The Science of Living Forms from Linnaeus to Kant* (Baltimore: Johns Hopkins University Press, 1994), 132–33.

34. Caspar Friedrich Wolff, *Theoria Generationis (1759)*, ed. and trans. Paul Samassa. 2 vols. (Leipzig: W. Engelmann, 1896); In German, *Theorie von der Generation: in zwo Abhandlungen erklärt und bewiesen* (Berlin: Friedr. W. Birnstiel, 1764).

35. C. F. Wolff is famous for having brought forward epigenesis and its arguments for the entirely new creation of organisms. After he moved to Petersburg, however, where he was confronted with a large collection of deformities, Wolff wrote a manuscript on inheritance laws. Raijkov, still the only scholar to have worked extensively on this, claims that Wolff draws conclusions pointing well beyond what was believed in his time. Cf. Boris E. Raijkov [Rajkov], "Caspar Friedrich Wolff," *Zoologische Jahrbücher Systematik* 91 (1965):

51–65. See also Shirley A. Roe, *Matter, Life, and Generation: Eighteenth-Century Embryology and the Haller-Wolff Debate* (Cambridge: Cambridge University Press, 1981); and Larson, *Interpreting Nature*.

36. Johann Friedrich Blumenbach, *Über den Bildungstrieb und das Zeugungsgeschäfte* (Göttingen: Dieterich, 1781, repr. 1791). ———. *An Essay on Generation* (London: T. Cadell, Strand, 1792).

37. Helmut Müller-Sievers, *Self-Generation: Biology, Philosophy and Literature around 1800* (Stanford: Stanford University Press, 1997), 183.

38. See Helmut Müller-Sievers discussion of Kant's use of the concept (ibid., 48–64).

39. Abraham Trembley, *Mémoires pour servir à l'histoire d'un genre de polypes d'eau douce: à bras en forme de cornes* (Leide: Jean & Herman Verbeek, 1744).

40. "Hoffentlich ist für die mehresten Leser die Erinnerung überflüssig, daß *das Wort* Bildungstrieb . . . zu nichts mehr und nichts weniger dienen soll, als eine Kraft zu bezeichnen, deren constante Wirkung aus der Erfahrung anerkannt worden, deren *Ursache* . . . für uns *qualitas occulta* ist. . . . Das Verdienst beym Studium dieser Kräfte ist nur das, ihre Wirkungen näher zu bestimmen und auf allgemeinere Gesetze zurück zu bringen" (Blumenbach, Über den Bildungstrieb, 32–33). [It is to be hoped, that there is no necessity for reminding the reader, that the expression Formative Nisus, like that of Attraction, serves only to denote a power, whose constant operation is known from experience, but whose cause, like the causes of most of the qualities of matter, is a qualitas occulta to us. . . . But the great merit in the study of these powers, is to ascertain more accurately their effects, and to reduce them under general laws (Blumenbach, *An Essay on Generation*, 21–22).]

41. See also Rheinberger and Müller-Wille, "Introduction," 4 and Larson, *Interpreting Nature*, 61–65.

42. While *Zedler* (1732) lists the word "Anlage," the entry does not discuss the meaning of an innate talent at all (Johann Heinrich Zedler, *Großes vollständiges Universal Lexicon aller Wissenschafften und Künste* [Leipzig and Halle: Zedler, 1732], 351–52).

43. Johann Christoph Adelung, *Grammatisch-kritisches Wörterbuch der hochdeutschen Mundart* (Leipzig: Breitkopf und Compagnie, 1793), 1:329.

44. We can observe a similar interpretation in the *Grimm* (1854), but with a clearer reference to the medical discourse. According to *Grimm*, "Anlage" in the sixteenth century is understood almost exclusively in financial terms (taxes, costs, later also in the sense of investment). At the time of the 1854 publication of the first volume of *Grimm*, however, the term has become understood as an attachment (to a letter etc.), a plan (for a garden etc.), and then also as "natürliche anlage: der mensch hat gute anlagen, eine anlage zur krankheit" [natural predisposition (Anlage): the man has good predispositions, a predisposition for sickness]. Hence, by the mid-nineteenth century, the connection of "Anlage" to the medical-biological discourse seems to have been established. "Anlage" in Jacob Grimm and Wilhelm Grimm, *Deutsches Wörterbuch*, 15 vols. (Leipzig: S. Hirzel, 1854–1913), 1:389.

45. "Anlage" in *Deutsche Encyclopädie oder allgemeines Real-Wörterbuch aller Künste und Wissenschaften von einer Gesellschaft Gelehrten*, ed. Heinrich Martin Gottfried Köster and Johann Friedrich Roos (Frankfurt am Main: Varrentrapp Sohn und Werner, 1778), 1:506–7.

46. *Allgemeine Encyclopädie der Wissenschaften und Künste*, ed. Johann Samuel Ersch and Johann Gustav Gruber (Leipzig: Johann Friedrich Gleditsch, 1819), 4:154.

47. Ibid., 4:155.

48. Ibid., 4:156.

49. Klügel, *Enzyklopädie*, 6.

50. Carus, *Nachgelassene Werke*, 101.

51. The increased frequency of encounters with people of different skin colors in the wake of colonial enterprises and the sight of children from interracial couples had led to an extensive literature on race and the pattern of race transmission of which Kant's theoretical deliberations were a part. Cf. Renato Mazzolini, "Las Castas: Inter-Racial Crossing and Social Structure (1770–1835)," in *Heredity Produced: At the Crossroads of Biology, Politics, and Culture, 1500–1870*, ed. Staffan Müller-Wille and Hans-Jörg Rheinberger (Cambridge, MA: MIT Press, 2007), 349–74.

52. Like Kant, Johann Friedrich Blumenbach assumed one human race, distinguished from apes. Blumenbach quoted Haller, Linnaeus, and Buffon, which all, despite their differences, agreed in this point: "Alle dreye hielten den Menschen vom Orangutang himmelweit verschieden, und hingegen alle wahre Menschen, Europäer, Neger etc. für blose Spielarten einer und derselben Stammgattung." [All three considered the human being tremendously different from the Orangutan, but on the other side (they considered) all true humans, Europeans, Negroes etc. just variations of one and the same species.] (*Beyträge zur Naturgeschichte Erster Theil* [Göttingen: Dieterich, 1790], 58.)

Georg Forster, "Noch etwas über die Menschenrassen. An Herrn Dr. Biester," *Teutscher Merkur* (October 1786): 57–86; and "Beschluss der im vorigen Monat angefangenen Abhandlung des Herrn G. R. Forsters über die Menschenrassen," *Teutscher Merkur* (November 1786): 150–66. Samuel Thomas von Sömmering, *Über die körperliche Verschiedenheit des Negers vom Europäer* (1785), ed. Sigrid Oehler-Klein (Stuttgart: G. Fischer, 1998). See also Wolfdietrich Schmied-Kowarzik, "Der Streit um die Einheit des Menschengeschlechts. Gedanken zu Forster, Herder und Kant," in *Georg Forster in interdisziplinärer Perspektive*, ed. Claus-Volker Klenke (Berlin: Akademie Verlag, 1994), 115–32.

In the course of the nineteenth century, we can observe growing support for the polygenesis thesis. This is true for both, theories on the human race (see August Wilhelm Schlegel, "De l'origine des Hindous," in *Essais littérέraires et historiques* [Bonn: E. Weber, 1842], 441–518); and on animal races, see, for example, Joachim Friedrich Christian Dieterichs: "Der Stammvater der Hunde läßt sich nicht mit Bestimmtheit angeben, und es ist zu gewagt, den Hirten- oder Schäferhund dafür halten zu wollen, indem manche Rassen, wie Blumenbach es auch schon gesagt, zu viel Eigenes haben, als daß diese Eigenschaften für zufällige Folgen der Ausartung gehalten werden sollten. Es müssten meiner Ansicht nach alle Eigenschaften, die man von den verschiedenen Gattungen oder Rassen von Hunden aufzählen kann, alle in dem Schäferhund vereinigt seyn, wenn man ihn für den Vererber so vieler besonderer Eigenschaften halten wollte. . . . Daher ist es wohl zuverlässig, daß es schon in den ältesten Zeiten verschiedene Rassen oder Gattungen von Hunden gegeben habe." [The progenitor of dogs cannot be determined with certainty, and it is too daring to want to consider the sheepdog or the German shepherd as such, because some races, as Blumenbach has also already said, have too many peculiar features to consider these qualities as accidental results of degeneration. In my opinion, all features, which can be named from the different species or races of dogs, would have to be united in the shepherd, if one wanted to consider it the progenitor of so many peculiar features. . . . Therefore, it is probably reliable (to consider) that different races or species existed already in the oldest times.] (*Von der Zucht der Hunde, ihren vorzüglichen Rassen und den wichtigsten Krankheiten derselben* [Leipzig: Baumgärtner, 1831], 5).

Similar Röttger von Veltheim: "Uebrigens darf ich hier nicht übergehen, dass Wollstein zuerst wagte, öffentlich die Hypothese aufzustellen: dass ursprünglich wenigstens zwei wesentlich verschiedene Pferdegattungen, d.h. das südliche und nordische Pferd, deren jedes seine verschiedenen Nüancen habe, erschaffen seyn müssten" [By the way, I shouldn't ignore that Wollstein was the first who dared to formulate the hypothesis that originally there must have been created at least two fundamentally different horse species, namely the Southern and the Northern horse, of which each has its own peculiar nuances.] (*Abhandlungen über die Pferdezucht* . . . [Braunschweig: Druck und Verlag von Friedrich Vieweg, 1833], 379).

53. Carl Linnaeus had given the following definition of species and variety already in 1737: "There are as many species as the Infinite Being produced different forms in the beginning; which forms afterwards produced more forms, but always similar to themselves according to inherent laws of generation, so that there are no more species now, than came into being in the beginning. Hence, there are as many species as different forms or structures of plants occurring today, those rejected, which place or accident exhibits to be less different (varieties)," quoted in Staffan Müller-Wille and Vítezslav Orel, "From Linnaean Species to Mendelian Factors: Elements of Hybridism, 1751–1870," *Annals of Science* 64.2 (April 2007): 175.

54. Georges Louis Leclerc Buffon, *Buffon's Natural history, containing a theory of the earth, a general history of man, of the brute creation, and of vegetables, minerals, &c. &c.* (London: J. S. Barr, 1792), 2: 265.

The exceptional status of the mule had been noted before and the animal had been considered an "unnatural" being early on. Winter in his 1670 *Gründliche und kurtz durchgehende Beschreibung, wie eine Stutherey anzurichten* stresses the artificiality of the mule when using the word "invent" (erfinden) for the mule's assumed first appearance, as mentioned in the Bible: "Die Maulthier oder Maul-Esel, wie man dieselbe zu nennen pflegt, sind nicht allererst für etlich 100. Jahren/ oder in unserm *Seculo* auf und in Brauch kommen, sondern ihr erster Ursprung komet *ex Arabia deserta*, und sind von *Ana* dem Sohne *Sebeon*, einem Enkel *Esau*, als er seines Vatters Esel in der Wüsten hütete, erfunden worden/ *vid. Genes. C. 36*. . . ." [The mules or mule-donkeys, as one usually calls them, have become common not just in the past several 100 years or in our time, but their first origin comes from the Arabian desert, and they were invented by Ana, the son of Sebeon, a grandson of Esau, when he kept watch over his father's donkeys in the desert. . . .] (86). See also Joseph Gottlieb Koelreuter, *D.J.G.K.'s Vorläufige Nachricht von einigen das Geschlecht der Pflanzen betreffenden Versuch und Beobachtungen* (Leipzig: Gleditsch, 1761–1766); and Thomas L. Hankins, *Science and the Enlightenment* (Cambridge: Cambridge University Press, 1985), 137.

55. Buffon is often quoted as one of the first who enabled a developmental notion of nature. Cf. Buffon, *Natural history*. See also Staffan Müller-Wille, "Genealogie, Naturgeschichte und Naturgesetz bei Linné und Buffon," in *Genealogie als Denkform im Mittelalter und Früher Neuzeit*, ed. Kilian Heck and Bernhard Jahn (Tübingen: Niemeyer, 2000), 109–19.

56. Immanuel Kant, *Kants Gesammelte Schriften*, ed. Herausgegeben von der Königlich Preußischen Akademie der Wissenschaften (Berlin: Georg Reimer, Walter de Gruyter, 1912–1983), 2:429.

57. Immanuel Kant, "Of the Different Races of Human Beings (1775)" in *The Cambridge Edition of the Works of Immanuel Kant* (Cambridge: Cambridge University Press, 2011), 84–85.

58. After a long passage on the terms for different degenerations from the original form (Spielarten, Abartungen, Nachartungen, Schlag), Kant defines "Rasse" as that variety which persists even under different climatic conditions (like the black race in Europe or the white race in Africa) (Kant, *Gesammelte Schriften*, 2:430).

59. Already Linnaeus had been convinced: "Varietas est planta mutate a causa accidentali" [a variety is a plant that mutated because of accidental reasons], quoted in Larson, *Interpreting Nature*, 62.

60. Again, it might have been Buffon who gave Kant the idea of several germs composing an organism. Buffon writes: "an individual is a whole, uniformly organized in all its parts; a compound of infinity of resembling figures and similar parts, and assemblage of germs, or small individuals of the same kind, which can expand in the same mode according to circumstances, and form new bodies, composed like those from whence they proceed" (*Natural history*, 2:276–77). The idea that the organism is composed of a number of germs was also discussed in the context of freshwater hydra research in order to explain why individual body parts (not only entire organisms) could be regenerated.

61. By introducing "Anlage" and "Keim" as the site of modification, Kant pursued an original path. Blumebach, whom Kant followed in many places, assumed that it is the "Zellgewebe" [cellular fiber] which reacts to stimuli and which is most open for modification in the human being: "Da aber auch zugleich dieses beym Menschen so ausnehmend geschmeidige Zellgewebe . . ., die erste und wichtigste Hauptwerkstätte des Bildungstriebes ist, so begreift sich aus allem diesen zusammengenommen, warum der Mensch folglich auch in Bildung seines Körpers und der Theile desselben so mannichfaltiger Degeneration in Spielarten, ausgesetzt ist." [Since the cellular fiber, which is so extraordinary malleable in the human being . . ., is the first and most important workshop of the formative drive, we can understand from all that that the human being in the formation of his body and body parts consequently is exposed to so many degenerations and variations.] (Blumenbach, *Beyträge*, 50–52).

62. Carus, *Nachgelassene Werke*, 100.

63. Winter, *Gründliche*, 15, 35, 19. In the early eighteenth century, qualities like skin color were mostly understood as a result of climatic conditions, which are considered temporally and, therefore, contingent forces whose influence on the organism only exist as long as they are present. However, in older breeding manuals, as that by Winter, we still find an explanation of skin color via cosmic forces: "Der Pferd Farben sind vielerley, welche fast nit alle zubeschreiben; diese haben nun ihre *Tinctur* auß den *tribus principiis*, als nemlich, *ex Sale*, *Sulphure & Mercurio*, und entstehen entweder, wann der Beschäler und Stuthe gleich von Farben sind, oder, wann *in ipsa horâ congresius maris cum foemellâ* eine gewisse *Constellation* einfällt, auch die Sonn und Mond in gewissen Himmelshäusern sich befinden, oder durch eine starke *impression* einer gewisse Farb beschihet" [Horses have many colors, one can hardly describe them all; they have their tint from the tribus principiis, namely, ex Sale, Sulphur & Mercurio, and (they) emerge either when the stud and mare are of the same color or when in ipsa horâ congresius maris cum foemellâ a certain constellation happens, also when sun and moon are in certain houses in the sky, or by means of a strong impression of a certain color] (55).

64. Joachim Christian Bergen, *Anleitung für die Landwirthe zur Verbesserung der Viehzucht* (Berlin and Stralsund: G. A. Lange, 1781).

65. Graf von Pfeil, "Ueber Voll-Veredlung der Schafe durch spanische Race (mit Tabellen)," in *Annalen des Ackerbaus 1807/6*, ed. Albrecht Daniel Thaer (Berlin: Realschulbuchhandlung, 1807), 222.

66. Wood and Orel, *Genetic Prehistory*, 13. See also Harold B. Carter, *His Majesty's Spanish Flock: Sir Joseph Banks and the Merinos of George III of England* (London: Angus & Robertson, 1964), 7.

67. Immanuel Kant, "Bestimmung des Begriffs einer Menschenrasse. Berliner Monatsschrift, Nov. 1785," in *Pädagogik*, ed. Wilhelm von Weischedel (Frankfurt am Main: Suhrkamp, 1993), 11:66.

68. Immanue Kant, "Determination of the Concept of a Human Race (1785)," in *The Cambridge Edition of the Works of Immanuel Kant* (Cambridge: Cambridge University Press, 2011), 146.

69. The close connection between sheep wool and human skin is also made by Caleb Hillier Parry: "There is another respect in which the skin of the Merino race differs from that of our native sheep. It is of a fairer hue, with a vivid tint of what is called carnation or flesh color; bearing the same relation to that of our English breeds, as the skin of women with red or auburn hair does to that of those whose hair is dark brown, or black. This tint is particularly conspicuous on those parts which are naturally free from wool, as the eye-lids and lips." ("Part II. History of the Merino-Ryeland Breed of the Author," in *Communications to the Board of Agriculture; on subjects relative to the husbandry, and internal improvement of the country* [London: Bulmer and Co., 1806], Eighteenth Century Collections Online: http://quod.lib.umich.edu/e/ecco [accessed November 5, 2011], 5:343).

70. Wood and Orel, *Genetic Prehistory*, 44, 46.

71. Johann Heinrich Fincke, *Schreiben über ungekünstelte und sichere Verfeinerung aller groben Wolle aus eigener Erfahrung und Thathandlungen aufgestellt* (Celle: Ernst August Richter, 1790), 5–9.

72. Von Veltheim, *Abhandlungen*, 380.

73. The change from horse to sheep as the preferred animal for breeding experiments coincided with a change in class of the breeders. Horse breeding traditionally was associated with aristocratic families. Winter reports: "Es haben nicht allein von vielen unerdencklichen Jahren/ grosse Herren und *Potentaten* sich der Pferde-Zucht in allem beflissen, sondern daneben auch auf dieselbe grossen Unkosten gewendet; wie man dann von *Alexandro Magno* schreibt, daß sein *Bucephalus* auß deß *Philonici* Gestüth/ in *Thessalia* umb 13. biß 16. *Talenta* seye erkaufft worden/ deß *Equi Sejani*, welches 100000. *Sestertios* gekostet/ und viel anderer zugeschweigen" [For many unfathomable years great lords and rulers have occupied themselves not only with horse breeding, but have spent great amounts of money on it; as it is written from Alexandro Magno that his Bucephalus was bought from the stud farm of Philonicus in Thessaly for 13. to 16. talents, not to speak of horse Sejanus which costed 100,000 Sesterces and of many others] (*Gründliche*, "Preface," n.p.). Bakewell, however, was yeoman. Wood and Orel suggest that Bakewell was highly aware of these class issues and that they informed his breeding praxis: "Part at least of Bakewell's reluctance to speak in detail of his breed's origin probably resulted from his impatience with the concept of pedigree (so dear to horse breeders). He gave every evidence of judging a sheep for breeding purposes more by its qualities, and those of its immediate relatives, than its ancestry" (Wood and Orel, *Genetic Prehistory*, 69).

74. Johanne Ernesto Francisco Kraus, *Der Edle Gestütt-Garten, oder aufrichtige Anleitung, zur Gestütt- und Pferd-Zieglung, auch Maulthier-Zucht* (Nürnberg: Joh. Fried. Rüdiger, 1724), 3–4.

75. Buffon, *Natural history*, 5:128.

76. Ibid., 5:128–30.

77. Kant seems to take over Buffon's idea of the perfect form as a middle in his *Kritik der Urteilskraft*. In sec. 17 with the title "Vom Ideale der Schönheit," he writes on such a "Normalidee":

> Jemand hat tausend erwachsene Mannspersonen gesehen. Will er nun über die vergleichungsweise zu schätzende Normalgröße urteilen, so läßt . . . die Einbildungskraft ein große Zahl der Bilder (vielleicht alle jene tausend) auf einander fallen; und . . . in dem Raum, wo die meisten sich vereinigen . . . da wird die mittlere Größe kenntlich, die sowohl der Höhe als Breite nach von den äußersten Grenzen der größten und kleinsten Staturen gleich weit entfernt ist; und dies ist die Statur für einen schönen Mann. . . . Diese Normalidee ist nicht aus von der Erfahrung hergenommenen Proportionen, als bestimmten Regeln, abgeleitete; sondern nach ihr werden allererst Regeln der Beurteilung möglich. Sie ist das zwischen allen einzelnen, auf mancherlei Weise verschiedenen, Anschauungen der Individuen schwebende Bild für die ganze Gattung, welches die Natur zum Urbilde ihren Erzeugungen in derselben Spezies unterlegte, aber in keinem einzelnen völlig erreicht zu haben scheint. (Kant, *Kritik der Urteilskraft*, ed. Wilhelm Weischedel [Frankfurt: Suhrkamp, 1974], 152–53).
>
> [Say, for instance, a person has seen a thousand full-grown men. Now if he wishes to judge normal size . . . then imagination (to my mind) allows a great number of these images (perhaps the whole thousand) to fall one upon the other, and . . . in the space where they come most preponderantly together . . . one gets a perception of the average size, which alike in height and breadth is equally removed from extreme limits of the greates and smallest statures; and this is the stature of a beautiful man . . . This normal idea is not derived from proportions taken from experience as determinate rules: rather it is according to this idea that rules for judging first become possible. It is something intermediate between all singular intuitions of individuals, with their manifold variations—a floating image for the whole genus, which nature has set as an archetype underlying those of her products that belong to the same species, but which in no single case she seems to have completely attained.] (Kant, *Critique of Judgment*, transl. Nicholas Walker [Oxford: Oxford University Press, 2007], 64–65.)

78. Wood and Orel, *Genetic Prehistory*, vii-viii. Bakewell's openness to inbreeding was possibly prepared by Natural Law theorists who rejected the incest taboo as a cultural norm (see chap. 3).

79. As a response to changing tastes and a new preference for fine wool, Saxony started its own breeding experiments with Merino sheep in the 1760s. The first Merino sheep arrived in Saxony in 1765. "They came as a flock of the famous Escurial *cabaña*, a personal gift to the Elector from his cousin King Carlos III of Spain. . . . a second major shipment arrived from the same Escurial source in 1776–8" (ibid. 153).

80. Cf. Albrecht Daniel Thaer, *Einleitung zur Kenntnis der englischen Landwirthschaft und ihrer neueren practischen und theoretischen Fortschritte in Rücksicht auf Vervollkommnung deutscher Landwirthschaft für denkende Landwirthe und Cameralisten* (Hannover: Bey den Gebrüdern Hahn, 1798); and his *Handbuch für die feinwollige Schafzucht* (Berlin: Ville and Krammer, 1811).

Bakewell and his breeding methods are often quoted in German breeding manuals around 1800, so also in (Dieterichs, *Von der Zucht der Hunde*, 2). On the knowledge of Bakewell's breeding methods in Germany see also (Wood and Orel, *Genetic Prehistory*, 118). Fink, Baumann, and John note in their commentary to the *Journeyman Years* that among the works on agriculture from which Goethe drew his information was Thaer's *Grundsätze der rationalen Landwirtschaft* (1812) (Gonthier-Louis Fink, Gerhart Baumann, and Johannes John, "Kommentar," in Johann Wolfgang Goethe, *Sämtliche Werke nach Epochen seines Schaffens*, ed. Karl Richter et al. (München: Hanser, 1985–1998), 17:1131.)

81. Thaer, *Einleitung*, 623, see also 680.

82. Thaer, *Handbuch für die feinwollige Schafzucht*, 48–49.

83. For Thaer's rejection of the idea that inbreeding would lead to degeneration, see also Thaer, *Einleitung*, 20, 30, 36, 48, and 49.

84. We find a similar assessment in Südeckum, but without the explanatory power found in Thaer: "Man hat hiebey gewiß nicht zu befürchten, daß die Begattung in zu naher Verwandtschaft eine Entedelung hervorbringt, weil diese—bloß auf alte Vorurtheile beruhende Besorgniß dadurch sattsam als ungegründet erscheinet, daß man, auf den meisten Schäfereyen, seine selbst gezogene Böcke zum Belegen braucht, ohne je üble Folgen davon zu bemerken" [One certainly does not have to fear here that intercourse among too close relatives leads to imperfection, because this worry—based only on old prejudices—appears totally ungrounded considering the fact that one uses in most sheep farms rams from one 's own breed for mating, without ever noticing bad consequences]. (Friederich Südeckum, *Practische Bemerkungen über die Veredelung der Schafzucht auf Niedersächsischen Schäfereyen, sowohl über deren Hindernisse als Beförderungsmittel* [Braunschweig: Friedrich Vieweg, 1800], 27.)

85. Karl Christian Gottlieb Sturm's most important written contributions to animal husbandry and breeding were his *Ueber die wichtigsten Rassezeichen der Hausthiere* (Jena 1812) and *Die Viehrassen auf den großherzoglich weimarischen Kammergütern* (1820). Sturm's *Lehrbuch der Landwirthschaft* (2 vols., 1819–1821) made him known to a larger audience (cf. Carl Leisewitz, "Sturm, Karl Christ. Gottl." In *Allgemeine Deutsche Biographie* [Leipzig: Duncker & Humblot, 1894], 37:40–41).

86. Karl Christian Gottlieb Sturm. *Ueber Raçen, Kreuzungen und Veredlung der landwirthschaftlichen Hausthiere* (Eberfeld: Büschler, 1825), 30.

87. Ibid., 3–4.

88. Ibid., 26.

89. Ibid., 30.

90. A similar assessment of the breeding process can be found in Dieterichs. In his *Von der Zucht der Hunde* from 1831, Dieterichs still bemoans the removal of the dog from its natural station by means of breeding: "Wenn gleich durch die Zähmung und außerordentliche *Entartung* der Hunde, wodurch sie zu *weit von ihrem natürlichen Zustande* abgezogen worden, ihr Geschlechtstrieb nicht an so bestimmte Zeiten gebunden ist" [Even though their sexual drive is not bound to specific times, because of the taming and extraordinary degeneration of the dogs, by which they are removed too far from their natural condition]. (6, italics mine).

91. Ibid., 110.

92. In Sturm we do not only find references to Bakewell and Thaer (Sturm, *Ueber Raçen*, 22), but we find also references to the importance of breeding experiments in Germany and in Goethe's immediate surrounding when referring to Tieffurt: "Des Großherzogs von Weimer K. H. ließen über die Paarung verschiedener Gattungen von Vögeln früher Versuche anstellen" (19).

93. "Veränderlichkeit der Racen," in Goethe, *Sämtliche Werke nach Epochen seines Schaffens*, ed. Karl Richter, 13.2:312. See Sturm (1824). See also Goethe, "Die Faultiere und die Dickhäutigen": "wir glauben auch an die Mobilität aller Formen in der Erscheinung. Hier kommt jedoch zur Sprache, daß gewisse Gestalten, wenn sie einmal generisiert, spezifiziert, individualisiert sind, sich hartnäckig lange Zeit durch viele Generationen erhalten und sich auch deshalb bei den größten Abweichungen immer im Hauptsinne gleich bleiben" [we also believe in the mobility of all forms in the appearance. Here, however, comes up the fact that certain forms, once they are generated, specified, and individualized, remain stubbornly stable for a long time through many generations and, therefore, they remain in principle true to themselves even under the greatest variation] (*Sämtliche Werke nach Epochen seines Schaffens*, ed. Karl Richter et. al., 12:245).

94. Robert J. Richards, *The Romantic Conception of Life: Science and Philosophy in the Age of Goethe* (Chicago: University of Chicago Press, 2002); and Astrida Orle Tantillo, *The Will to Create: Goethe's Philosophy of Nature* (Pittsburgh, PA: University of Pittsburgh Press, 2002).

95. Charles Darwin, *The Origin of Species by Means of Natural Selection; or, The Preservation of Favoured Races in the Struggle for Life*, 6th ed., with additions and corrections (London: John Murray, 1891), 281. Refining Goethe and Geoffrey's suggestions, Darwin contains the idea of economy in animal organization under a broader law: "I suspect, also, that some of the cases of compensation which have been advanced, and likewise some other facts, may be merged under a more general principle, namely, that natural selection is continually trying to economise every part of the organisation." (183)

96. Another example is Carus. In his *Nachgelassene Werke*, Carus formulated his evolutionary approach in the following way: "Die Materie ist die Mutter, in welcher sich die edlere Organisation ursprünglich verschloß; die Erde ist älter als die Thierheit, wie diese älter als die Menschheit" [Matter is the mother, in which the nobler organization was originally locked; earth is older than animality, as this is older than humanity] (90). While Carus's acknowledgment of the youth of humankind and the implied transition from animal to human being is innovative, his insistence that future, nobler forms are locked ("verschloß") in matter betrays his lack of comfort with the idea that matter itself could have evolved to more complex forms. Although Carus recognizes development (from animal to human being), he is also already certain about the outcome of this development: "Der erste geborne Mensch war aber eben darum noch immer nicht vollständiger und vollkommner Mensch, sondern nur halber werdender Mensch. Es gab demnach nie einen ersten Menschen im strengen Sinne, wie es keinen gebornen Menschen, d.i. einen mit der Geburt gleich fertigen vollendeten Menschen gab (wie es auch jezt noch keine eigentlichen Menschen-Geburten sondern mehr Menschenwerdungen geborner Thiere gibt). Die Natur producirte blos Menschenkeime, die sich hernach selbst entwickelten, und producirte auch diese nur als thierische Natur. War nun die Thierart, aus welcher der Mensch stammt, eine edlere Race, so auch ihr Product,—edlere Thierjungen—Menschen" [The first born human

being was (just) because of (being first born) not at all yet a complete and perfect human being, but only a half, developing human being. Hence, there was never a first human being in the strict sense, as much as there is no born human being, that is a human being that is at the moment of birth complete and perfect (similarly, we don't have real human-births now but rather the development of human beings form born animals). Nature produced only germs of humans, which developed themselves afterwards, and (nature) produced these too only as animal nature. If the kind of animal from which the human being originates was a nobler race, so was its product,—nobler animal offspring—humans] (200). See also (Olaf Breidbach, "Einleitung." In *Über Entwicklungsgeschichte der Thiere: Beobachtungen und Reflexion*, by Karl Ernst von Baer [Hildesheim, Zürich and New York: Olms-Weidmann, 1999], v–xxiii, in particular xvi).

97. See Hankins, *Science and the Enlightenment*, 137.

98. Carl Friedrich Kielmeyer, *Ueber die Verhältniße der organischen Kräfte unter einander in der Reihe der verschiedenen Organisationen, die Geseeze und Folgen dieser Verhältniße* (New unaltered reprint. Tübingen: Christian Friedrich Osiander, 1814), 18.

99. Johann Wolfgang Goethe, *Sämtliche Werke*, ed. Dieter Borchmeyer (Frankfurt a.M.: Deutscher Klassiker, 1987), series 1, 24:233–34.

100. Johann Wolfgang Goethe, *Goethe's Collected Works*, ed. Douglas Miller (New York: Suhrkamp), 12:121.

101. Goethe, *Sämtliche Werke nach Epochen seines Schaffens*, ed. Dieter Borchmeyer, 235.

102. For a different account see Astrida Tantillo who has recently argued that in Goethe "[n]ature is not spontaneous, but will driven. Goethe's principle emphasizes the will and the drive of individual entities to change their own forms and functions" (*The Will to Create*, 9).

103. This observation differs from claims by scholars such as Sabean and Nora who stress a new focus on horizontality in kinship relations.

104. Sturm. *Ueber Raçen*, 3.

105. Johann Gottfried Herder, *Auch eine Philosophie der Geschichte zur Bildung der Menschheit*, ed. Hans Dietrich Irmscher (Stuttgart: Philipp Reclam, 1990).

106. Trembley, *Mémoires*, 153–54; 175, italics mine.

107. Jacob Christian Schäffer, *Die Armpolypen in den süßen Wassern um Regensburg.* Ev. Prediger in Regensburg und Mitglied der Königl. Deutschen Gesellschaft in Göttingen. Nebst 3 Kupferplatten mit Figuren in Farben. (Regensburg: E. A. Weiss, 1754).

108. Ibid. 35, italics mine.

109. Blumenbach, Über den Bildungstrieb, repr. 1791, 88–89.

110. Erasmus Darwin, *Zoonomia; or, The Laws of Organic Life* (London: Printed for J. Johnson, in St. Paul's Church-Yard, 1794–1796), 491–501, italics mine.

111. Ibid., 480. E. Darwin's critique of language ("Owing to the imperfection of language") stands in a long tradition of language critique in English empiricism, which start with Francis Bacon, Locke, and Berkeley. See Karl-Otto Apel, *Die Idee der Sprache in der Tradition des Humanismus von Dante bis Vico* (Bonn: H. Bouvier, 1963).

112. E. Darwin's conceptualization of heredity becomes possible because he imagines the parent-child relationship analogue to the stem-branch relationship within one organism. Although Erasmus Darwin is indebted in many ways to Enlightenment notions of nature, this early "organic" approach connects him closely to Romantic biologists. The following is an early example of this Romantic approach from Kielmeyer (1793):

> Kindheit, Jugend, Alter und Tod bieten einander bei jedem wechselsweise die Hände; und in jedem dieser Zustände stehen die Wirkungen dieses Individuums mit den Wirkungen anderer Individuen der nehmlichen Gattung wieder von neuem in ein größeres System von Wirkungen verbunden, Kindheit des einen hängt an dem Alter des andern, und Jugend des einen an der Jugend des andern, und auch diese Verbindung ist so enge, daß wir nach unserer Redens- und Vorstellungsart glauben sollten, die Natur hätte die Nerven des Individuums mit denen des andern in ein Nez verschlungen, und die Eindrücke des einen würden im Sensorium des andern gefühlt.
>
> [Childhood, youth, old age, and death offer each other the hand in each person; and in each of these circumstances the effects of one individual are connected in a greater system of effects with the effects of other individuals of the same species, childhood of the one is connected to the old age of the other, and youth of the one is connected to the youth of another, and this connection too is so close that we should believe—in our manner of speaking and imagining—that nature had intertwined the nerves of the individual with the nevers of the other in a net, and the impressions of the one would be felt in the sensory apparatus of the other.] (Kielmeyer *Ueber die Verhältniße*, 6–7)

113. Fincke, *Schreiben*, 56.

114. Lorenz Oken, *Die Zeugung* (Bamberg and Würzburg: Joseph Anton Goebhardt, 1805), 55. A decade earlier, Blumenbach had already referred to Koelreuter's experiments ("Bildungstrieb," 1791, 74–76). Joseph Gottlieb Koelreuter had started his experiments on the tobacco plant in the 1760s, first results were published in 1761. His experiments were motivated by the attempt to prove the fixity of species. "He found the answer when his tobacco hybrids proved to be sterile. . . . It was for him, 'one of the most wonderful of all events that have ever occurred upon the wide field of nature.' Koelreuter's hybrid was, as he said 'the first botanical mule which has been produced by art.'" (Hankins, *Science and the Enlightenment*, 137). Koelreuter's experiments were highly regarded in the scholarly community and are still quoted in Charles Darwin's *Origins of Species* (10).

115. Sturm, *Ueber Raçen*, 16.

116. Sturm, *Ueber Raçen*, 25–26. We find a very similar assessment in Oken: "Es ist auffallend, wie hier der Vater mehr beiträgt als die Mutter. . . ." [It is noteworthy how the father contributes here more than the mother] (*Die Zeugung*, 53).

117. However, Thaer states, in contrast to Sturm, "Es ist jetzt wohl außer allen Zweifel gesetzt, daß bei der Erzeugung das männliche und weibliche Geschlecht gleich viel beitragen, und daß im Allgemeinen das junge Thier zur Hälfte die Eigenschaften des Vaters, zur Hälfte die der Mutter besitze. Die Meinung, daß der eine Theil auf die äußere Gestalt, der andere auf die innere Konstitution mehr Einfluß habe, ist völlig ungegründet" [Probably, it has now been shown beyond doubts that in reproduction male and female contribute the same, and that in general the young animal has half of its qualities from the father and half from the mother. The opinion that one part has influence on the outside form, the other on the inner constitution is totally unfounded] (Thaer, *Handbuch für die feinwollige Schafzucht*). An older formulation of the belief in the influence of the male contribution can be found in Kraus: "ein gesundes nobles *Hengst*-Roß, oder *Stallon*, kan ehender eine Stutt in der *Nobilität corrigi*ren, als eine Stutte das Roß oder *Stallon*" [a healthy and noble stallion . . . can rather correct a mare into nobility, than a mare the stallion . . .] (*Der Edle Gestütt-Garten*, 56).

118. Sturm eventually refers to Christian mythology to illustrate his claim that the female contribution is connected to materiality and animality: "Moses läßt Gott den Herrn als Element des Weibes nicht einen Theil des Gehirn's von Adam nehmen, noch ein Stückchen vom Herzen, was unsere mystischen Poeten würden gerathen haben, sondern eine Rippe als das Materiellste und eine Bedeckung der rein thierischen Functionen" [Moses does not let God, the Lord, take a piece of Adam's brain as element of the woman, neither a little piece of the heart, which our mystical poets would have chosen, but a rip as the most material and a covering of the plain animalistic functions] (*Ueber Raçen*, 102).

119. Ibid.,7, 17–18.

120. For the significance of gender on the conceptualization of nature in the long eighteenth century see Londa Schiebinger, in particular "Were Women on the Chain?"in *Nature's Body: Gender in the Making of Modern Science* (New Brunswick, NJ: Rutgers University Press, 2004), 145.

121. Similarly, Gottfried Reinhold Treviranus speaks of changes by means of "Einwirkung eines ungleichartigen männlichen Zeugungsstoffs" [influence of a different (different from the female) male seminal fluid] (*Biologie, oder Philosophie der lebenden Natur für Naturforscher und Aerzte* [Göttingen: Johann Friedrich Röwer, 1803], 0:499).

122. Ibid. 3.

123. Similar Dieterichs, *Von der Zucht der Hunde*, 5.

124. Friedrich Karl Gustav Gericke, *Praktische Anleitung zur Führung der Wirthschafts-Geschäfte für angehende Landwirthe. Erster Theil. Von der Viehzucht. Herausgegeben und mit einer Vorrede begleitet von Albrecht Thaer*, ed. Albrecht D. Thaer (Berlin: Realschulbuchhandlung, 1804), 284–85.

125. The idea of an organic unit which carries information, but does not necessarily express this information in each generation did not yet exist. Kant had contributed the notion of "Anlage" which allowed already for units smaller than the organism, but since its expression was considered to be dependent on climatic circumstances it was considered to be expressed whenever the climatic circumstances were present. There was one area where the idea of small units transmitted from parent to child was brought up and the possibility of skipping generations was considered. However, the idea did not emerge directly in the context of heredity, but in the context of parasites. Lorenz Oken, in his treatise on reproduction, considered the transfer of small organic units from one generation to the next in the context of parasitic diseases: "[D]ass sie [Eingeweide-Würmer] von Thier zu Thier übergehen, ist die noch häufig angenommene, . . . allein wegen der Erklärung ihrer Entstehung durch Forterbung von den Aeltern zum Kinde, . . . nicht durchgängigen Beifall finden konnte. . . . Aber warum hat das Kind die Würme des Vaters und der Mutter nicht, die hingegen wieder bei dem Enkel hervorkommen? Sollten die Eier durch mehre Generationen im Blute umhergeschwommen sein ohne Zersetzung, um sich in der Leber des spätern Enkels ausbrüten zu lassen?" [Most often assumed is still that they (worms in the entrails) are transferred from animal to animal, but (this) couldn't meet universal approval because of the explanation of their development by means of hereditary transmission from parents to child. . . . But why does the child not have the worms of father and mother, but they reappear again in the grandchild? Is it perhaps so that the eggs have swum in the blood for several generations without decomposition in order to be hatched in the liver of the later grandchild?] (*Die Zeugung*, 32–33). Although Oken's text refers to parasites and not to germs or similar units, Oken's hypothesis is fascinating because it—in this respect very different from Erasmus

Darwin's approach—suggested that there are units smaller than the organism which can be transferred from one generation to the next without being developed in each generation. By way of analogy, Oken's thought on parasites could have become a potent model for the question of feature transmission—but it did not.

126. A similar idea of potentiality we find in Carus:

> Anlage ist etwas Uebersinnliches und Bleibendes, zwar modificabel, aber eigentlich nicht modificirt. Sie besteht aber nicht in dem Bleibenden an sich, sondern nur in demjenigen Bleibenden, welches allgemein ist, und trotz besonderer, sogar erblicher Verschiedenheiten, dennoch immer hervorstechend seyn und sich immer verrathen, nie aber ganz verlieren musste. Anlage ist sonach der (für jede Wesen-Gattung) angelegte Keim, die Möglichkeit das ursprünglich Gegebene als das Bleibende unter allen angenommenen Modificationen zu erhalten; namentlich bei den Menschen: die Möglichkeit, das Göttliche (mithin das Genie) unter allem Wechsel des Irdischen, das Uebersinnliche unter allen Schranken des Sinnlichen sich innig und immer inniger anzueignen. Es gibt eigentlich nur Anlage für das Allgemeine (also für Körper und Geist) und sie geht auf das unzertrennte, fest in einander verschlungene, Ganze. Daher gibt es eigentlich nur Anlagen (für das Unendliche) für den allgemeinen Gebrauch der Kräfte, nicht für den besonderen, Einer Kraft.
>
> Auf dieser Ur-Anlage—(eine zufällige oder abgeleitete gibt es streng genommen nicht) beruht die Möglichkeit unter allen Veränderungen bestehen zu können, als Gattung zu beharren. Die Anlage enthält demnach die Bestandstücke der Gattung (z.B. der Erdwesen, oder namentlich der Thierheit, Menschheit). Von den ersten Menschen müssen wir daher alle [. . .] noch Etwas haben, da wir sogar Manches noch von den Thieren besizzen, und wäre es auch nicht in der zufällig äussern Form, so doch in der Innern des Gemüths. In dem ersten einzigen Urstamme lagen die Anlagen zu allen Classen-Verschiedenheiten. (Carus, *Nachgelassene Werke*, 96–97)
>
> [A trait is something metaphysical and something that stays; modifiable but actually not modified. However, it is not defined by the actual thing which stays (Bleibenden an sich), but by that staying which is general, and which—despite particular, even hereditary differences—nevertheless always needed to stand out and to reveal itself, and which, however, was never totally lost. Hence, trait is (for each species) the innate germ, the potential to maintain the originally given as that which stays in all assumed modifications; in the human being: the potential to acquire more and more the divine (the genius) in all changes of the earthly being, the metaphysical in all limits of the sensual. There is actually only one trait for the general (that is for body and spirit) and it is directed towards the undivided, deeply entangled, whole entity. Therefore, there are actually only traits (for the infinite) for the general use of forms, not for (the use) of one particular, of one force.
>
> The potential to exist in all changes, to stay as species, is based on this first trait (in the strict sense, there is no accidental or derived one). Hence, the trait contains the components of the species (for example of earth beings, or of the animality, humanity respectively). Therefore, we must have still something from the first human beings, since we have even still something from the animals, and even if it were not in the accidental outer form, it still exists in the interiority of the mind. The traits for all class differences were contained in the original and only stem].

127. E. Darwin, *Zoonomia*, 505.

128. However, the question of whether change should be considered degeneration and a temporary phenomenon or a mode of progress continued to be debated. In the first decade of the nineteenth century, research on hybridity moved to the center of agricultural breeding. Three essay competitions show how the interest in this research developed. Still in the tradition of the eighteenth century, in 1819/1822, the *Prussian Academy of Sciences* announced an essay competition on the question: "Is there hybrid fertilization in the realm of plants?" In 1830, however, the *Dutch Academy of Sciences* was already asking: "What does experience teach us concerning new species and anomalies via the artificial insemination of one flower with the pollen of another, and which agricultural crop or ornamental plants can be produced and reproduced in such a way?" (Hans-Jörg Rheinberger, "Naudin, Darwin, Nägeli: Bemerkungen zu den Vererbungsvorstellungen des 19. Jahrhunderts," *Medizinhistorisches Journal* 18, no. 3 [1983]: 198–212).

129. For a consideration of the concept of breeding before its biologization see Jenny Davidson, *Breeding: A Partial History of the Eighteenth Century* (New York: Columbia University Press, 2009).

130. See Schiebinger, *Nature's Body*; and Ernst Mayr, *The Growth of Biological Thought: Diversity, Evolution, and Inheritance* (Cambridge, MA: Belknap of Harvard University Press, 1982), in particular 97–99; and 370 respectively.

131. To what extent such hierarchies started to be questioned in the early nineteenth century can be seen in Baer's *Entwicklungsgeschichte der Thiere*, in which the author makes the remark: "Es ist überhaupt der Mensch wohl nur in Hinsicht seines Nervensystems und dem, was zunächst damit verbunden ist, die höchste Form der Thiere . . . wenn man nicht den Magen des Rindviehes, der das Gras in Chylus umwandelt, für vollkommner halt, als den Magen des Menschen" [Probably the human being is the highest form of animals only with respect to the nervous system and that which is primarily connected to it . . . unless one considers the stomach of the cattle, which transforms grass in chylus, for less perfect than the stomach of the human being] (241–42).

132. Sturm, *Ueber Raçen*, 80–81. In his work on the *Über den Bildungstrieb*, Blumenbach had already suggested that habit could become hereditary: "Die merkwürdigsten aber von solchen Veränderungen und Abweichungen der Bildung sind . . . die, so anfangs durch die Kunst aus Gewohnheit, Landessitte etc. veranlaßt, nach und nach haftend, gleichsam zur andern Natur und erblich worden sind" [The strangest of such changes and modifications of form are . . . those that—first triggered artificially (literally "by art") through habit, regional customs etc.—become fixed and become so to speak second nature and hereditary] ([1781], 260).

133. Ibid., 8.

134. Ibid, 107–9.

135. For an illustration of the energy that early modern thinkers devoted already to maintain a non-material beginning of the human being see Albrecht Koschorke's discussion of William Harvey (1578–1657) ("Inseminationen, Empfängnislehre, Rhetorik und christliche Verkündigung," in *Kunst—Zeugung—Geburt: Theorien und Metaphern ästhetischer Produktion in der Neuzeit*, ed. Christian Begemann and David E. Wellbery [Freiburg im Breisgau: Rombach, 2002], 89–100).

136. Kant, *Kants Gesammelte Schriften*, 2:438.

137. Kant, "Of the Different Races," 92–93.

138. In the beginning of the nineteenth century, Treviranus still worked with the climate theory when discussing racial differences. In order to demonstrate the difficulty of reversing forms by means of climate once the form has achieved some stability, he referred to the Jewish population: "Einen Beweis, wie schwer die Ausartung bey dem Menschen von statten geht, geben die Juden. Schon seit so vielen Jahrhunderten aus Palästina verbannt, und in alle Weltgegenden zerstreut, behauptet dieses Volk, selbst unter den verschiedensten Zonen, noch immer seine eigenthümliche Bildung" [A proof of how difficult modification works in the human being is provided by the Jews. Banned from Palestine since so many centuries and dispersed througout the world, this people maintains its particular formation even in the most diverse zones] (Treviranus, *Biologie*, [1803], 496).

139. Kant, *Kants Gesammelte Schriften*, 9:316. Kant, *Natural Science*, ed. Eric Watkins, trans. Lewis Whitebeck, Jeffrey B. Edwards, Olaf Reinhardt, Martin Schönfeld, and Eric Watkins (Cambridge: Cambridge University Press, 2012), 576.

140. Kant, *Kants Gesammelte Schriften*, 6:60.

141. Kant, *Kants Gesammelte Schriften*, 4:239. See Wimmer, 124–128.

142. Today, we seem to have settled the question of equality by assuming that the person who enters the legal realm is equal to other persons who enter this space and by acknowledging that this equality says little about our biological differences (among others). But the question of the strict line between these two spaces haunts us when we try to account for biological difference *within* the legal system, a question which is most prominently asked in the context of reliability.

143. Kant, *Kants Gesammelte Schriften*, 9:442.

144. Kant, "Lectures on Pedagogy (1803)," 438.

145. Kant, *Kants Gesammelte Schriften*, 9:441, italics mine.

146. Kant, "Lectures on Pedagogy (1803)," 437, italics mine. I have replaced "Discipline and training" given in this translation with "Discipline and breeding," since this stays closer to the biological connotation of the word "Zucht." I have also replaced the word "intelligence" with "reason" as a translation for "Vernunft."

147. Kant, *Kants Gesammelte Schriften*, 9:441. Kant, "Lectures on Pedagogy," 437.

148. Kant, *Kants Gesammelte Schriften*, 9:443. Kant, "Lectures on Pedagogy," 439.

149. Kant, *Kants Gesammelte Schriften*, 9:443–44.

150. Kant, "Lectures on Pedagogy," 439.

151. Kant, *Kants Gesammelte Schriften*, 9:445.

152. Kant, "Lectures on Pedagogy," 440–41.

153. Immanuel Kant, "Beantwortung der Frage: Was ist Aufklärung? Berliner Monatschrift. Dez. 1784," in *Schriften zur Anthropologie, Geschichtsphilosophie, Politik und Pädagogik*, ed. Wilhelm von Weischedel (Frankfurt am Main: Suhrkamp, 1993), 11:481–97.

154. In *Two Treatises on Government* (1689), John Locke states: "For every man's children being by nature as free as himself [their father], or any of his ancestors were, may, whilst they are in that freedom, choose what society they will join themselves to, what commonwealth they will put themselves under" (*Two Treatises on Government*, ed. Mark Goldie [London: J. M. Dent; Rutland, VT: C. E. Tuttle, 1993], 151–52). Paine formulated similarly in the *Rights of Man* (1791–1792): "Every age and generation must be as free to act for itself, *in all cases*, as the age and generations which preceded it. The vanity and presumption of governing beyond the grave is the most ridiculous and insolent of all tyrannies. Man has no property in man; neither has any generation a property in the generations which are to

follow" (*Rights of Man. Being an Answer to Mr. Burke's Attack on the French Revolution* [London: J. S. Jordan, 1791] 11; emphasis in original).

155. Carus, *Nachgelassene Werke*, 81.

156. Carus, "Abstammung," in *Nachgelassene Werke*, 87.

157. Ibid.

158. "Erbe," 3:708 and 3:710, in Jacob Grimm and Wilhelm Grimm, *Deutsches Wörterbuch* (Leipzig: S. Hirzel, 1854–1913).

159. Ibid., 3:710.

160. Similarly, in Zedler's *Universial-Lexikon* heritage is defined as that which is not feud: "Erbe, ist, das nicht Lehen ist" (8:1480).

161. Grimm, "Wörterbuch" in *Deutsches Wörterbuch*, 3:710.

162. Besides the possibility of inheriting land and objects, the Grimms also mention the possibility of inheriting certain character qualities as well as hereditary diseases.

163. Von Grolman in Koselleck, "Landrecht," 62.

164. Ibid., § 5, I, 1.

165. Ibid., § 42, I, 1:

166. Ibid., §§ 1 and 2, II, 3.

167. Ibid., § 2, II, 67: "Pflicht der Mütter, ihre Kinder selbst zu säugen." Ludmilla Jordanova has shown how the new relationship between child/individual and mother/nurturing entity also found its expression on a visual level. While in the beginning of the century, mother and unborn are depicted as forming one unity. In the end of the eighteenth century, the fetus is depicted independently, thereby, as Jordanova argues, reducing the mother to a mere vessel for the child's growth. ("Interrogating the Concept of Reproduction in the Eighteenth Century," in *Conceiving the New World Order: The Global Politics of Reproduction*, ed. Faye D. Ginsburg and Rayna Rapp [Berkeley: University of California Press, 1995], 369–86.

168. See for example Peter Uwe Hohendahl who speaks of "Die Krise der Männlichkeit im späten 18. Jahrhundert. Eine Problemskizze." *Zeitschrift für Germanistik* 12, no. 2 (2002): 275–86.

169. This new autonomy also found its legal expression. In the Prussian law, it is for example stated that children above fourteen have to be given free choice over their confession (§ 84, II, 2); that a young person can freely chose her profession and can go to courts, if the father opposes this choice (§ 112, II, 2); and that parents are not allowed to force children in a marriage (§ 119, II, 2). (Cf. Koselleck, "Landrecht," 62–63) For the literary discussion of the subject see Schindler (1994).

170. Johann Peter Frank, *System einer vollständigen medicinischen Policey. Erster Band. Von Fortpflanzung der Menschen und Ehe-Anstalten, von Erhaltung und Pflege schwangerer Mütter, ihrer Leibesfrucht und der Kind-Betterinnen in jedem Gemeinwesen* (Mannheim: Schwan und Götz, 1784), viii, translation mine.

171. See for example Giorgio Agamben's discussion of Foucault in *Homo sacer: Die souveräne Macht und das nackte Leben* (Frankfurt am Main: Suhrkamp, 2002), 13–15.

172. The other center of this interest was certainly the site of epidemics. It would be interesting to ask in what relationship the concept of inheritance and that of epidemic stand. In many respects they build an opposite pair, with inheritance being inherent and transmitted vertically and epidemics being a threat from outside transmitted horizontally. There are, however, also a number of common features as for example the moment of

transfer from one person to the other and the vocabulary of pollution and purity shows. For nineteenth-century questions see Jean-Paul Gaudillière, and Ilana Löwy, *Heredity and Infection: The History of Disease and Transmission* (London and New York: Routledge, 2001).

173. Johann Christian Majer, *Teutsche Staatskonstitution* (Hamburg: Carl Ernst Bohn, 1800), 460, quoted in Sara P Eigen's "A Mother's Love, a Father's Line: Law, Medicine, and the 18th-Century Fictions of Patrilineal Genealogy," in *Genealogie als Denkform in Mittelalter und Früher Neuzeit*, ed. Kilian Heck and Bernhard Jahn (Tübingen: Niemeyer, 2000), 102–3.

174. Wilhelm Gottfried Ploucquet, *Ueber die physische Erfordernisse der Erbfähigkeit der Kinder* (Tübingen: Heerbrandt, 1779), 125–26.

CHAPTER TWO

Und das Übel häuft sich von Generation zu Generation!—Denn nicht genug, dass wir Sünden unserer Väter zu leiden haben, sondern wir überliefern auch diese geerbten Gebrechen, mit unseren eigenen vermehrt, unsern Nachkommen. —Goethe to Eckermann, March 1828

[*And the evil accumulates from generation to generation! It is not enough that we have to suffer for the sins of our fathers; we hand down to posterity these inherited vices increased by our own.*]

Incest and Inbreeding

HEREDITY AS DETERMINATION

IN SEPTEMBER 2012, the European Court of Human Rights made a final decision in the case of Patrick Stübing. For months, even years, the fate of Patrick and his sister Susan Karolewski had occupied the international press. As the German weekly *Der Spiegel* put it, the key issue was "whether the protection of a powerful moral taboo is sufficient justification for punishment. And whether there are reasons beyond that taboo for locking someone up, for depriving children of their father, a woman of her partner."[1] The moral taboo in question was sibling incest. Stübing, who had been adopted as a young child, and his biological sister, who was born after he had left his birth family, got to know each other when Patrick was a young adult and Susan a teenager. After their mother's death half a year later, they developed a consensual sexual relationship and now have four children together. All but one of these children were taken away from them and Stübing has served a two-year prison sentence for being in the incestuous relationship. Patrick Stübing—who meanwhile has undergone vasectomy—has fought the prosecution in front of German and European courts with the argument that the law violates his human right to respect for private and family life. In September 2012, the European Court has made the final decision that his human rights have not been violated by German court decisions. However, the case continues to dominate discussions in Germany regarding the legal anchoring of what is considered, in the words of *Der Spiegel* article quoted above, a "moral taboo."

By placing the prohibition of incestuous relationships in the framework of morals and culture, the article implicitly questions the existence of consequences of incestuous cohabitation other than moral—for example biological. It goes on to question whether even if such consequence did exist—in fact some of Stübing's children are considered slightly handicapped—a prohibition is justified considering that marriage and cohabitation of non-incestuous couples is generally not prosecuted although many of them carry statistically similar hereditary risk for the offspring. The article's implied assumption is that culturally grounded morals do not suffice to justify prosection, but neither do potential biological risks. The article closes by quoting Stübing's lawyer, stating "One can't put this poor person in jail again."

There can be no doubt that the couple's suffering deserves our greatest sympathies. However, in focusing on the prosecuted parents by means of identification and empathy, *Der Spiegel* article overlooks that what is at stake in the Stübing case is not also the unjustified intrusion of the law in our private lives, but also how to protect such a space for private and family life while at the same not shying away from addressing biological challenges. The problem is not only that an anachronistic "moral law" has transgressed into the legal area, but the challenge is also how the legal framework could respond once it assumes that the potential problem for the offspring exists—regardless of whether or not incest should have social or moral consequences for the parents. What is ultimately at stake is how, or whether, moral, biological, and legal regulations relate to each other. The Stübing case and the arguments made in its context share, as this chapter will demonstrate, important aspects with the Mignon episode in Johann Wolfgang Goethe's *Wilhelm Meister*. In this episode, Goethe experiments with possible answers to the difficult questions surrounding incest and the incest-child in a historical moment in which for the first time biology (as object and discipline) is thought independently from legal and cultural inscriptions.

In Goethe scholarship on the *Lehrjahre* (Apprenticeship), much attention has been devoted to the difference between Mignon and Wilhelm in terms of their relationship to the Tower Society. One persistent line of argument has been that Mignon, as a poetic figure, has to perish in the prosaic and economically driven world of the society, while Wilhelm enters the association to become a useful member as a surgeon (Schlechta, Schlaffer). Little attention has been paid to the difference between Mignon and Wilhelm in terms of their ability to establish a family and to find their place in a genealogy. The novel, however, highlights that the end of Wilhelm's years of apprenticeship coincides with the establishment of a genealogy, with Wilhelm's recognition of his son Felix:

> [Wilhelm] sah die Welt nicht mehr wie ein Zugvogel an. . . . Alles, was er anzulegen gedachte, sollte dem Knaben entgegen wachsen, und alles, was er herstellte, sollte eine Dauer auf einige Geschlechter haben. In diesem Sinne waren seine Lehrjahre geendigt, und mit dem Gefühl des Vaters hatte er auch alle Tugenden eines Bürgers erworben.[2]

> [(Wilhelm) surveyed the world around him, but not like a bird of passage . . . Everything he planned was now to mature for the boy, and everything he built was to last for several generations. His apprenticeship was therefore completed in one sense, for along with the feeling of a father he had acquired the virtues of a solid citizen.][3]

For Wilhelm, the foundation of genealogy becomes the precondition for the termination of the formation process and for the attaining of adult selfhood, which allows him to enter the bourgeois society as a useful member. Paternity, citizenship, and adult selfhood coincide.

The opposition between Wilhelm and Mignon is stressed by the image of the migratory bird, whose nomadic existence Wilhelm wants to leave behind. Mignon, however, longs for the bird's migratory life: "Nun, sagte sie, Mignon klettert und springt nicht mehr, und doch fühlt er noch immer die Begierde über die Gipfel der Berge wegzuspazieren. . . . Wie beneidenswert sind die Vögel, besonders wenn sie so artig und vertraulich ihre Nester bauen."[4] ["Well," she would say, "Mignon can't jump and climb anymore, but she still feels the urge to walk over the tops of mountains . . . How I envy the birds, especially when they are building their nests nicely and quietly."][5] While Wilhelm's end of apprenticeship is marked by the wish to build something that can "last for several generations," Mignon continues to envy the birds for which even the building of nests is an integral part of the never-ending migratory journey.

Considering the close connection between the foundation of genealogy and adult selfhood that is proposed in the description of Wilhelm's entrance into the Tower Society, the question of why Mignon is unable to inscribe herself into a genealogy becomes crucial for an understanding of selfhood in the *Lehrjahre*. This chapter will examine the reasons for which Mignon has to stay outside genealogy by paying close attention to Mignon's incestuous origin. The following chapter will inquire into eighteenth-century knowledge on the transmission of hereditary defects in incestuous reproduction and their effect on genealogy in order to find answers to the question of why Mignon cannot reach adult selfhood.

In order to demonstrate the significance of Mignon's incestuous origin, it will be first necessary to inquire into eighteenth-century paradigms of genealogy and adoption. We will look at these paradigms by focusing on conceptionalizations of the relationship between paternity, citizenship, and selfhood. In the second section, this chapter focuses on dominant scholarship that assumes that Mignon is a girl and the questions that remain unanswered in the framework of this assumption. In the third section, the argument widens to engage with Mignon's incestuous origin and contemporary experiments in animal husbandry that used the method of inbreeding to obtain desired traits in the offspring. Against the backdrop of the examinations of debates in natural law and animal husbandry, the fourth section of this chapter will ask whether these debates can be made productive for a reading of the Mignon figure. Investigating this question will lead us in a fifth section to contemporary discourses on anatatomical collections and the representation of the monstrous body. The sixth section will provide an exploration of the iconographic possibilities which are opened once the Mignon figure is read as a body that is shaped by its incestuous origin. The chapter will close with a discussion of whether Goethe's take on the Mignon episode allows for a reading of Goethe as a thinker who anticipates the biologization of human life.

PATERNITY, CITIZENSHIP, SELFHOOD

It has become a standard assumption that around 1800 the paradigm of genealogy is replaced by a paradigm of adoption. Drawing on a Freudian model, Lynn Hunt has argued in her influential *The Family Romance of the French Revolution* (1992) that the model of society propelled in the wake of the French Revolution was that of a community of brothers which was established and maintained by reference to the dead (royal) father.[6] This new community of brothers was ideologically not any longer based on blood and kinship, but understood in terms of the adoptive family. Studying this model of community, scholars have pointed out the paradox that despite the preoccupation with paternity and its importance for the stability of society, eighteenth-century thinkers are unable to find biological certainty for it. In discussions of the *Lehrjahre*, authors such as Elisabeth Krimmer and Heidi M. Schlipphacke have highlighted the fact that Wilhelm's fatherhood remains in question and that instead of biological paternity homosocial bonding in the name of fraternity and adoption ultimately serves as a model for society in Goethe's novel and in eighteenth-century German literature more generally.[7] Krimmer observes that "Goethe is not only acutely aware that fatherhood is not naturally given but socially created, he is also convinced that it is this social reality that counts." Consequently, in these interpretations of the novel becoming a father does not mean finding biological certainty but to accept paternal responsibilities. Citing the example of the Harper and of Mignon, Krimmer concludes that in the *Lehrjahre* "great unhappiness and suffering arise from a father's unwillingness to accept his paternal responsibilities."[8]

Krimmer's reading convinces through its careful attention to the differences between male and female gender construction in a time when motherhood was seldom doubted but genetic paternity tests—as Krimmer notes—were not available yet. However, the focus on adoption also leaves open puzzling questions: if the Harper's problem arises from his father's refusal to accept paternal responsibilities, thereby obfuscating the sibling relation between Harper and Sperata, why does the Harper's problem start only after his incestuous relationship has been banned by family and church? And if Mignon's tragic fate originates from the Harper's own refusal to accept paternal responsibilities, why is Mignon's "sonderbare Natur"[9] [strange nature] mentioned well before she displays any signs of trauma? Even more puzzling is the case of Therese: in the *Lehrjahre*, the Tower Society inquires carefully into Therese's lineage in order to establish whether she should be allowed to marry Lothario. This marriage seems first impossible because Lothario had at some point in the past an affair with Therese's mother. However, once it turns out that the woman who raised Therese and with whom Lothario was involved is not Therese's biological mother the Tower Society decides that Therese can marry Lothario. If it were only "social reality that counts," the inquiry into Therese's parenthood could not bring greater clarity into the question of whether or not she can become Lothario's wife. Socially, the parents with which Therese grew up remain the same regardless of their biological connection to her. The care with which the Tower Society inquires into Therese's lineage—in the narrative in close proximity of the history of Mignon's incestuous origin—and the decision that Therese can marry Lothario because his former lover is not her biological mother suggests that the problem at stake goes beyond social reality. Krimmer's article points us in an important direction. It is one of the few works which notes a differentiation of cultural and biological registers in the understanding of gender and selfhood in Goethe's *Lehrjahre*, by arguing that motherhood around 1800 is defined via the body while fatherhood—in absence of biological proof—is defined via cultural parameters. However, questions remain regarding the novel's obsession with incest, paternity, and lineage beyond their cultural signification. While there can be no doubt that the *Lehrjahre* text provides moments of uncertainty and "ambiguities of patrilineal descent,"[10] the fascination with the model of adoption has made us overlook the renewed importance of biological fatherhood, reproduction, and the genealogical paradigm in which the conditions of selfhood are discussed in the *Wilhelm Meister*.

One author who is often quoted in scholarship on eighteenth-century ideas about fatherhood and citizenship is Kant student and friend Theodor Gottlieb von Hippel (1741–1796). In his 1774 publication *Über die Ehe* [On Marriage], Hippel finds strong words for the connection between paternity and citizenship: "Das Wort Vater ist ein großes Wort, das größte im Staate; wer nicht Vater ist, verdient auch den Namen Bürger nicht, und, um freigebig zu sein, nur halb den Namen Mensch"[11] [The word father is a grand word, the most important in the state. Whoever is not a father does not deserve the title "citizen" either; indeed, it is being generous to say he even halfway deserves to be called a human being].[12] As we have just seen, the exemplarity of such apodictic judgments on the coincidence of fatherhood, citizenship, and even humanity for eighteenth-century understandings of individual and society have been questioned in the scholarship of the last decades. Indeed, with its strong focus on the significance of education for individual and society, Hippel's text seems to support in many ways a non-biological, constitutional understanding of the state. He states at some point that "nichts ist wahrer, als jener . . . Gedanke, daß man seinem Erzieher mehr, als seinem Erzeuger zu danken habe"[13] [nothing is truer than that . . . thought that one has to thank one's educator more than one's genitor]. However, a more comprehensive reading of Hippel is helpful here. It is important to recognize that Hippel connects the coincidence of fatherhood and citizenship explicitly to *biological* paternity. Hippel's defense of education is based on the conviction that biology provides a necessary condition which cannot be ignored despite the force of social constructs. This conviction is expressed nowhere more clearly than in those passages in which Hippel discusses the fate of individuals who cannot "father." He illustrates this situation with the figure of the eunuch:

> Es soll nie ein Kastrat ein großer Bösewicht gewesen seyn; wenn es wahr wäre, so siehe dich um, ob er Fähigkeiten genug besaß, ein Bösewicht zu werden. Einem Menschen, der nur einen halben Körper hat, fehlt es auch an Seele; er konnte nach dem Mosaischen Gesetzbuche nicht in die Gemeine des Herrn kommen, und der Name Mensch steht ihm nur als Schmutztitel zu. Das Leiden eines Verschnittenen ist größer, als das Leiden des Kreuzträgers Hiob, der nach ausgestandener Versuchung Söhne und Töchter zeugte. . . .Er kann nicht an Kindesstatt annehmen, weil die Adoption eine Fiktion, eine schöne Kunst ist, die sich nach dem Lineal der Natur richtet; und kurz, er muß sich auf das Singen, so wie der Kapaun auf schönere Federn, einschränken.[14]

> [It is said that there never has been a castrato who was a heinous criminal; if that is true, we ought to ask ourselves whether a castrato in fact even has the wherewithal to become a criminal at all. A person with only half a body will also be lacking in soul; according to Mosaic Law, such a person was not permitted to "enter the congregation of the children of Israel," and the designation "human being" could never be more than a bastard title for him. The suffering of a eunuch is greater even than that of the cross-bearer Job, who at least was able to beget sons and daughters after he had borne his trials . . . He cannot adopt, for adoption in truth is merely a fiction, a beautiful art regulated by the rules of Nature. In short, he must limit himself to his singing, as the capon to his pretty feathers.][15]

Hippel's acknowledgment of the biological precondition of selfhood and citizenship climaxes in a statement which connects the possibility of freedom of will—and by extension, the possibility of a soul and of humanity—intrinsically to biology.[16] According to Hippel, only that person which fulfills certain bodily conditions has the ability to decide freely between good and evil and has, therefore, the potential to become a villain. The bodily shortcomings extend beyond the inability to procreate; they render it also impossible for the eunuch to adopt. Adoption might be a "fiction," but this fiction is intrinsically conditioned by nature. Hence, the condition of possibility of adoption is, paradoxically, the ability to procreate. Far from superseding corporeal conditions, adoption is dependent on them. If we follow Hippel's argument and look at the *Wilhelm Meister*, the question of whether or not Wilhelm is indeed Felix's biological father is not as important as his potential of becoming a biological father, and, therefore, an adoptive one. Even the fictional adoption needs biology as its reference point. His potential to sire children also allows Wilhelm to become a "Mensch" [human]. The inability to prove biological fatherhood in the eighteenth century opened the "fiction" of adoption to virtually everybody. However, where the inability to procreate seemed to be certain as in the case of sexual "degenerations" humanity and selfhood were in question.[17]

Hippel's assessment of the coincidence of fatherhood, citizenship, and humanity thus raises the question of whether Mignon's problem in the *Lehrjahre* can be explained sufficiently by reference to social conditions. The question arises whether Mignon's problem is not so much that she does not have a father, but that she is not able to "father" herself.[18] Searching for the reasons behind Mignon's inability to reach adult selfhood, we need to consider whether Mignon might have been born with a hereditary burden connected to her incestuous origin which makes it impossible for her to reproduce.

IT'S A GIRL!

A significant part of scholarly discussions on the Mignon figure can be summarized under two connected topics: first, Mignon's androgyny and, second, her inability to overcome this androgyny as the result of a traumatic childhood—in contrast to other androgynous figures in *Wilhelm Meister*, like Therese and Nathalie, who eventually embrace their assigned gender roles.[19] Thomas W. Kniesche interprets Mignon's problems as a hysterical neurosis connected to the social "Zwang, ein Mädchen werden zu müssen"[20] [coercion to become a girl]. And Friedrich A. Kittler shows that the Tower Society explains Mignon's androgyny as an unfortunate result of her traumatic childhood and lack of proper *Bildung*.[21]

Throughout the literature on Mignon, it is mentioned again and again that she does not want to wear woman's clothes. The assumption in these articles remains, however, that woman's clothes are those which fit her sex and that Mignon's refusal is the result of a psychological problem or predisposition. In short, in the dominant scholarship Mignon has been read as a female figure and has been listed among the female figures that Wilhelm encounters along his way to *Bildung*. Typical assessments are: "Die überirdische Mignon und die unterirdische Melusine . . . können aus feministischer Perspektive als kulturelle *Präsentationsformen des Weiblichen* . . . gedeutet werden"[22] [From a feminist perspective, the celestial Mignon and the subterranean Melusine . . . can be read as *cultural representations of femininity*]; or "Vor allem an den Figuren der Mariane und Mignon stellt sich die Bearbeitung des *weiblichen Körpers* dar . . ."[23] [(p)articularly, the figures of Mariane and Mignon represent the discussion of the *female body*]; and "[w]ith Mignon in the novels *Wilhelm Meisters Lehrjahre*, Ottilie in *Die Wahlverwandtschaften*, and to some degree with Makarie in *Wilhelm Meisters Wanderjahre*, Goethe created three *female characters*."[24] Even for readers who have offered comprehensive and careful accounts of androgyny in Goethe's *Meister*, it has remained challenging to accept Mignon's liminal stage. In her detailed study *Embodying Ambiguity* (1998), Catriona MacLeod assumes that Mignon is a girl. In reference to Natalie's suggestion that Mignon wears female clothes, MacLeod remarks "it is Natalie who attempts to have Mignon dress in a manner *befitting her sex*" and in reference to Mignon's preference for the angelic dress, MacLeod continues "[b]ut in keeping the long white dress, she [Mignon] is *not acknowledging her female sex*."[25] Astonishing is the fact that no scholar seems to question that Mignon *is* a girl. Despite the abundance of discussions on Mignon's androgyny, the two-sex model which Thomas Walter Laqueur has described so convincingly prevails even in recent scholarship on the Mignon figure, thereby demonstrating the model's force well beyond the time of its conception.[26]

Only a few scholars have tried to turn the tide. Robert Tobin in his article "The Medicinalization of Mignon" questions the Tower Society's attempts to normalize Mignon within established gender norms. Rejecting the dichotomy that the Tower Society establishes and drawing on the fact that the word "mignon" was used in the late eighteenth century to refer to male homosexual prostitutes, Tobin highlights the homoerotic quality of Mignon's relationship to Wilhelm.[27] He rejects the Tower Society's medicinalization of Mignon as an inability to accept erotic desires outside the heterosexual paradigm.[28] Tobin is correct in stressing the Tower Society's inability to accept that Mignon does not fit the gender dichotomy; a dichotomy which is perhaps most clearly articulated by Therese whose embrace of Wilhelm arguably leads to Mignon's death.[29] However, Tobin's wish to rescue a space for homoerotic desire within a system that considers gender a cultural construct makes it also impossible to (re)consider Mignon's body. Providing a critical reading of the Tower Society and its attempts to medicinalize Mignon, Tobin states that the "novel's initial presentation of Mignon has nothing to do with a medical perspective: She is simply 'ein junges Geschöpf.'" Tobin further supports his argument against a corporeal, medical, explanation of Mignon's otherness by stating that the "novel's neutrality at this point with regard to her health condition is especially striking because of the androgyny which is practically the first characteristic attributed to her."[30]

Tobin's careful observations are important and his assessment that the novel presents us with a critique of the Tower Society's insistence on a heterosexual culture convincing. However, Tobin's reading of Wilhelm's uncertainty concerning Mignon's sex and the claim of underlying homoerotic desires remain confusing since Tobin at the same time continues to argue that Mignon is biologically a girl. This confusion is highlighted when Tobin refers to Kurt Robert Eissler's interpretation of Mignon's seizure in book 2 of the *Lehrjahre*. Eissler reads this seizure as a detailed and realistic description of female orgasm.[31] Following Eissler, Tobin summarizes that of all erotic scenes in the novel, the scene of Mignon's seizure "portrays most explicitly a woman's feelings, sensations, and sexuality." It is "the only scene in the novel portraying a *woman's orgasm*."[32] Much like scholars before him, Tobin assumes a split between Mignon's body, her androgynous feelings, and Wilhelm's homoerotic desires. By declaring the described seizure as a female orgasm and Mignon as a girl, many scholars so far have cemented the two-sex model that they want to criticize.

Despite the abundance of scholarly accounts that work with the assumption that Mignon is a girl and that her problems are of traumatic nature, these assumptions leave a number of questions unanswered: first, if Mignon is a girl, why does the text continue to switch between male and female pronouns even after Wilhelm has declared her a girl (this oscillation is even more pronounced in the *Theatralische Sendung*)? Second, if Mignon's problem is located on the psychological-social level, why does the Tower Society inquire so deeply into Therese's genealogy? Third, if Mignon is an androgynous girl, why is this androgyny depicted as so problematic, while Natalie's and Therese's "Mannweiblichkeit" [literally manwomanhood] is depicted as unproblematic and eventually overcome? Fourth, how is it possible to reconcile the assumption that Mignon has a "woman's orgasm" with the claim that Wilhelm has homoerotic desires for her? Fifth, if Mignon's problems are the effect of trauma, why is she described as a "sonderbare Natur" already before traumatic events such as the separation from the mother have taken place?

We learn of Mignon's "strangeness" only after her death in the context of the narration of her childhood:

> [Speratas] Kind wuchs heran, und zeigte bald eine *sonderbare Natur*. Es konnte sehr früh laufen, und sich mit aller Geschicklichkeit bewegen, es sang bald sehr artig, und lernte die Zither gleichsam von sich selbst. Nur mit Worten konnte es sich nicht ausdrücken, und es schien das Hindernis mehr in seiner Denkungsart als in den Sprachwerkzeugen zu liegen.[33]

> [(Sperata's) child grew and soon revealed *strange characteristics*. It began very early to run and develop great skill in bodily movements, it would sing very pleasingly, and soon learned by its own efforts to play the zither. But it could not express itself in words, and the obstacle seemed to be in its mind rather than in its speech organs.][34]

The report seems relatively vague regarding the reasons for Mignon's "strange characteristics" (literally "strange nature") and attributes her inability to speak to her "mind." However, an inquiry into this strangeness might hold important clues for an understanding of the Mignon figure. We will need to dig deeper into her biography to find out about the possible reasons for Mignon's strangeness and for her inability to speak.

MIGNON'S INCESTUOUS ORIGIN

In the center of Mignon's childhood story, stands her incestuous origin. As is revealed at the end of the *Lehrjahre*, Mignon is the daughter of siblings, Augustin and Sperata, who did not know about their kinship at the time that Mignon was conceived. When, after their father's death, it turns out that Augustin and Sperata's relationship is incestuous, the brothers urge Augustin not to see the sister any longer. Although Mignon has already been conceived, the family feels a social and moral necessity that brother and sister separate. Augustin, the brothers are convinced, "[Augustin] solle überlegen, daß er nicht in der freien Welt seiner Gedanken und Vorstellungen, sondern in einer Verfassung lebe, deren Gesetze und Verhältnisse die Unbezwinglichkeit eines Naturgesetzes angenommen haben"[35] [(Augustin) should remember . . . that he was not living in the free world of his own thoughts and ideas but in a state whose laws and customs had the inviolability of natural law].[36] The brothers argue that the historical norm, the tradition, has reached validity comparable to the law of nature. Rejecting the brothers' plea, Augustin insists that his incestuous inclination is justified precisely because it follows nature:

> Nichts schien ihm [Augustin] heilig als das Verhältnis der Natur und der Religion, der sittlichen Rechte und der bürgerlichen Gesetze wurden von meinem Bruder aufs heftigste durchgefochten. Nichts schien ihm heilig als das Verhältnis zu Sperate, nichts schien ihm würdig als der Name Vater und Gattin. Diese allein, rief er aus, sind *der Natur gemäß*, alles andere sind Grillen und Meinungen. Gab es nicht edle Völker, die eine Heirat mit der Schwester billigten? Nennt eure Götter nicht, rief er aus, ihr braucht die Namen nie, als wenn ihr uns betören, uns von dem *Wege der Natur* abführen, und die edelsten Triebe, durch schändlichen Zwang, zu Verbrechen entstellen wollt.[37]

> [My brother vehemently opposed all he (the confessor) said about the demands of nature, of religion, morality and social order; nothing was sacred to him save his relationship to Sperata, no names more worthy than those of father and wife. "Such designations," he said, "are *natural*, all else is fancy or opinion. Haven't there been great nations that have sanctioned marriage with one's sister? Don't talk about your gods, you only refer to them when you want to fool us, lead us *away from nature*, distort our noblest instincts into crimes by infamous coercion, committing your victims to utter distraction of mind and disgraceful misuse of their bodies, burying them alive . . ."][38]

Augustin's argument is closely aligned with what natural law philosophers had to say about the incest taboo.[39] In the context of the rejection of religious institutions and theological justifications, Enlightenment authors reconsidered moral rules and strove to base these rules on nature. Many scholars were of the opinion that the incest taboo cannot be justified in the context of natural law.[40] This argument was made with reference to indigenous populations that were thought to live in the natural state and for which incest was not a taboo[41] and it was further supported by the claim that the reproduction of the first people had to be necessarily incestuous, if one assumed with the Bible that there was *one* couple in the beginning. In some instances, for example in Montesquieu's *Histoire d'Aphéridon et d'Astarté* within the *Lettres persanes* (1721), incestuous relationships were even depicted as a bourgeois ideal of a private and egalitarian union against the public sphere reigned by hierarchies. Using the argument of "double retour," there were also discussions whether incestuous relationships and the offspring of such relationships were more perfect than others since they combined innate blood inclination with erotic attraction. By arguing for such a "voice of the blood," Enlightenment authors revived an opinion propagated by non-Christian authors like Ovid, who declared incestuous relationships particularly strong, since they combined kinship and love: "Jupiter allowed what pleases / and the marriage between sister and brother legitimizes everything. / This kinship relationship is tightened by a strong chain, / which Venus herself has forged." Concerning incest, Enlightenment authors generally free the incestuous individual of the accusation of guilt and they do so in the framework of natural law.[42]

Like natural law philosophers, Augustin considers the incest taboo as unjustifiable by reference to nature and as an expression of human-made laws and conventions. In the Enlightenment tradition, he refers to indigenous, "edle" [noble], people who do not know the taboo. He turns against Christian religious and moral tradition and refers to nature, which he equates with the heart, to legitimize his relationship with the sister: "Fragt nicht den Widerhall eurer Kreuzgänge, nicht euer vermodertes Pergament, nicht eure verschränkten Grillen und Verordnungen; fragt die Natur und euer Herz. . . ."[43] [Don't listen to the echoes of your cloisters, don't consult your musty parchments, your crotchety and quirky regulations: ask Nature and your hearts].[44]

Augustin's reference to nature and heart, to a place outside of social and clerical limitations echoes a Roussian understanding of nature as an entity before all civilization, but it also strives for a grounding of this anti-clerical impetus in the Christian tradition. Revisiting biblical rhetoric and imagery, Augustin says to his brothers: "Seht die Lilien an, entspringt nicht Gatte und Gattin auf Einem Stengel? verbinde [sic] beide nicht die Blume, die beide gebar, und ist die Lilie nicht das Bild der Unschuld und ist ihre geschwisterliche Vereinigung nicht fruchtbar?"[45] [Consider the lilies: Do not husband and wife grow on one and the same stem? Does not the blossom they bear unite them? And is not the lily the image of innocence? Is not its sibling union fruitful?][46]

Augustin's lily imagery alludes to the Immaculate Conception and it is rhetorically closely related to Matthew 6:28–29: "And why are you worried about clothing? Look how the lilies of the field grow; they do not toil nor do they spin, yet I say to you that not even Solomon in all his glory clothed himself like one of these."[47] Furthermore, for eighteenth-century readers Augustin's evocation of the lily invited connections to Linnaean taxonomy, which drew close connections between plant morphology and human marriage.

The image of the lily then is used here in four different ways. As in the Christian iconography where the lily refers to the Immaculate Conception, it is used symbolically to illustrate Augustin's and Sperata's innocence (and the innocence of Mignon's conception). Second, the rhetorical form that Augustin employs alludes in structure and vocabulary to a Biblical verse that stresses the greatness of human security in God's protection. Third, as in the Linnaean system of plants, the lily is anthropomorphized and—in an inversion of Linnaean's application of human marriage to plant reproduction—becomes for Augustin a metaphor for the legitimacy of the couple's union. Finally, Augustin takes the comparison between humanity and plants literally: he compares the process of reproduction in flowers and humans in order to argue that both of them follow a law which is natural rather than human. The intensity of Augustin's rejection of the legal framework becomes even clearer when we consider the verse in Matthew which succeeds the verse to which Augustin refers (the statement on the lilies). Solomon is traditionally viewed as the highest incarnation of human justice, yet even his "glory" pales besides the perfection of the divine law of nature, embodied in the lily. This super-human rightness is claimed by Augustin for his love to the sister.[48]

What is the significance of Augustin's reference to nature for an understanding of incest and the Mignon figure? Nonnenmacher has observed that there is basically no literary work before the *Lehrjahre* in which incest is connected to botany.[49] However, despite Nonnenmacher's brief consideration of botany and biology in the context of Mignon's incestuous origin, he comes to the conclusion that in the Mignon episode: "Die Handlungsführung läßt letzlich offen, ob die Natur den Inzest oder nicht doch dessen Verbot rechtfertigt. Diese Unklarheit darüber, welche Norm von der Natur gestützt werde und welche nicht, rückt Goethes Inzestgeschichte in deutliche Distanz zur aufklärerischen Motivtradition, in der über die Naturrechtskonformität des Inzest weitgehend Einvernehmen bestanden hatte"[50] [In the end, the plot line leaves it open whether nature justifies incest or, after all, its prohibition. The uncertainty regarding the question of which norm is supported by nature and which is not, separates Goethe's incest story significantly from the Enlightenment motive tradition, in which there had been broad agreement that incest is conformed with natural law]. While Nonnenmacher acknowledges the differences between the standpoint taken by the *Lehrjahre* and natural law philosophy with its clear rejection of the incest taboo, he also stops short of reading the Mignon episode as an argument for the justification of the taboo. Considering *both* the tragic outcome of Augustin's and Sperata's relationship *and* Mignon's exclusion from the Tower Society, it is understandable that scholars hesitate to read any normative statements in Goethe's treatment of incest. I want to suggest that this uncertainty in the assessment of the Mignon episode is largely based on the fear that an inquiry into, or an acknowledgment of, Mignon's problematic "nature" would lead to her parents', or for that matter Mignon's, moral condemnation. However, it is precisely the differentiation between nature and morality that is decisive for Goethe's take on the incest question.[51]

Despite his demonstrative distinction between nature and culture, Augustin himself confuses the two realms—and the novel's exposure of this confusion can be read as an implicit critique of the lack in precision in the notion of "nature" which natural law philosophy used as its basis. Confronted with the brothers' accusations, Augustin invites them to visit the place where he and the sister meet: "[B]esucht uns an jenen Spalieren, wo die Zitronen und Pomeranzen neben uns blühn . . . und dann wagt es, uns mit euren trüben, grauen von Menschen gesponnenen Netzen zu ängstigen"[52] [(V)isit us in those groves where lemons and pomegranates surround us . . . and then try to frighten us with your dismal, gray, man-made entrapments!].[53] In using the reference to the garden, Augustin aims to highlight nature's justification of his incestuous relationship. However, it is remarkable that the fruits that he names are trained on a trellis ("Spalieren"). As a product of both nature and culture, they do not support the dichotomy between the two spheres that Augustin wants to evoke when rejecting the coercion ("gray, man-made entrapments") of human laws.

The doubts regarding Augustin's justifications climax with Augustin's claim that not only the couple's love, but also the product of their incestuous union demonstrates nature's justification of the incestuous relationship: "Wenn die Natur verabscheut, so spricht sie es laut aus; das Geschöpf, das nicht sein soll, kann nicht werden, das Geschöpf, das falsch lebt, wird früh zerstört. Unfruchtbarkeit, kümmerliches Dasein, frühzeitiges Zerfallen, das sind ihre Flüche, die Kennzeichen ihrer Strenge."[54] [Nature clearly indicates what it abhors: a creature that should not exist, cannot exist, develops wrongly, or is soon destroyed. The marks of her curse, the signs of her severity are: barreness, stunted growth, premature decay.][55] Augustin's justification of the incestuous relationship via the reference to nature appears in the narrative after Mignon's death and—considering Mignon's sad existence and early demise—turns into his accusation. Nature, it seems, has condemned Mignon who was born with a strange nature, has problems to speak, and dies early.

Mignon's strangeness is located in her "nature" and is observed already before Mignon is removed from her mother and before she could experience any trauma, psychological or educational harm. We have observed before that the reasons for Mignon's strangeness and for her inability to speak are difficult to ascertain. However, it is important to note that in connecting the inability to speak to the child's "Denkungsart" rather than to the "Sprachwerkzeug"[56] ("the obstacle seemed to be in its mind rather than in its speech organs"),[57] the novel establishes a connection to the case of a woman which was quoted in medical literature of the time to illustrate that the ability to speak was located not in the "Sprachwerkzeug" but in some yet to define organ related to the brain. When in 1805, in the midst of the excitement surrounding the beginning of phrenology, Christian Heinrich Ernst Bischoff (1781–1861) published the second edition of his *Darstellung der Gallschen Gehirn- und Schädel-Lehre* (Presentation of Gall's Brain and Skull Theory)—together with an attached treatise on the subject by Goethe's physician Christoph Wilhelm Hufeland—Bischoff attributed the discussion of the woman's case to the Strasbourg professor for surgery and anatomy Johann Friedrich Lobstein (1736–1784):

> Daß manche Kinder so schwer und so spät erst sprechen lernen, liegt in unvollkommener Entwicklung dieses Organs, und nicht der eigentlichen Sprachwerkzeuge, wie man gewöhnlich glaubt. Letzteres wird dadurch bewiesen, daß oft Menschen bei den unvollkommensten Sprachwerkzeugen sprechen lernen (wie Lobsteins Dissertation, betitelt: *Feminae elinguis historia* beweist), und bei der größten Integrität derselben oft sehr schwer oder gar nicht.—Kömmt dieses Organ gar nicht zur Entwicklung, so ist Blödsinn die Folge davon. [58]

> [The reason for which some children learn to speak with such difficulty and so late is to be found in the underdeveloped organ (related to the brain), and not in the actual speech organs, as one usually believes. The latter can be proven by the fact that frequently individuals learn to speak although their speech organs are underdeveloped (as Lobstein's Dissertation, entitled *Feminae elinguis historia* shows), and that they often learn only with difficulty or not at all although their speech or-

gans are perfect.—If this organ (related to the brain) is not developed, debility is the result.]

Both Bischoff and the narrator in the *Lehrjahre* use the vocabulary of the "Sprechwerkzeug" [speech organ] and distinguish it from some ability located in the organ responsible for intellectual capacities, stating that the ability to speak is more closely aligned with that capacity than with the actual tools of speech. Johann Friedrich Lobstein, to whom Bischoff referred to make his case, was professor for surgery and anatomy in Strasbourg where the young Goethe had participated in the lectures and exercises on anatomy offered by "dem dicken Chirurgen"[59] [the corpulent surgeon]. These lectures and exercises instilled in Goethe a life-long fascination with anatomy. Forty years later, he still followed with great interest Gall's phrenological theories which Bischoff described in his publication.[60] Reading Mignon's inability to speak in the context of the case that Bischoff attributes to Lobstein, suggests that her difficulty predates any trauma she could have experienced in her childhood or later and is located in the body. Mignon's early death and her "disability" suggest that Goethe is interested in a biological explanation of Mignon's "strange nature." Goethe seems interested in following Augustin's conviction that nature establishes an unbreakable rule that reigns over man-made laws, but he also seems to draw a different conclusion: the law of nature does not justify incest, it condemns it.

Unlike the nature to which Augustin refers, this nature is neither in necessary harmony with emotional needs nor with ethical norms, instead it is governed by strictly biological laws. If incest is problematic in the *Lehrjahre*, it is not problematic because it violates moral laws, but because it leads to problems of the mind, early death, and infertility in the offspring. The determinism which comes to light once we read Mignon in the context of anatomy and heredity might be shocking, but it might also explain why so much of Mignon's figure which has continued to puzzle scholars cannot be explained with reference to "Poesie." Mignon's "Eiertanz" [egg dance] is not an expression of some notion of autonomous art, it rather develops, as Hellmut Ammerlahn has noted, with mechanical precision.[61] If art is evoked here, it is the artistry of the machine which had served the French Enlightenment as model to explain nature's determinism. However, where Enlightenment thinkers such as La Mettrie (1709–1751) had totalized this understanding of nature by declaring it the governing principle of life, Goethe—whose opposition to French materialism is well known—is interested in both its force and its limit (more on this below).

Augustin's reference to nature's strict laws—even though he might be mistaken in his assumptions about the content of these laws—raises the question of the degree of contemporary knowledge on incestuous reproduction. Interestingly, but perhaps not surprisingly, the assessment of incest varied widely, depending on the discipline and discourse. Possibly under the influence of natural law philosophy, a liberal tendency in the understanding of incest is recognizable in German court decisions around 1800.[62] In what can be called the most modern German law of the time, the *Allgemeine Landrecht* from 1794, regulations concerning the prohibition of marriage among relatives cover a great variety of relationships:

> §. 3. Ehen zwischen Verwandten in auf- und absteigender Linie sind gänzlich verboten. §. 4. Auch Ehen zwischen voll- und halbbürtigen in oder außer der Ehe erzeugten Geschwistern sind unzuläßig. §. 5. Stief- oder Schwieger-Aeltern dürfen sich mit ihren Stief- oder Schwieger-Kindern ohne Unterschied des Grades, nicht verheirathen. §. 6. Diese Eheverbote (§. 5.) dauern fort, wenn gleich die Ehe, wodurch die Verbindung zwischen Stief- oder Schwieger-Aeltern und Kindern entstanden war, durch Tod der richterlichen Ausspruch wieder getrennt worden. §. 7. In allen übrigen Graden der Verwandtschaft und Schwägerschaft ist die Ehe erlaubt, und bedarf es dazu keiner Dispensation.[63]

> [§. 3. Marriages among relatives in descending and ascending line are strictly prohibited. §. 4. Marriages among sanguineous and half-sanguineous siblings, whether born in or outside wedlock, are also prohibited. §. 5. Stepparents or parents-in-law are not allowed to marry their stepchildren or children in law without difference of degree. §. 6. These prohibitions of marriage (§. 5.) continue even if the marriage through which the relationship between stepparents or in-laws and children developed, is ended through death or legal sentence. §. 7. In all other degrees of kinship and affinity the marriage is allowed, and no dispensation is necessary.]

This passage from the *Allgemeine Landrecht* allows for two observations: first, in the *Allgemeine Landrecht* the prohibition of marriage was not limited to blood relationships. Marriage between stepparents and in-laws was as much prohibited as marriage between biologically related siblings. Secondly, the prohibition had a relatively limited range regarding blood relationships. It is true that marriage between relatives in descending or ascending lines as well as marriages among siblings was prohibited. But while marriage between uncle and niece was not allowed, the law did not prohibit cousin marriages.

The lack of distinction between blood and other kinship connections shaped the rules that the *Allgemeine Landrecht* formulated. This does not mean that legal reformers were completely unaware of the possibility to differentiate between biological and socio-historical kinship definitions. Gottlieb Carl Svarez (1746–1798), the leading eighteenth-century reformer of the Prussian law which resulted in the *Allgemeine Landrecht*, stated in his lectures on law and state:

> Die Gründe, warum Ehen unter nahen Verwandten verboten werden, sind teils physische, teils moralische. Die physischen gründen sich auf Erfahrung, die man bei allen Arten der Tiere gemacht hat, daß aus der Vermischung zu nahe verwandten Blutes Rassen entstehen, die, wenn besonders die Vermischung durch mehrere Generationen fortgesetzt wird, durch Schwäche, Dummheit, Verdorbenheit sich sehr nachteilig auszeichnen. Die weit stärkeren moralischen Gründen sind a) daß, wenn nahe verwandte Personen . . . die Erlaubnis hätten, einander zu heiraten, die Aussicht hierzu und die gemeine Vertraulichkeit . . . die Ausschweifungen der Wollust zu reizen und Sittsamkeit und Scham . . . aus den Familien verbannen würden; b) daß, wenn die Familien untereinander heiraten dürften, daraus ein moralischer Egoismus und eine dem Staat höchst nachteilige Absonderung der Familien voneinander entstehen würde, anstatt daß nach der entgegengesetzten Einrichtung die Familien sich durch Heirat miteinander verbinden müssen und so am Ende der Staat eine einzige große Familie wird.[64]
>
> [The reasons for which marriages among close relatives are prohibited are partly physical, partly moral. The physical (reasons) are based on experience, which one has made with all kinds of animals, (namely) that from the mixing of too closely related blood, races develop which—in particular when the mixing is continued through several generations—are marked very unfavorably by weakness, stupidity, depravity. The much stronger moral reasons are a) that, if closely related individuals . . . had the permission to marry each other, the prospect of it and the common intimacy . . . would stimulate the excesses of lust and ban demureness and chastity . . . from families; b) that, if family members were allowed to marry among each other, a moral egotism would develop, and

> a segregation of families from each other, which were very disadvantageous for the state, instead of the opposite arrangement according to which families have to unite in (exogamous) marriages and this way the state eventually becomes one big family.]

By referring to experiences made in animal husbandry, Svarez clearly distinguishes between the physical and moral effects of incest.[65] The quote, however, makes the biological considerations secondary and focuses on the moral and social effects of the rejection of the incest taboo. Svarez supports the incest taboo because it guarantees the mixing of society and, by extrapolation, helps to avoid the accumulation of wealth and influence in certain social subgroups. Svarez imagines the state as one big family in which all members are increasingly related by blood. The rejection of the incest taboo is for him not based on biological hesitations, but on the social and ethical consequences.[66] Svarez is typical for Germany. German authors in the eighteenth century agreed that biological effects of incest ultimately could not have any legal significance. It is at this exact moment of a beginning awareness for, but indecision about, the significance of the biological impact that Goethe joins the discussion.

CONTEMPORARY KNOWLEDGE IN ANIMAL HUSBANDRY

In discussions of eighteenth-century German literature in general and the Mignon episode in particular, the incest motive has been discussed frequently. However, the question of a biological dimension of the motive has hardly ever been touched. A notable exception is David W. Wilson's 1984 article, in which he addressed this dimension without exploring it. Wilson concluded this article with the remark that contemporary thinkers "did not have the benefit of Darwin, Mendel, and recent studies. . . . without a 1980s knowledge of genetics, they have nothing on which to base their suspicion, so that suspicion took on the status of superstition."[67] A close look at the literature shows, however, that documented knowledge about the effects of inbreeding in animals predates Darwin and Mendel by several decades and that Goethe is at the forefront of this consideration.

The cultural rejection of the incest taboo in the context of natural law philosophy possibly prepared the openness to the use of inbreeding in animal husbandry. Robert Bakewell's success with inbreeding was well-known throughout Europe (see chapter 1). However, although the reports by Culley, von Brühl, Thaer, and others on Bakewell's breeding methods and successes were published in Germany, German breeders were cautious in applying inbreeding methods in their agricultural practices. A popular book on breeding in Germany was Georg Stumpf's *Versuch einer pragmatischen Geschichte der Schäfereien* [Essay on a pragmatic history of sheep farming] (1785). In this publication, the author strongly recommended a different ram for every generation of crossing.[68] Drawing on comprehensive research, Wood and Orel conclude that the "[o]pposition to breeding in-and-in was commonplace among German writers . . . [Rees 1819]."[69]

The rejection of inbreeding as a method in animal husbandry might have been connected to traditional religious sentiments in German lands vis-à-vis incest. The incest taboo was less pronounced in Germany than in other European countries—especially in France where anti-clerical feelings were particularly strong—in part because a loosening of the incest taboo had been underway since Luther had pleaded for the continuation of marriages which only later were revealed to be incestuous. As a result, there was widespread acceptance of relationships that developed without knowledge of their incestuous quality (as in Augustin and Sperata's case). Increasingly, the rejection to inbreeding was *not* any longer based on religious and social conventions, but based on experience with inbreeding experiments themselves.[70] Already Buffon had alluded to potential problems of degeneration (see chapter 1), but in the late seventeen and early eighteen hundreds British breeding experts like Caleb Hillier Parry (1755–1822), John Saunders Sebright (1767–1846), and John Sinclair (1754–1835), who had witnessed the extension of inbreeding practices, started to issue warnings about the degenerative effect of inbreeding.[71] These warnings referred in particular to loss of sexual differentiation and increasing infertility among the offspring of incestuous breeding in livestock, particularly sheep.

In 1806, Parry, member of the Royal College of Physicians of London and honorary member of the Physical Society of Göttingen, contributed a long article on the history of the Merino-Ryeland sheep breed to the British *Communications to the Board of Agriculture*.[72] Among other aspects like wool quality, Parry reflected upon the options for the sheep's improvement by means of fattening (the latter had been one of Bakewell's prime concerns). Operating initially within a framework which stressed the influence of environmental factors, he observed that the fattening of the Merino-Ryeland "may probably be accelerated by early and uniform luxuriance of keep." But he added:

> Perhaps also the same end may be promoted by *breeding in and in*. This has been suggested to me by Mr. Davis, who thinks the early fattening of

> the New-Leicester breed to be chiefly owing to this cause. He says, that this constant *incestuous intercourse* produces, in both sexes, a *deficiency of the powers of generation*, and that of nursing in the female; reducing them to a state approaching to that of *eunuchs* . . . If this opinion be well founded, it shows that the Divine Law against incest has a physical, as well as moral end.[73]

Here then, we can observe that the author draws a close connection between incestuous reproduction and degeneration, in particular the loss of clearly distinguishable sexual features which coincides with infertility. What shapes the given quote is a strange tension between a rehabilitation of inbreeding as a means to improve certain flock qualities and a rejection of incestuous propagation as reason for sexual degeneration in the offspring. That both statements are possible at the same time is closely related to the fact that the criteria for the assessment of incestuous propagation are biological and economical. The results are judged according to these criteria and therefore the question of whether the outcomes are desirable depends solely on the market and purpose for which the sheep is bred. The explanation of the anomaly in sexual determination and reproductive abilities does not need the reference to a divine order any longer, however, it is still considered as conforming to it. Biology increasingly becomes independent from moral discourses.

HERMAPHRODITISM

Mignon's first introduction in the *Wilhelm Meister* is as "ein junges Geschöpf"[74] [a young creature].[75] Mignon's declaration as girl is presented as the result of a rather arbitrary decision process by Wilhelm: "Er sah die Gestalt mit Verwunderung an und konnte nicht mit sich einig werden, ob er sie für einen Knaben oder für ein Mädchen erklären sollte. Doch entschied er sich bald für das letzte"[76] [He looked at the figure with amazement, uncertain whether it was a boy or a girl. But he finally decided in favor of the latter].[77] Despite Wilhelm's decision and despite the dominant reading of Mignon as a girl, doubts about Mignon's sex remain alive throughout the text and textual evidence pointing to this ambiguity is abundant. When Melina wishes that Mignon wears "Weiberkleider" [clothes for women], Mignon clings to Wilhelm and responds with agitation: "Ich bin ein Junge, ich will kein Mädchen sein"[78] [I am a boy, I don't want to be a girl].[79] Similarly—as many commentators have noted—throughout the novel and in even more so in the *Theatralische Sendung*, the narrator switches between male and female pronouns and articles. And even after Mignon's death, the boys' choir still switches back and forth between pronouns when referring to Mignon: "Ach! wie ungern brachten wir *ihn* her! Ach! und *er* soll hier bleiben! . . . als wir mit Rosen kränzten *ihr* Haupt, blickte *sie* hold und freundlich"[80] [Oh! How reluctantly did we bring *him* here! Oh! And *he* ought to stay here! . . . when we garlanded *her* head, *she* looked dear and friendly].[81] In her last song, Mignon wishes for a state where the gender question is mute: "Und jene himmlische Gestalten, / Sie fragen nicht nach Mann und Weib"[82] [For all those glorious heavenly forms, / They do not ask for man or wife].[83] Wolfgang Riedel, puzzled by the changing pronouns, attributes the confusion to a mistake by Goethe or one of his aides: "Maskuliner Gebrauch von 'Mignon' lag im Sprachgebrauch der Goethezeit . . . durchaus nahe und könnte also sehr leicht auch als Fehlleistung (Goethes, seines Schreibers oder auch des Setzers) in den Text gekommen sein"[84] [Masculine use of "Mignon" suggested itself in the language use of Goethe's time . . . therefore, it could have easily come into the text by a mistake (of Goethe, his scribe or also of the typesetter)]. The textual indecision concerning Mignon's sex, which has irritated commentators, invites thoughts that Jarno's rejection of Mignon as "ein albernes, zwitterhaftes Geschöpf"[85] [a silly, hermaphrodite creature][86] is more than an annoyed remark about Wilhelm's company with the theatrical nomads and his hesitation to enter a more productive life. Against the backdrop of contemporary animal husbandry, we can conclude that Mignon is presented as deeply shaped by her incestuous origin displaying a "sexual degeneration" that makes it impossible to assign one sex to her body. Drawing on Jarno's statement, I want to suggest that Mignon is biologically a hermaphrodite (or intersex person as we would say today).[87]

Only if we understand Mignon's "problem" as one that is conditioned by her body, it becomes clear why the androgyny which Therese and Natalie display earlier in the *Lehrjahre* is never depicted as problematic.[88] In contrast to Mignon, Therese and Natalie derive their masculinity from clothes that they wear. Their double gender is the result of masquerade, not of the body.[89] Since they can replace their androgynous display with clearly defined gender roles, they can become adults.[90] Mignon's body, however, is not replaceable within a system marked by contingency and, therefore, allows her neither to reproduce nor to become a member of society, which is based on classificatory systems that her body resists.[91]

In her book on comic androgyny in Jonson and Shakespeare, Grace Tiffany has observed that Renaissance authors used the terms hermaphroditism and androgyny still interchangeably and that the terms "almost always described a psychological and behavioral condition."[92] Ruth Gilbert comes to the same conclusion for the early modern period.[93] While Ann Rosalind Jones and Peter Stallybrass observe that in Renaissance Europe "gender is never grounded: there is no master discourse which is called upon to fix the essence of gender,"[94] in the eighteenth century the body obtained an unprecedented normative force. In the new paradigm, it was thought possible to distinguish between the body and its cultural inscription, between biology and culture. In the *Lehrjahre*, Mignon's "Zwitterhaftes" [hermaphroditism] is bound to the body, while Therese and Natalie's "Mannweiblichkeit" [androgyny] remains a cultural and psychological phenomenon. By negotiating androgyny and hermaphroditism, Goethe's *Lehrjahre* is one of the works in which the split between the body and its cultural inscription is newly formulated and negotiated.

By the eighteenth century, hermaphroditism had already a long history in the cultural imagination of Western Europe, in which biological aspects were included but not separated from mythical-religious or aesthetic accounts. From antiquity to medieval times, the figure's otherness had opened up a space for the interpretation of the hermaphrodite's possible meaning. In *The City of God*, Saint Augustine expressed the challenges that the hermaphrodite poses for classificatory attempts and formulated one of the most influential descriptions: "[A]n age can seldom be without an hermaphrodite, though they be not ordinary—persons I mean that are so perfect in both sexes that we know not what to term them, man, or woman."[95] As a solution to the classification problem, Augustine declared the hermaphrodite a monstrosity.[96] As such, the hermaphrodite became in the Christian tradition a sign and message of God. As Ruth Gilbert states, monsters "were, in fact, signs and inherently connected to representation (demonstration). To prove the point he [Augustine] noted that the etymology of the word monster 'evidently comes from *monstrare*, »to show« because they [monsters] show by signifying something.'"[97]

Reading the hermaphrodite as monster and, therefore, as a sign of God, shaped the medieval understanding of the hermaphrodite and still informed early modern accounts, like Jacob Rueff's account of the famous "Monster von Ravenna" (see fig. 2.1). He says:

> [I]m Jar als man zehlet 1512. ward zu Rauenna eine Wundergestalt geboren / mit einem Menschenhäupt / vnnd wolgeformierter Brust vnnd Leib / hatt auff dem Häupt ein Horn / keine Arm / an statt der Armen zween Flügel / Weib vnd Mans Glieder neben einander . . . an

der Brust aber zwey Zeichen Y vnd + . . . Solcher vnd anderer Wundergeburten / können wir wol gedencken / dass viel mehr gesehen vnd an tag kommen seyn / denn wir hierinnen begreiffen können / an solche aber sollen wir deß ewigen Gottes Wunderwerck erlehrnen / auch seine Straffe vnd gerecht Urtheil erkennen / vnd die zu vnserer Besserung auffnemmen.[98]

[In the year 1512, a miraculous figure was born in Ravenna, with a human head and with well formed breast and body. It had a horn on top of the head, no arms, instead of the arms two wings. Female and male organs next to each other . . . at the breast, however, two signs Y and + . . . Of such and many other miraculous births, we can think that many more were seen and came to light than we can grasp here. They are supposed to teach us the eternal wonder works of God, also his punishment and just judgment and we are supposed to perceive them to our improvement.]

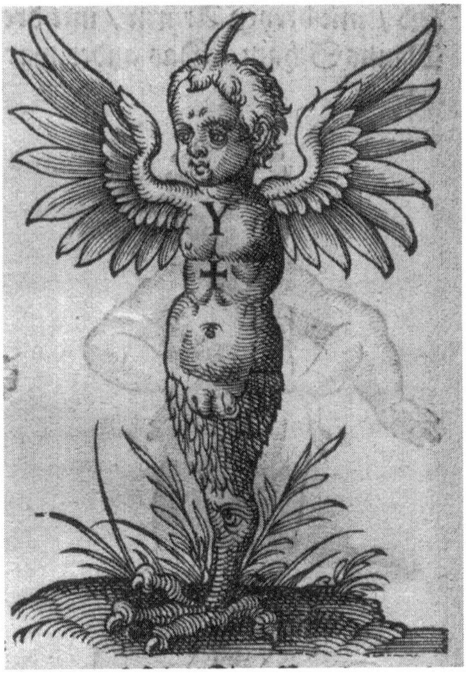

fig. 2.1. The Monster of Ravenna. Courtesy of Thomas Fisher Rare Book Library, University of Toronto. Jacob Rueff. Hebammen Buch *(Midwife Book). Gedruckt zu Franckfort am Mayn, 1580.*

Although the hermaphrodite challenged existing classificatory systems, for Rueff there was no doubt that ultimately the figure could be read as a sign and message of God. It was precisely the inability to place the hermaphrodite in available explanatory systems—it was precisely the abundance of meaning—that turned the figure into a prominent site of reading of God's message.

At the same time, the early modern period experiences the onset of a decisive shift in the understanding of the hermaphrodite. Medical practitioners pointed increasingly to the fact that the hermaphrodite's body can neither be read within traditional cultural systems nor as a sign of a divine order. Even more, we can observe a growing conviction that as mere body the figure of the hermaphrodite resists classifying interpretation in general. As Kathleen Long reports, the hermaphrodite became the center of an intensive debate in which surgeons, who learned their trade by observing the living body, were facing doctors and professors who taught medicine referring to classical philosophy, particularly Aristotle. Long arrives at the conclusion that the figure of the hermaphrodite in the early modern period can be associated "with the rise of empirical science and the valorization of clinical practice over bookish learning."[99]

Despite the new focus on an empirical approach to the hermaphrodite body, the drive to classify this body did not subside. The evidence, however, was now differently produced. The seventeenth and eighteenth century, as Maximilian Schochow has described in his comprehensive study on *Die Ordnung der Hermaphroditen-Geschlechter* [The order of the hermaphrodite sex], experienced not only a heightened interest in hermaphroditism, but also brought forward a new way of observing the hermaphrodite body by focusing on his/her sexual organs.[100] Where the hermaphrodite monstrosity was read by Augustine as a divine sign, physicians in the seventeenth century started to observe, touch, and vivisect specific body regions to decide the sex of the hermaphrodite.[101] In the search for a sexual classification of the hermaphrodite body, seventeenth- and eighteenth-century physicians like Jacques Duval (1555–1615?), and in particular surgeons like Georg Arnaud (1698–1774) and Jean Jacques Louis Hoin (1722–1772), did not read the signs of the body any longer for a divine message. For the first time, they examined sexual parts with their fingers and tried to find the unseen truth by entering the openings of the body to discover the hidden sexual organ.[102]

Goethe's physician Christoph Wilhelm Hufeland (1762–1836) operated within this methodological framework when, in 1801, he described the often discussed body of the hermaphrodite Maria Dorothea Derrier whose unusual sexual organs had been discovered when she visited the Berlin Charité for an unrelated reason (see fig. 2.2).[103] Hufeland reports:

> Die Ruthe ist von ansehnlicher Größe, sehr gut und vollkommen formirt, bis auf einen kleinen Theil freystehend, mit einer völlig gebildeten Vorhaut versehen, die sich über die Eichel vor- und zurückschieben lässt, auch bekommt sie zuweilen, besonders gegen Morgen, schwache Erektionen. . . . Die weiblichen Geburtstheile sind übrigens in folgender natürlichen Beschaffenheit vorhanden. Die großen Lefzen ganz natürlich, die kleinen nur schwach gedeutet, die Oeffnung der Harnröhre unter der Ruthe, der Mutterscheide, welche aber so enge ist, daß man kaum einen Federkiel hineinbringen kann, auch das Becken ist ganz weiblich gebildet.[104]
>
> [The rod (penis) is of remarkable size, very well and perfectly formed, freestanding except a small part, equipped with a perfectly formed foreskin, which can be glided back and forth over the glans; she also has from time to time, particularly in the early morning, weak erections. . . . By the way, the female sexual organs are present in the following natural constitution. The great lips totally natural, the small ones are indicated only faintly, the opening of the urethra under the penis, under the vagina, which, however, is so narrow that one can hardly stick a quill in it, the pelvis is also perfectly formed.]

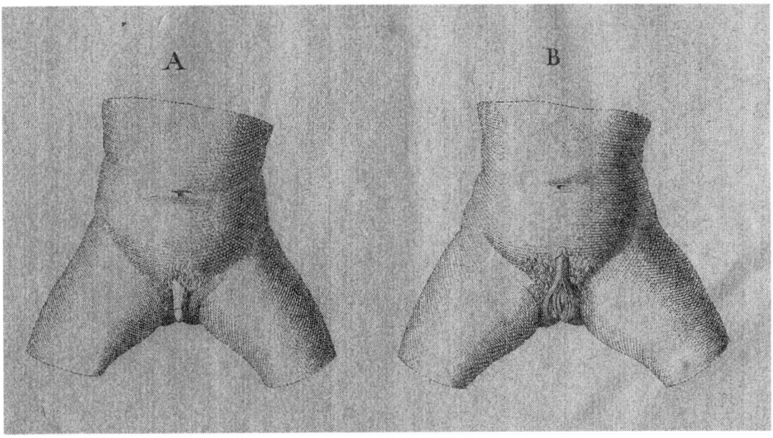

fig. 2.2. Maria Dorothea Derrier (Mursinna, 1801). Universitätsbibliothek Leipzig, Allg. med. Lit. 1527-n (1.1801), Abb. Christian Ludwig Mursinna. "Von einer besonderen Naturbegebenheit (Of a particular natural event)." Journal für die Chirurgie, Arzneykunde und Geburtshilfe. *Erster Band, drittes Stück. Berlin: in der Felischischen Buchhandlung, 1801, 555-59.*

Focusing on the sexual organs, Hufeland's surgical eye and hand discovered both sexes. Like Hufeland, contemporary physicians were confronted with a variety of bodily formations. In contrast to Saint Augustine, for whom the hermaphrodite presented a perfect unity of male and female features, eighteenth-century physicians were overwhelmed by an abundance of forms which seemed to evade any norm. But the urge for deciding for one of the two sexes persisted. Hufeland ended his report on Maria Dorothea Derrier with the reassuring words: "Da nun offenbar die Haupttheile und Eigenschaften der Weiblichkeit vorhanden sind, und die wesentlichen Theile der Mannheit, die Testikel fehlen, so ist diese Person für nichs anders, als für ein weibliches Geschöpf mit einer monströsen Clitoris zu halten"[105] [Since apparently the main parts and features of femininity are present, and the main parts of masculinity, the testicles, are missing, the person is to be considered for nothing else than a female being with a monstrous clitoris].

Hufeland's contemporary Johann Christian Stark (1753–1811) examined Derrier again, but came to a somewhat different conclusion. In particular he denied that there had been any menstrual flow as previously observed. In his opinion, the opposite was true: "Dagegen hat sie öfter einen nach ihrer Beschreibung wahren Saamenschleim unter Erektionen während des Schlafs abgehen sehen, nach dem ich und andere sie auf diese Erscheinung aufmerksam gemacht haben"[106] [In contrast, according to her (Derrier's) description—after I and others have pointed out this phenomenon to her—she has seen more often true ejaculation combined with erections during sleep]. While Stark made different observation than Hufeland, his conclusion was similar in the sense that he also denied real hermaphroditism and tried to solve the challenge within the two-sex model. For Stark, Derrier is a "missgestaltete Mannsperson"[107] [degenerated man].

As Hufeland's and Stark's observations illustrate, hermaphroditism in the eighteenth century cannot be understood without considering the redefinition of gender under the paradigm that emerges around 1800 and that Laqueur has called the "two-sex model."[108] For centuries, scholars and medical doctors had followed Aristotle's gender theory, in which woman was understood as lesser man.[109] This model became questioned in the eighteenth century and more and more replaced by a model which emphasized the difference between man and woman and located this difference in the body. This new model, in which the body obtained a previously unknown normative power, enforced gender divisions and hierarchies—and was bound to increase interest in, or concern about, the body of the hermaphrodite. Where the medieval age recognized the hermaphrodite's body as a sign of "deß ewigen Gottes Wunderwerck"[110] [the miracle work of the eternal God], practitioners in the eighteenth century followed a new two-sex model and attempt to categorize each individual hermaphrodite as either (rather) male or female.[111]

The zeal for sexual categorization within the binary model eventually led to the wish for surgical "corrections" of the hermaphrodite's sexual "degeneration." An often discussed case in contemporary medical treatises is that of the hermaphrodite "B.*M.*" This intersex person was born in 1745 and first declared a girl. Only in puberty, B.*M.* realized that her/his sexual organs were not typical for a girl. After the parents' death, B.*M.* decided never to tell anyone about her/his condition and to take the knowledge of it to the grave.[112] The case came only to light when the then forty-nine year old B.*M.* intended to marry. A medical and legal battle ensued. Although doubts remained regarding B.*M.*'s ability to sire children, B.*M.* was eventually declared a man and allowed to marry. What is interesting in this case is not only the act of categorization, but also that one of the medical practitioners concerned with the case, Johann Feiler, considered the possibility of surgical intervention with the goal to enable the hermaphrodite to have intercourse. Although Feiler apparently did not perform the operation, he concluded that "Eine solche Operation ist nicht nur denkbar, sondern auch sehr einfach"[113] [such operation is not only conceivable, but also very simple] and provided two illustrations supporting his claim (see fig. 2.3).

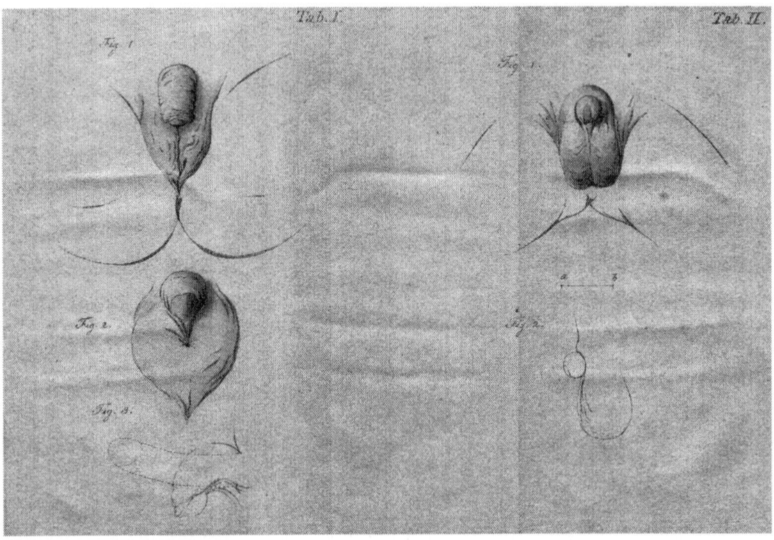

fig. 2.3. Surgical intervention. Universitätsbibliothek Leipzig, Anat. 4283, Abb. Johann Feiler. Über angeborene menschliche Missbildungen im Allgemeinen und Hermaphroditen insbesondere *(On congenital human deformations in general and hermaphrodites in particular). Landshut: Philipp Krüll, Universitätsbuchhändler, 1820.*

The medical profession had entered the path to finding ways to correct hermaphroditism to better match the two-sex paradigm by means of surgical interventions—interventions which are so intimately linked to a history of humiliation and pain by those who underwent them.

Not surprisingly, the figure of the surgeon plays an important role in the context of the Mignon episode. Mignon's first encounter with the surgeon happens after the theatrical group has been surprised by a band of robbers. Like Wilhelm, Mignon has been hurt in the violent event: "Mignon war einige Tage sehr still gewesen, und als man in sie drang, gestand sie endlich, daß ihr rechter Arm verrenkt sei. . . . Man schalt sie, daß sie das Übel nicht eher entdeckt habe, doch merkte man wohl, daß sie sich vor dem Chirurgus gescheut, der sie bisher immer für einen Knaben gehalten hatte"[114] [Mignon had for several days been very quiet, and when asked why, she finally admitted that she had sprained her right arm. . . . They scolded her for not telling them sooner that she was hurt, but they had noticed that she was afraid of the surgeon who all this time had taken her for a boy].[115] We can read Mignon's hesitation vis-à-vis of the surgeon as a sign of general shyness, but we can also read it as sign of her unwillingness to disclose her sexuality and the discovery of the "evil" as the discovery of Mignon's hermaphroditism (Mignon's arm can be read metonymically for another dislocated "member" of her body).

On the basis of these observations regarding the hermaphrodite body, we can now have a new look at the seizure scene. Drawing on Eissler's and Tobin's perceptive recognition of the erotic potential of the scene, I want to suggest that its closure has been misread. The narrator describes the end of the scene with the following words: "[Mignons] starren Glieder wurden gelinde, es ergoß sich ihr Innerstes"[116] [(Mignon's) rigid members became gentle, her innermost flowed over].[117] It is, of course, possible to read this over pouring of the inner as the flow of those tears which are mentioned shortly before. However, associations surrounding the transition from stiffness to relaxation accompanied by an outpouring of liquid rather suggest an ejaculation. In fact, as we have seen, when considering the hermaphrodite body, surgeons of the time consider ejaculations one of the main criteria in the assessment of the hermaphrodite body.[118] Mignon's seizure has also to be read within this context of arousal and ejaculation.[119]

By looking at the history of cultural imagination and medicalization of hermaphroditism, we have observed that the figure of the hermaphrodite emerged in the eighteenth century as a body which escaped existing classificatory systems and which was thought in close connection to new empirical methodologies and the profession of the surgeon. We have also observed that despite the challenges to systematization and despite new empirical methodologies, efforts to subsume the hermaphrodite body under culturally sanctioned paradigms did not subside. The urgent wish to classify this body as precisely as possible led to a new interest in the hermaphrodite corpse. While Hufeland seems to have been satisfied with examining the living body, Stark exclaimed: "Doch habe ich immer gewünscht! wünsche es noch sehr, dass man die Zwitter nach dem Tode möchte secirt haben, oder die noch existenten nach ihrem Tode öfnen möge, um die Uebereinstimmung der äussern und innern Theile zu sehen"[120] [However, I have always wished! And still wish it very much, that one dissected a hermaphrodite after death or that one might open the ones which still exist after death, in order to see the correlation of outer and inner parts]. It is precisely such dissection which takes place at the end of the *Lehrjahre*.

ANATOMICAL COLLECTIONS

Scholarship so far has claimed that the preparation of Mignon's corpse for the Hall of the Past belongs in the context of mummification practices in reference to Egyptian rituals.[121] While such references are not excluded, Mignon's corpse also belongs in the context of a contemporary craze for the preservation of dead bodies, in particular teratological ones, for anatomical collections. It is astonishing that in scholarship so far, it has not been noticed that the preparation of Mignon's corpse is staged as the preparation of a biological, teratological, specimen. Immediately after Mignon's death the doctor announces in eerily objectifying language what he and the surgeon—from the first moment fascinated with the riddle of Mignon's body—have planned all along:

> Halten Sie sich von diesem traurigen *Gegenstande* entfernt, und erlauben sie mir, dass ich den Resten dieses *sonderbaren Wesens*, soviel meine Kunst vermag, einige Dauer gebe. Ich will die schöne Kunst, einen Körper nicht allein zu *balsamieren*, sondern ihm auch ein lebendiges Ansehn zu erhalten, bei diesem geliebten *Geschöpf* sogleich anwenden. Da ich ihren Tod voraussah, habe ich alle Anstalten gemacht, und mit diesem Gehilfen [der Chirurgus] hier soll mir's gelingen.[122]

> [Stay away from this mournful sight (*object*) and let me use my art to give some permanence to the remains of this unusual person (*strange being*). I (the doctor) will start immediately to employ the delicate art

> of *embalming*, and also preserve an appearance of life in this beloved *creature*. Since I foresaw that she was dying, I have made all preparations, and my assistant (the surgeon) and I will see that we succeed.][123]

The doctor refers to the dead Mignon here as an "object," a "strange being," and a "creature." He mentions that he has anticipated Mignon's death and his words suggest that he has eagerly awaited it, since he has already prepared everything for the procedures which are supposed to preserve the dead body ("I have made all preparations"). This objectifying language together with the uncontained eagerness with which the doctor seems to have awaited Mignon's death reminds us not only of Johann Christian Stark's wish to dissect a hermaphrodite, but also of the so called "Materialmangel"—that is the lack of dead bodies—which in the eighteenth century was felt in the medical sciences regarding the need of corpses for the training of doctors. Goethe had great contempt for the resulting run for dead bodies and the lack of respect for the individual who just died.[124] We can feel some of this contempt when the topic of dissection resurfaces again in the third book of the *Wanderjahre*. There, the suspicion is raised that even murders have been committed to obtain corpses for the training of doctors.[125] The lack of respect for the dead and their bereaved families is a topic that occupied Goethe throughout his life. In the third book of the *Wanderjahre*, but also in a letter written in the last weeks before his own death, Goethe still worked on suggestions to alleviate the need for dead bodies in the medical training by replacing them with wax models.[126]

In the *Lehrjahre*, however, wax is employed in a different way. Embalming ["balsamieren"], the word that the doctor uses to describe the preparation of Mignon's body, was a term frequently employed for the description of a preparation method of corpses for the *Naturalienkabinett* [anatomical collection]. "Embalmed bodies" form a distinct group within the systematic of such collections. In the printed index of the Leipzig *Museum Linckianum* from 1783, we find the following rubrics: "A. Unzeitige Geburten. B. Mißgeburten. C. Gerippe. D. *Einbalsamierte Körper*. E. Einzelne anatomische Theile" [127] [A. Premature births. B. *Monstrosities*. C. Skeletons. D. *Embalmed Bodies*. E. Individual anatomical parts, italics mine]. Similarly, Caspar Friedrich Jenckel, also known as C. F. Neickelio, used the term when presenting such anatomical collection in his influential *Anleitung zum rechten Begriff und nützlicher Anregung der Museorum oder Raritätenkammern* [Instructions for the right definition and useful idea of museums and anatomic collections], published in 1727:

> Etliche Curiöse, und vornemlich Medici, legen sich auch auf ein Anatomie-Cabinet, oder in grossen Städten findet man gantze Anatomie-Kammern, in welchen . . . allerley anatomirte oder zerschnittene, zubereitete und *balsamirte* Cörper und Glieder gezeigt werden. Je zierlicher nun die Sceleta aufgerichtet seyn . . . , daß nicht allein das Auge, sondern auch vornemlich *das Gemüth und Gedächtniß* den meisten Nutzen in Betrachtung derselben empfinden kan; um desto schöner ist ein solches Cabinet bestellet. Die besondere Curiosität in Anatomicis aber besteht auch darinnen, wenn man *monströse oder ungemeine Objecta* vorzeigen kan. . . .[128]
>
> [Many scholars, and particularly doctors, establish an anatomical cabinet, or in big cities one can find entire anatomical collections, in which . . . are shown all kinds of dissected or cut, prepared and *embalmed* bodies and members. The finer the skeletons are positioned . . . , so that not just the eye but particularly also the *mind and memory* can draw the most advantage in observation of it, the more beautiful such a cabinet is organized. Particular curiosity in such anatomical cabinets consists also in the ability to present *monstrous or immense objects*. . . .]

Jenckel's vocabulary reminds us of the vocabulary that is used in the context of the preparation of Mignon's corpse for the Hall of the Past. Not only are the bodies embalmed for display, but this display convinces in particular through its effect on emotions and memory, as it seems intended for the Hall of the Past. Jenckel also stresses that the particular pride of the anatomical collection is its wealth in teratological exhibition pieces, among which Mignon's body could be counted.[129]

Once the connection of Mignon's embalmed body to the anatomical collection and the medical-biological context is established, it becomes clear that the Abbé does not simply describe a mummification process when presenting Mignon's corpse to the mourners, but a modern preservation practice:

> [D]ie Geschicklichkeit des Arztes konnte das schöne Leben nicht erhalten. . . . Aber wenn die Kunst den scheidenden Geist nicht zu fesseln vermochte, so hat sie alle ihre Mittel angewandt, den Körper zu erhalten und ihn der Vergänglichkeit zu entziehen. Eine *balsamische Masse* ist durch alle Adern gedrungen und färbt nun *an der Stelle des Bluts* die so früh verblichenen Wangen. Treten Sie näher meine Freunde und sehen Sie das *Wunder der Kunst* und der Sorgfalt![130]

> [The doctor's skill could not preserve the beauty of her life. . . . But if art could not give permanence to her spirit, it could employ every skill to preserve her body and save it from decay. *Balsam* has been introduced into all her veins and, *instead of blood*, this colors those cheeks that faded so early. Draw near, my friends, and observe the *wonders of art*, the sum of solicitude!][131]

The balsamic mass which fills Mignon's blood vessels is not connected to Egyptian practices (where the body is drained from fluids) as Kniesche would have it, but quotes injection techniques widely used for representation and preservation of blood vessels. If a connection to Egyptian practices is possible, then this connection can only be established via the self-understanding of contemporary surgeons. As Barbara Stafford notes, the "French surgeon [Garengeot (1688–1759)] located its [surgery's] venerable origins in the Egyptian custom of embalming."[132] To illustrate the presence of Egyptian embalming practices in eighteenth-century understandings of surgical practices, Stafford points to an illustration in the *Physique Sacrée* (1732–1737) by Johann Jakob Scheuchzer (1672–1733) which presents the embalming of Israel (see fig. 2.4).[133]

Injections used to present blood vessels belonged to the highlights of anatomical preparations in the seventeenth and eighteenth century. Many discoveries of vessels and organs became only possible through these injections.[134] While the technique for injection of vessels dates back at least to Leonardo da Vinci (1452–1519), decisive progress was made by Dutch researchers, in particular Jan Swammerdamm (1637–1680) who introduced the wax injection—a method which was in practice until the twentieth century. Since the success of the technique depended on size and condition of the vessels, the best results were obtained with corpses of children. The preparation of the body was a delicate and time consuming process which needed careful preparation. In a treatise, *Anweisungen zur praktischen Zergliederungskunst* [Instructions for the practical art of dissection] (1791) which describes the necessary procedures, Johann Leonhart Fischer (1760–1833) recommended having everything ready at the time of death: preparing a sufficient amount of water in order to warm the organ which is supposed to be injected, warming the syringes, and making a strong enough fire in order to melt the injection mass.[135] It is for this reason that the doctor in *Wilhelm Meister* announces that he has, in anticipation of Mignon's death, prepared everything in order to guarantee a successful procedure. Further stressing the length of the process, the doctor adds: "Erlauben Sie mir noch einige Tage Zeit, und verlangen Sie

fig. 2.4. The embalming of Joseph. Herzog August Bibliothek Wolfenbüttel: Tb2° 33. Johann Jacob Scheuchzer. Kupfer-Bibel *(Copper-Bible). Vol. II. Ulm: Wagner, 1731.*

das liebe Kind nicht wieder zu sehen, bis wir es in den Saal der Vergangenheit gebracht haben"[136] [Grant me but a few days and don't ask to see her until we have brought her into the Hall of the Past].[137] The complexity of the process is mentioned a third time, when the narrator comments on Wilhelm's impatience and the doctor's rejection of his wish to see Mignon: "Man wollte ihm [Wilhelm] ein für allemal Mignons Körper nicht zeigen"[138] [They had refused to allow him to see Mignon's corpse].[139] The injection technique when used for the bodies of older children, as Mignon would be, required the opening of the thorax and belly covers. The body was then laid for an hour in water and the blood removed via the vena portae and other vessels. After this work, the (colored) wax mass was injected in the vessels (see fig. 2.5). The fact that Wilhelm is not allowed to see Mignon's body hints to the fact that the corpse does not look very pretty while undergoing the preparation.

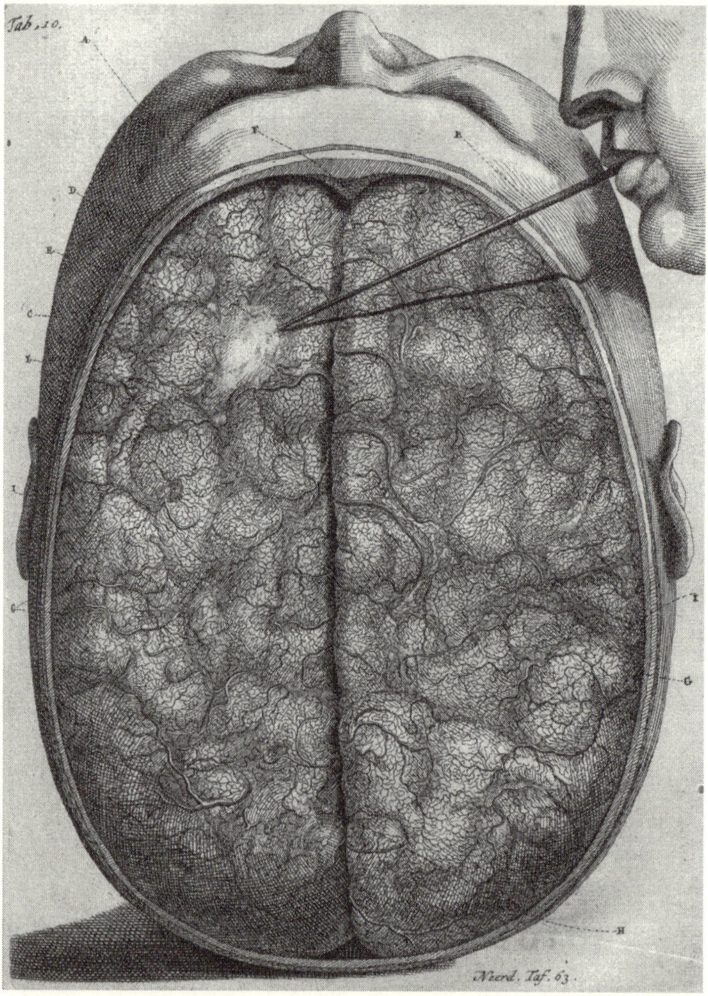

fig. 2.5. Preparation technique. Courtesy of Thomas Fisher Rare Book Library, University of Toronto. Frederik Ruysch. Opera Omnia Anatomico-Medico Chirurgica. *Amsterdam: Janssonio-Waesbergios, 1737.*

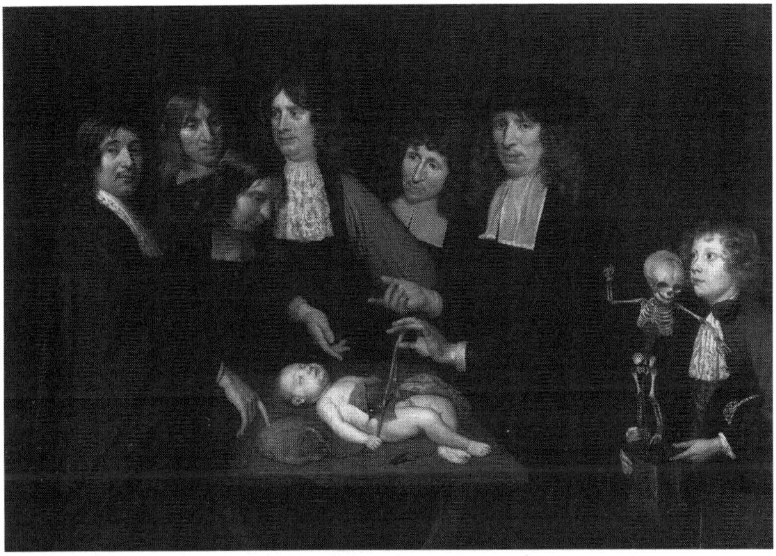

fig. 2.6. The Anatomy Lesson of Dr. Frederick Ruysch (Jan Van Neck, 1683). Collection Amsterdam Museum.

The epistemological questions at the center of the embalming practice and science are perhaps nowhere clearer than in the work of the Dutch anatomist Frederik Ruysch (1638–1731), one of the most talented and famous producers of anatomical preparations in the seventeenth and eighteenth century (see fig. 2.6).[140]

Drawing on Swammerdamm's inventions, Ruysch was known as having experimented with the injection of a number of liquids in blood vessels and his models were praised for their faithfulness. As suggested in Caspar Friedrich Jenckel's description of the anatomical collection, the organization of Ruysch's cabinet was less systematic than moral-aesthetic. Skeletons of children or fetuses were displayed as weeping or mourning figures, which dried their "tears" with handkerchieves from lung tissue. Ruysch's work experienced enormous popularity, which can be estimated from the kind of visitors which his collection attracted. Among the many visitors was Peter the Great from Russia who saw the collection several times and eventually bought significant parts of it. (As a proof for the faithfulness of Ruysch's specimens, the anecdote was told that Peter the Great kissed one of Ruysch's mummified children because he thought it was alive.[141]) Other segments of the collection were auctioned after Ruysch's death in 1731 by King Stanislaw who later gave the collection as a royal present to the University Wittenberg. For decades after Ruysch's death, his mummies and their eerie mimicry of life remained part of the popular imagination. In 1827, almost one hundred years after Ruysch's death, the fascination with his work still surfaced in the "Dialogo di Frederico Ruysch e delle sue mummie," in Giacomo Leopardi's *Operette morali.*[142]

Ruysch was not indifferent to his reputation. He actively promoted his name as the most talented anatomist of his time in Europe via extensive publications with elaborate illustrations (see fig. 2.7).

fig. 2.7. Opera Omnia, Frontispiece. Courtesy of Thomas Fisher Rare Book Library, University of Toronto. Frederik Ruysch. Opera Omnia Anatomico-Medico Chirurgica. *Amsterdam: Janssonio-Waesbergios, 1737.*

As Margócsy observes, the "dissemination of knowledge about his museum soon became an obsession for Ruysch."[143] As a result, Ruysch's publications were widely distributed and quoted. The exhibition catalogue was referenced in detail in Michael Bernhard Valentini's *Musaeum musaeorum*, a guide to famous collections for wealthy travelers interested in scientific matters (as Wilhelm Meister might be one) which came out in Frankfurt in 1714.[144] Margócsy has shown that an important point of Ruysch's publications was that they insisted on a hierarchy between object and text. In his publications, Ruysch advertised that through his method dead bodies would remain "very beautiful, well-shaped, and full of lively colour." This insistency on the actual object obviously had economic motivations since Ruysch charged for the visit of his collection. However, as Margócsy argues convincingly, it also stressed the "epistemological primacy" of the object over the text.[145]

Margócsy's point is well taken and focuses our attention on the relationship between body and text in the seventeenth and eighteenth centuries. However, it glosses too easily over the fact that the body to which Ruysch's text refers is itself not alive nature but its remnant and as such postpones the problem of access to nature and body without ever being able to solve it. The treatment of Mignon's preparation for the Hall of the Past demonstrates that Goethe was acutely aware of the problem. He brings this moment of absence to our attention by highlighting that the preparation and preservation of the body can only take place as a process of replacement. The Abbé's words demonstrate awareness for the fact that Mignon's embalmed body is representation *not* nature, and that the wax that fills the vessels of her corpse is there "an der Stelle des Bluts" [*in place* of blood].

The complex relationship between the live body and its replacement, nature and its representation, which is brought to our attention by reference to Ruysch becomes a crucial aspect of the Mignon episode. The fact that Mignon's embalmed body can only be a replacement where the living body is missing or a reference to the absent life is stressed by the recurrence of a motto from Ruysch's catalogue in Goethe's text.[146] The motto that Ruysch advertised in the foreword to *Opera Omnia* was "venite et videte."[147] It is this "come and see" that reappears in the context of the presentation of Mignon's embalmed body. At Mignon's funeral, the Abbé invites the friends: "*Treten Sie näher*, meine Freunde, *und sehen Sie* das Wunder der Kunst und Sorgfalt!"[148] [*Come* closer, my friends, *and see* the miracle of art and care!].[149] When the Abbé invites the audience to look at the embalmed corpse, the "come and see" promises presence, but the Abbé refers to a body which is only a reference to a living body and, even more, the reference to a text

(Ruysch's) which promises life-likeness, but refers to other corpses. The condition of possibility of this circulation of references is a living body which can only be grasped in its absence.[150] The absence of the living body instead of being overcome by the life-likeness of the embalming process is highlighted by it: if life would go on, the body would decay. Only in its absence which is referenced in the embalmed and aestheticized corpse can Mignon's hermaphrodite body be accepted in the culture of the Tower Society.[151] The phrase "gedenke zu leben"[152] [Remember to live],[153] the "Grundmaxime"[154] [principle] of the Tower Society which can be read from the entrance of the Hall of the Past, replaces, as Schiller has already noted in a letter to Goethe, the baroque "Memento Mori."[155] However, it does so by aestheticizing and halting the decay which is life itself.

We can observe that several features of the preparation of Mignon's corpse for the Hall of the Past connect this preparation to the ones taking place for anatomical collections. There is, first, the word and process of "embalming." Then, second, there is the stress that Jenckel puts on the fact that a particular pride of the anatomical collection is the number of monstrosities or strange objects. Interesting is further, third, that there is mentioning of the emotional and moral components which are highlighted in Jenckel's text on the *Naturalienkabinett*, in Ruysch's arrangement of his displays, and in the context of the Hall of the Past. Fourth and finally, the connection to the *Naturalienkabinett* is stressed by the words that the Abbé employs to describe the preparation of Mignon's corpse. "Wunder" [miracle] and "Kunst" [art] resonate with the term *Wunder- und Kunstkammer*, a title that was used to describe the agglomeration of objects assembled in anatomical collections of the early modern period and well into the eighteenth century.[156] The fact that the uncle (Oheim) possesses such a collection is mentioned in the "Confessions of a Beautiful Soul" (the sixth book of the *Lehrjahre*)[157] and then again shortly before Mignon's death when it is described how Wilhelm, in the early morning hours, discovers in the uncle's house where the members of the Tower Society convene "eine Bibliothek, eine Naturaliensammlung, ein physikalisches Kabinett"[158] [a library, a collection of natural history specimens, and another of stones and metals].[159]

The spatial and conceptual proximity of anatomical collection and Hall of the Past can shed further light on the Tower Society's ideology vis-à-vis Mignon—and Goethe's critique thereof. As Marion Ruisinger has observed, from the early modern period until well into the eighteenth century anatomical collections were places that combined research and wonder and that served both scientific curiosity and worship. As in Ruysch's displays of weeping child-skeletons, the vanitas thought and its implied moral teaching was one of the thematic foci of these presentations which tried to leave an impact on mind and memory. The Abbé alludes to the vanitas motive when he describes the function of the Hall of the Past:

> Nach ihm [dem Oheim], dem Erbauer dieses Hauses, dem Errichter dieser Stätte, haben wir zuerst einen jungen Fremdling hierher gebracht, und so faßt schon dieser kleine Raum zwei ganz verschiedene Opfer der strengen, willkürlichen und unerbittlichen Todesgöttin. Nach bestimmten Gesetzen treten wir ins Leben ein, die Tage sind gezählt, die uns zum Anblick des Lichts reif machen, aber für die Lebensdauer ist kein Gesetz. Der schwächste Lebensfaden zieht sich in unerwartete Länge, und den stärksten zerschneidet gewaltsam die Schere einer Parze, die sich in Widersprüchen zu gefallen scheint.[160]
>
> [The designer and builder of this hallowed place came first; now we have brought here a young stranger, so that this one room encloses two very different victims of the solemn, arbitrary, and inexorable goddess of death. Fixed laws govern our entry into life and the number of our days, our maturing in countenance of the light, but there is no law that prescribes the length of our life. The feeblest lifethread may stretch into unexpected length, and the strongest may be forcibly severed by Fate, which seems to delight in inconsistency.][161]

Jenckel conceptualized the cabinet as the place where "Naturalibus, oder denjenigen Dingen, so die Natur hervor bringet" [Naturalibus, or those things that nature produces] and "Artificiosis, oder was die Kunst durch Menschen verfertiget" [Artificiosis, or that which art produces through the human being] are united.[162] For him, the cabinet is the place "worinnen gleichsam die menschliche Vernunfft und Sinnen ob dem Anschauen der wunderbaren Geschöpffe GOttes und der künstlichen Hände-Werck in verwunderliche Bestürtzung gesetzet werden"[163] [wherein, as it were, human reason and senses are put in strange amazement, confronted with the miraculous creations of God and of the artificial craft]. The natural and artificial objects displayed in the cabinet were always also understood as part and representation of the divine creation which demanded the visitor's reverence.[164] As such, the anatomical collection and cabinet predated the clear split between biology, aesthetics, and morality and, at the same time, it was the place where this split was prepared.[165] By placing the Hall of the Past in close proximity to the anatomical collection, Goethe highlights the biological curiosity of Mignon's corpse and, at the same time, points out to what extent the hermaphrodite body and its incest origin is meant by the Tower Society to provide a moral tale which—much as the specimen in the "Wunderkammer"—it cannot deliver.

By placing the Hall of the Past as kin to the "Kunst- und Wunderkammer," Goethe also suggests that the Tower Society is not the progressive union as which it understands itself, but utterly anachronistic. Barbara Stafford has considered anatomy and dissection "the eighteenth-century paradigms for any forced, artful, contrived, and violent study of depths."[166] By privileging what Stafford has called "an absolutist and inflexible meta-discourse," the Enlightenment rendered invisible "the unsystematic, irreducible characteristics of the specific phenomena under study."[167] Goethe highlights the extent to which the Tower Society controls meaning by sanitizing the hermaphrodite's monstrosity.

For the funeral ritus, Mignon wears the winged garb that she has also worn in the last days before her death. This costume has often been read as the representation of a figure of the immaterial, as Psyche or an angelic being.[168] This reading seems legitimate in particular since Mignon in her last song, which contains the line "Und jene himmlische Gestalten, / Sie fragen nicht nach Mann und Weib" [For all those glorious heavenly forms, / They do not ask for man or wife][169] refers to Matthew 22:30: "For in the resurrection they neither marry nor are given in marriage, but are like angels in heaven."[170] The reference to the Biblical verse invites a reading in which Mignon's being transcends gender differences. In this context, Mignon's plea "Laßt mich scheinen, bis ich werde" [Let me appear until I become][171] has been discussed as a reference to the division of spirit and body, of which the body is read as "scheinhaft" [mere appearance] aggregate which Mignon transcends in the moment of death to enter the realm of real essence. Consequently, Mignon's death is read as the "Verklärtheit des überirdischen Seins von Mignon"[172] [transfiguration of the metaphysical being of Mignon]. However, another reading becomes possible, if we consider the winged appearance of Mignon's body in the context of the discussed hermaphroditism. This winged body might be a reference to angelic figures, but it also suggests a connection to the antique god Phanes. In Greek mythology, the god Phanes is described as a winged figure, at times monstrous or hermaphrodite, who is invisible but radiates pure light (see fig. 2.8). As the god's name suggests, Phanes (from phaino: bringing to light/make appear) makes things visible, but remains invisible him/herself.[173]

fig. 2.8. Phanes. Relief with Mithraic and Orphic symbols, Roman imperial art, Modena, Galleria Estense. Su concessione del Ministero dei Beni e delle Attività Culturali e del Turismo—Archivio Fotografico della SBSAE di Modena e Reggio Emilia.

Against the backdrop of the Phanes figure, we can read Mignon's song "Lasst mich scheinen"[174] [Let me appear] as a plea to let her be that source of light that makes the world visible, but remains invisible herself. In Benjamin Hederich's *Gründliches mythologisches Lexicon*, which was published in Schwaben's revised version in 1770 and which was in Goethe's possession, Goethe could find an even slightly more complex reading of the figure. The lexicon entry describes "Phanaeus" as an attribute name of Sol or Apollo, and adds:

> Ob er aber solchen Beynamen von dem Vorgebirge [phanaische Vorgebirge] oder von einem alten Könige, Phanäus, . . . oder nur von φάνεθαι, erscheinen, hat, weil er täglich *wieder erscheint* . . . *oder machete, daß alles erscheint* und gesehen werden kann . . . darüber ist man noch nicht einig.[175]
>
> [Whether he has such attributive name from the mountain range or from an old king, Phanaus, . . . or just from φάνεθαι, to appear, because he appears every day anew . . . or makes that everything appears and can be seen . . . has not been decided yet.]

The uncertainty and indeterminacy is increased in this lexicon entry by reentering the impossibility of a difference between appearing and to make appear in the figure itself. In the entry that follows "Phanaeus," the entry on "Phanes," Hederich's lexicon describes the figure as "soll der wahre und große Gotte seyn, den man den erstgeborenen nennet, weil vor ihm nichts und von ihm alles erzeuget worden. Er hat daher auch den Namen, weil er zuerst aus dem Unendlichen erschienen. . . . Im Grunde soll er nur ein Beynamen des Amors seyn, welchen er von φάνεθαι, erscheinen, bekommen, weil er zuerst aus dem Chaos hervorgetreten"[176] [is considered the true and great god, who is called the first born, because before him nothing was created and by him everything (is created). He also has this name, because he first appeared from eternity. . . . Basically it is considered only an attribute name of Amor, which he has received from φάνεθαι, to appear, because he appeared first from chaos]. If an allegorical reading of Mignon's hermaphrodite body is possible, then only as a figure who is the condition of possibility for interpretation to appear, but who has to remain outside the circulation of signs herself.[177] The plea of this figure remains unheard in the dissecting and classifying world of the Tower Society.[178]

Considering Mignon as Phanes and not as Psyche, we might read the Mignon figure as Goethe's consideration of questions that Schiller had addressed in the *Kallias* letter on beauty. Walter Müller-Seidel has shown how closely Schiller's essay on the beautiful from 1793 is connected to the beginning of Schiller's and Goethe's friendship in 1794.[179] In contrast to earlier scholarship, which focused on Goethe's account of the "glückliche Ereignis," Müller-Seidel has demonstrated that the beginning of the cooperation was as much shaped by a discussion on the metamorphosis of the plants (Goethe's memory) as it was defined by a debate on art and aesthetics (as Schiller's account in the letter to Körner suggests). One of the central topics in both cases, art and nature, was the question of autonomy.

In his letter to Körner, Schiller reported that Goethe had applied the "Idee: Schönheit sey Vollkommenheit mit Freyheit, auf organische Naturen"[180] [idea: beauty is perfection with freedom to organic natures]. However, as Müller-Seidel observes, it was Schiller who had already transferred Kantian terminology to organic nature, namely in the *Kalliasbrief*.[181] In the *Kallias*, Schiller declared beauty in nature (here animals and plants) as that moment "wo die Masse von der Form . . . und von den lebendigen Kräften (in die ich die Autonomie des organischen setze) völlig beherrscht wird"[182] [in which mass is completely dominated by form . . . and by living forces (in the autonomy of the organic)].[183] Schiller's example for the illustration of his thinking on autonomy in nature is the bird. Schiller writes:

> Unter den Thiergattungen ist das Vögelgeschlecht der beste Beleg meines Satzes. Ein Vogel im Flug ist die glücklichste Darstellung des durch die Form bezwungenen Stoffs, der durch die Kraft überwundenen Schwere.[184]
>
> [In the animal kingdom it is the birds which are the best proof of my claim. A bird in flight is the happiest depiction of matter dominated by form, of power overcoming weight.][185]

In the image of the bird, Schiller then expressed the overcoming of determinism. He concluded his remarks on the bird with reference to the symbolic power of the wing. And it is in this context that he brings up the figure of Psyche and the notion of freedom:

> Es ist nicht unwichtig zu bemerken, daß die Fähigkeit über die Schwere zu siegen oft zum Symbol der Freiheit gebraucht wird. Wir drücken die Freiheit der Phantasie aus, indem wir ihm Flügel geben; wir lassen Psyche mit Schmetterlingsflügeln sich über das irdische erheben, wenn wir ihre Freiheit von den Fesseln des Stoffes bezeichnen wollen.[186]
>
> [It is not unimportant to note that the ability to overcome heaviness is often used as the symbol of freedom. We express freedom of the imagination by giving it wings; when we want to describe Psyche's freedom from the bounds of matter, we let her soar above the world with the wings of a butterfly.][187]

In Schiller's *Kalliasbrief*, the winged figure of the bird and of Psyche—both images with which Mignon has been associated in scholarship on the *Lehrjahre*—stands in for the overcoming of nature's determinism. If Schiller stated that the "Schwerkraft verhält sich ohngefähr eben so gegen die lebendige Kraft des Vogels, wie sich . . . die Neigung zu der gesetzgebenden Vernunft verhält"[188] [The relationship of gravity to the living power of the bird is about the same as . . . inclination is related to law-giving reason],[189] then we need to read Mignon's wings at the end as a triumph of reason and freedom. However, such reading overlooks that it is in the context of the Tower Society that Mignon is cast in the angelic garb. If the enlightened Tower Society wants to see nature's determinism overcome by reason and celebrates in Mignon's death an apotheosis of this reason, Goethe is skeptical. He undermines this apotheosis of reason by providing a different context of reading Mignon, the winged hermaphrodite, Phanes. Furthermore, if Mignon has also Marianic attributes (lily, white dress, child, the cross tattoo), as Hannelore Schlaffer has noted,[190] then Goethe alludes to the fact that in both Christian and Schillerian thinking the body is only acceptable within dualistic structures which need to be overcome by spirit. Framing Mignon instead in the figure of Phanes, Goethe evokes a nature which is deterministic and creative. It is *appearing appearance*, or, to use Hederich's words on Phanes, "erscheint . . . oder machete, daß alles erscheint."[191] [appears . . . or makes that everything appears].

The tension between the determinism that shapes Mignon's life and her apotheosis as Phanes presents a sudden rupture in the interpretation which remains troubling without reliance on a dualistic model as Schiller offers it. However, we might be able to account for this tension if we draw on Spinozistic concepts. As Horst Lange has demonstrated brilliantly, the connection between Goethe and Spinoza is less obvious than often assumed and needs further exploration, an enterprise that is complicated by the fact that there are only few documents in which Goethe directly refers to Spinoza. Carefully searching the ways in which Spinoza might have influenced Goethe's thinking (and the ways in which Goethe misunderstood the Dutch philosopher), Lange encourages us to consider "the more indirect ways in which Goethe connected Spinoza to German Idealism."[192] The Mignon figure provides Goethe a space to think through Spinozistic concepts. We can read Mignon's two appearances as an illustration of Spinoza's concepts of natura naturata (produced nature) and natura naturans (productive nature). Both belong to a monistic, and deterministic, framework. In *Dichtung und Wahrheit*, Goethe confirms this conceptualization of nature as a deterministic entity: "Die Natur wirkt nach ewigen, notwendigen, dergestalt göttlichen Gesetzen, daß die Gottheit selbst daran nichts ändern könnte"[193]

[Nature operates according to eternal, necessary laws, which are so divine that the Divinity itself cannot alter them].[194]

While Goethe's indirect reference to Spinozistic concepts illustrates the distance to Schiller's dualistic model and to Schiller's belief in the possibility of "Freiheit von den Fesseln des Stoffes"[195] [freedom from the bounds of matter],[196] we will see in the next section of this chapter that Goethe does not conclude his thoughts on incestuous reproduction within the framework of Spinozistic determinism. As Lange observes, Goethe's "determinism ends where concepts such as development and personality enter the stage."[197] Each individual operates in a clearly delineated field within which freedom becomes possible. Although Goethe concedes that for some individuals, such as Mignon, this field might be extremely limited, he also demonstrates that such determinism will limit the possibility of selfhood only in the extreme case.

This chapter has considered the history of incest and has registered a new hesitation concerning incestuous reproduction around 1800. While the incest taboo before the eighteenth century had been understood in an ethical-religious context and the taboo's rejection was closely tied to a rejection of social conventions and laws, the new hesitations vis-à-vis incestuous reproduction can be understood in the context of new findings in animal husbandry. The hesitations in animal husbandry against inbreeding were connected to the (perceived) loss of sexual differentiation and the failure of inbred animals to produce fertile offspring. Without having yet certainty in terms of biology (in fact, we consider the case today biologically rather complex), Goethe explored in the Mignon episode what it would mean if the incestuous origin could constitute a determining factor for the offspring. Revisiting the Mignon episode, I have argued that Mignon's inability to reproduce and to enter a heterosexual society is not the result of a traumatic childhood or failed *Bildung* but the result of a biological problem (which also bars Mignon from becoming part of the cultural system of genealogy). Furthermore, Goethe's use of the figure of the hermaphrodite, in the specific historical moment in which he wrote, suggests that he placed Mignon's "monstrosity" as a body in opposition to the classificatory world of the Tower Society. Only as dead body is Mignon's nature represented and incorporated in the Hall of the Past.

The open problems thematized at the beginning of this chapter become answerable if Mignon is considered in the light of biology and hermaphroditism: the text switches between male and female pronouns, because Mignon is indeed neither one nor the other; Therese and Natalie's

"Mannweiblichkeit" is unproblematic, because it is tied to clothes and as such can be overcome; the source of Mignon's erotic "overflow" becomes readable as ejaculation; Mignon's "strange nature" is not the result of a trauma, but of a bodily condition; finally, the care with which the Tower Society inquires into Therese's lineage—in the narrative in close proximity of the history of Mignon's incestuous origin—and the decision that Therese can marry Lothario because his former lover was not Therese's biological mother becomes now understandable as an inquiry into Therese's biological genealogy and a sign of the Tower Society's wish to ban relationships that are biologically incestuous.

BIOLOGY AND KINSHIP

I would like to pause here for a moment and consider what we have observed so far and how this might relate to established and powerful assessments of kinship and incest in literary theory. This chapter was opened with the observation that in the *Lehrjahre* attaining selfhood is closely linked to inscribing oneself into genealogy. Mignon's inability to reach adult selfhood and her early death are closely linked to her biological condition originating in her incestuous conception which conditions the hereditary defect. By focusing on Mignon's biological condition, this chapter has offered an alternative reading to scholarship that locates Mignon's problem primarily in a socially constructed incest taboo, a childhood trauma, or the uncertainty of her descent. Focusing on the latter aspects, scholars have read the Mignon figure not only as an illustration of psychological problems, but even as the place where psychoanalysis is invented. Stefan Blessin has declared the Mignon episode the "Geburtsstunde der Psychoanalyse" [hour of birth of psychoanalysis]. This declaration is closely tied to a critique of that society which establishes such taboos. Blessin argues that it is "nicht der inzestuöse Vollzug als solcher, der die Gründe für Fehlentwicklungen liefert. Es ist das Inzesttabu, mit der unsere Kultur in die Partnerwahl eingreift. Dieses Tabu steuert die Ausbildung der Geschlechterrollen, und ähnlich wie beim Ödipuskomplex können Schäden die Folge sein"[198] [not the incestuous act as such, which provides the reasons for the wrong development. It is the incest taboo, with which our culture intrudes in our choice of partners. This taboo regulates the formation of gender roles, and damages might occur similar to (the damages in the case of) the Oedipus complex.]. Following this line of argument, scholars have stressed recently that in the *Lehrjahre* "identity is not determined by biology and fate but by the choice of the individual."[199] Such optimistic assessments if not of the reality but of the possibility of modern selfhood overlook both the extent to which the distinction between biological and adoptive relationship is impossible—and irrelevant—in the *Lehrjahre* and the arguably small, but eventually decisive precondition that biology provides for the individual's decisions and for society's regulation of them.

The most influential theoretical approach within the context of literary scholarship to questions of incest and its regulation in recent decades has probably been Claude Lévi-Strauss's and René Girard's theory of kinship. In *Structural Anthropology*, Lévi-Strauss has highlighted forcefully the arbitrariness of kinship systems and has stated the "essential way in which it diverges from nature."[200] Consequently, Strauss's approach rests on the assumption that there is no stringent connection between biological reproductive cycle and the symbolic structure of kinship. In *Violence and the Sacred*, René Girard further explores Strauss's suggestions and sets out to refine the relationship. He too highlights the significance of the distinction between "the *fact* of biological kinship" and the "*understanding* of these same procedures."[201] Pushing even further, Girard argues that the recognition of biological reproduction is only possible as an effect of the establishment of symbolic kinship structures. The incest taboo—as an arbitrary rule—becomes for Girard a condition of possibility for the recognition of biological laws. He concludes "if there were no incest prohibitions there would be no biological laws."[202]

Drawing on an anthropological methodology, Girard develops his model of kinship structures by turning to primitive societies. What remains puzzling in Girard's account of the emergence of biological laws is the fact that he introduces the distinction between biological fact and cultural interpretation in the analysis of societies that have no means to distinguish the two and are not aware of this distinction. In other words, the fact that the primitive cultures that he studies have introduced the incest taboo and exogamy is by no means a sign that these cultures have criteria to distinguish biological and other forms of kinship even though they might formulate what they consider biological laws. For the question that anthropologists like Girard and Lévi-Strauss want to answer and which aim for the formulation of anthropological constants (operating on the basis of astonishingly universalist assumptions), these differences might be irrelevant. However for projects that are interested in historical shifts such differences will matter. While anthropological models for understanding the behavior of group and individual remain a powerful tool for the analysis of literature, they also remain—and necessarily so—relatively insensitive regarding historical events and paradigm shifts.

Goethe articulates a split between biology and culture that marks a specific moment in the history of science and ideas which are unknown to the primitive societies that anthropology uses as case study and model. In contrast to Girard's claim that biological laws are an effect of the incest taboo, the discovery of biological laws becomes possible precisely at that moment in which animal breeders were willing to *ignore* the incest taboo—an ignorance which was made easier for these breeders because they assumed that cultural and moral laws apply to humans but not to animals. Only when animal breeders freely started to use inbreeding as a main form of reproduction, did they start to discover the biological laws of how traits are transmitted from one generation to the next. And only once the human being was considered also an animal could these laws become relevant for establishing new laws of reproduction. Until today—and much as Girard has formulated—these biological laws are often at odds with cultural laws which govern how marriage and kinship are regulated in our society. However, this does not make them mere effects of cultural structures. In stressing the contingency of social constructs, it is difficult to account for the challenges with which biology confronts individuals and societies. Unlike Girard who declares biological laws a function of the symbolic structure, Goethe—in the specific historical moment in which he is writing—acknowledges a fundamental precondition of natural conditions for notions like selfhood and society and thereby establishes a positive (in the sense of existing) notion of nature which defies cultural classification. At the same time, however, Goethe also stresses that this strong understanding of nature matters for notions like selfhood and love only to a very limited extent. With an acute awareness for the biology of his time, Goethe strives to define where nature ends and culture begins in order to come to a better understanding of the incest taboo.

A THEORY OF DESIRE AND LOVE

Assuming problematic consequences of incestuous reproduction for the offspring, the question arises: how can it be avoided? If incestuous reproduction is the consequence of incestuous desire, is—to use an often referenced Goethean concept—renunciation (Entsagung) the solution? In order to find Goethe's answer to these questions, we will have to investigate how desire develops and how it possibly can be channeled.

When studying eighteenth-century notions of desire, it is striking how often they are connected to incestuous inclinations. In fact, a convincing argument can be made that the incestuous quality of relationships is one of the most dominant features in the depiction of the emerging nuclear family around 1800. The ambiguous relationship between father and daughter in Gotthold Ephraim Lessing's *Emilia Galotti* (1772), or Kleist's *Marquise von O . . .* (1808), and the relationship between brother and sister in Lessing's *Nathan* (1779) or Tieck's *Eckbert* (1797) are but a few examples of this phenomenon. Goethe's early one-act play *Die Geschwister* (1776) is an interesting case in this series, because it reverses the usual plot-line. Instead of love followed by the lovers' recognition that they are relatives (the story of Mignons' parents and the story of Recha and the Templar in Lessing's *Nathan*), the two protagonists are here introduced as siblings and live like father and daughter because of the age difference. Only in the course of the play do the protagonists learn that they are not relatives and, therefore, can finally live their love.[203] However, what is most interesting about this inversion of the incest story is its suggestion that inclinations develop from (quasi) incestuous relationships. Goethe starts to articulate here a theory of desire which is fully formulated only in *Der Mann von funfzig Jahren*, the longest novella in the *Wanderjahre*. It is a story in which incest plays a necessary, albeit potentially problematic role.

If the *Lehrjahre* ends with the establishment of a genealogy by an individual (Wilhelm recognizes Felix as his son), *Der Mann von funfzig Jahren* begins with a family that is occupied with securing the continuity of such genealogy. If in the *Lehrjahre* the establishment of genealogy becomes impossible for Mignon because of her incestuous origin, in *Der Mann von funfzig Jahren* we are presented with the attempt to avoid incestuous affiliations. The novella starts out with a Major who returns from active duty to the family estate with the purpose of regulating and preparing the material and legal transfer of the inheritance for the family's youngest members, the cousins Hilarie and Flavio. All efforts of the Major and his sister, the Baroness, are geared towards preserving the family estate and to bring it in order for the cousins who are assumed to marry each other. Cousin marriages were not considered incestuous relationships according to German laws of the time.[204] In fact, as Elisabeth Joris has observed, the assumption that kinship became irrelevant for marriage patterns in Europe in the wake of the French Revolution, a dominant thesis in the historiography, cannot be confirmed by data on marriage behavior in the eighteenth and nineteenth century.[205] Marriages between first degree cousins became a particularly popular way to consolidate family estates in both aristocratic and bourgeois families.[206] Major and Baroness are eager to consult the family's genealogical table for the purpose of evaluating the size of their children's inheritance:

> [Der Major] fand seine Schwester vor dem Stammbaume stehen, den sie hatte aufhängen lassen, weil Abends vorher zwischen ihnen von einigen Seitenverwandten die Rede gewesen, welche, teils unverheiratet, teils in fernen Landen wohnhaft, teils gar verschollen, mehr oder weniger den beiden Geschwistern, oder ihren Kindern, auf reiche Erbschaften Hoffnung machten.[207]

> [(The Major) found his sister standing before the family tree, which she had hung up, prompted by conversation the previous evening about various collateral relations, who, because they were unmarried, or had gone to live in distant lands, or had even totally disappeared, raised in greater or lesser degree the hope of large legacies either for brother and sister or for their children.][208]

In this consideration of the genealogical table, kinship relations are noted with the goal to establish the transfer of inheritance. The genealogical table is here not a mimetic presentation of biological kinship relations but constructed under a specific perspective, namely the perspective of the children as the benefactors of the inheritance. With this focus in mind, Major and Baroness consider collateral lines not in their own rights, but only insofar as they might contribute to the children's inheritance. This subsumption of collateral lines becomes also possible because the wealth that Major and Baroness have in mind is based on land for which there is no exchange value so that those relatives who moved away cannot keep their possession nor sell it.[209] Instead, it will become part of Flavio's and Hilarie's inheritance.

Viewed through the eyes of Major and Baroness, the genealogical table resembles what Rheinberger and Müller-Wille have described as a premodern, not exclusively biological understanding of kinship. The mediality of the genealogical table highlights this constructive rather than mimetic function by techniques such as the translation of temporal succession into a spatial representation and by forcing a reading movement from bottom to top which runs counter to the naturalized reading movement from left to right and top to bottom.[210] Accordingly, this premodern understanding used the kinship criteria rather flexible. Such flexibility allowed the exclusion of close relatives and the inclusion of distant relatives when it seemed convenient for the purpose of succession and inheritance. For example, the focus on male succession often led to the exclusion of female relatives and collateral lines and the reduction of the genealogical table to a line.[211] Hence, while these genealogical tables promised to be mimetic in their representation of kinship relations, they also clearly constructed these relationships in specific ways with the aim to support specific legal claims. Such legal claims regarding inheritance and power were in fact one of the main motivators for the consideration of genealogical tables; Goethe himself, when working as a young lawyer at the Reichskammergericht in Wetzlar, would have encountered an abundance of legal disputes which often stretched over decades and which centered on genealogical evidence with the aim to establish questions of property and power.[212]

However, in *Der Mann von funfzig Jahren* the family's genealogical table receives a second consideration when the Major studies it again, this time together with his niece Hilarie:

> Die Baronin entfernte sich, der Major stand allein vor dem lakonischen Familiengemälde, Hilarie trat an ihn heran, lehnte sich kindlich an ihn, beschaute die Tafel und fragte: wen er alles von diesen gekannt habe? und wer wohl noch leben und übrig sein möchte? Der Major begann seine Schilderung von den ältesten, deren er sich aus seiner Kindheit nur noch dunkel erinnerte. Dann ging er weiter, zeichnete die Charaktere verschiedener Väter, die Ähnlichkeit oder *Unähnlichkeit* der Kinder mit denselben, bemerkte, daß oft *der Großvater im Enkel wieder hervortrete*, sprach gelegentlich von dem Einfluß der Weiber, die, aus fremden Familien herüber heiratend, oft den Charakter ganzer Stämme verändern.[213]

> [The Baroness withdrew, and the Major remained alone before this laconic family portrait. Hilarie approached him, and leaning against him in childlike fashion, gazed at the chart and asked him whom of all these people he had known and who might be alive and still left. The Major began his account with the oldest ones, whom he remembered but dimly from his childhood. Then he went on, sketched the character of various fathers, the ways in which their children *resembled* or *differed* from them, observed that *a grandfather often reemerged in a grandson*, spoke as the occasion warranted of the influence of the women, who, marrying in from other families, often altered the character of entire lines.][214]

In calling the genealogical table a "laconic" [lakonischen] picture of the family, the narrator highlights that the document's meaning depends on an act of reading. The reading which Hilarie initiates differs significantly from that which uncle and mother had provided. While Major and Baroness looked at the genealogical table under the aspect of inheritance and the transfer of property, Hilarie is interested in the relationship between people represented in the genealogical table. Triggered by her question "wen er alles von diesen gekannt habe?" [whom of all these people he had known],[215] the Major pays new attention to his own relationship to persons mentioned in the table, observes similarities between relatives, notes the influence of women, the significance of collateral lines, and mentions the qualities of individual members of the family. The Major also speaks of a number of ancestors and other relatives and their relation to each other. While before he and his sister were interested in virtually one line along which property travels, the words with which the Major now describes the relations among relatives have both vertical ("von den ältesten," "Großvater im Enkel") and lateral connotations ("weiter," "aus fremden Familien herüber").

In revisiting the genealogical table, the Major enlivens the laconic picture not only by highlighting similarities between the generations and by observing the influence of individuals on the genealogical transfer, but also by adding new spatial dimensions. These additions correspond to a new structure of the genealogical table that Ohad Parnes observes around 1800. Parnes describes how the genealogical table was reorganized with greater awareness for generational structures and by incorporating collateral lines.[216] While the premodern genealogical table was structured following the principle of succession and focused for the most part on the male (agnatic) line, in the late eighteenth and beginning nineteenth century we find increasingly genealogical tables that attempt to record kinship relations as comprehensively as possible.[217] In the second reading of the genealogical table, the focus is not any longer on economic ramifications, but on the structure and content of kinship relations. The criteria under which the genealogical table is considered anticipates a biological understanding of kinship by delineating similarities and dissimilarities between family members, including those similarities which are transmitted from grandfather to grandchild and, therefore, can hardly be attributed to culture ("that a grandfather often reemerged in a grandson").

Drawing on Parnes's observation, the changes from the first to the second consideration of the genealogical table can be read as the performance of a shift from a premodern to a modern understanding of genealogy. The novella's position in a moment of transformation is further supported by its presentation of a clash between two understandings of desire. The Major's careful plans regarding the transfer of inheritance, which are based on the assumption of the cousin marriage, are endangered by Hilarie's desire for another man. Shocked by the devastating news delivered by the Baroness, the Major exclaims: "Was sagst du?" rief der Major; "ist's möglich? indessen wir uns alle Mühe geben uns ökonomisch vorzusehen, so spielt uns die *Neigung* einen solchen Streich!"[218] ["What are you saying?" the Major exclaimed. "Is it possible? When we are taking such pains to make financial provisions for them, would inclination play us such a trick?"].[219] From the Major's perspective, the family's economic interests are endangered by the desires of the individual: love as passion and love in marriage are juxtaposed as following different regimes.[220] In presenting the opposition between economy and desire as a conflict between generations, the story is part of an often discussed historical shift from feudal household governed by the economy of the *oikos* and its male patriarch to the bourgeois family in which economies follow the laws of the exchange value.[221] This shift has consequences for the understanding of desire.

From the Major's perspective, desire is in tune with what he considers natural rhythms of generational succession. Relying on such framework, he declares Hilarie's feelings in a conversation with his sister as "nur scheinbar" [illusory] and is convinced "dass ein Selbstbetrug dahinter verborgen liegt" [they are based upon self-deception]. The Baroness, however, rejects this assessment of Hilarie's desires, "denn nach allen Symptomen ist es ein sehr ernstliches Gefühl" [because to judge by all the symptoms, it is a very serious feeling].[222] The Major's assessment of Hilarie's desires indicates the extent to which the distinction between true and false, natural and artificial is one of the core frameworks available to the Major. Bound to the idea of generational transfer as a representation of natural rhythms, it is impossible for the Major to account for feelings which occur outside this generational order. This remains even true after the Major has learned from his sister that he himself is the object of Hilarie's desire—an incestuous relationship in the understanding of the time. The Major can reconcile the individual's independence and the inclinations which contradict the planned succession only once he manages to explain them via larger cosmic and cyclical structures. Warming up to the idea that Hilarie loves him, the Major tries to place this love in an organic framework:

> Die Empfindung war für ihn höchst angenehm, ob er sich gleich solche weder gestehen noch erlauben wollte. . . . Der Garten war in seiner vollen Frühlingspracht, und der Major, der so viele alte Bäume sich wieder belauben sah, konnte auch an die Wiederkehr seines eignen Frühlings glauben. Und wer hätte sich nicht in der Gegenwart des liebenswürdigsten Mädchens dazu verführen lassen![223]

> [The sensation was very pleasant, although he was inclined neither to admit it nor indulge it . . . The garden was in full spring glory, and the Major, seeing so many old trees leafing out anew, could believe in a similar return of his own springtime. And who would not have been seduced into such thoughts in the presence of such a charming girl!][224]

By drawing analogies to nature, the Major finds his late love justified as natural and comes to believe that his aging body will rejuvenate like his mind. Much like Augustin in the Mignon episode—and not unlike the figures in the *Wahlverwandtschaften*—the Major tries to explain and justify his love by referring to nature. However, as in Augustin's incest justification, the text makes visible that the metaphor of nature does not match what it is supposed to describe.[225] As the following narrative suggests, the Major's body sets him limits where his mind and emotions still roam freely and these limits question the metaphorical connection that he has made. Hence, it is remarkable but not surprising how much doubt the narrator sheds on Hilarie's love and the Major's reaction to it by using subjunctive and verbs like "seemed" [schien], "to believe in" [Glauben beizumessen], "to seduce" [verführen]. By means of lexical choices and grammatical devices, the narrator undermines the Major's belief in what he considers natural love.

In contrast to the Major's belief, it is remarkable how from the novella's beginning emerging desires are exposed as results of manipulative and contingent situations. When the Baroness informs the Major about Hilarie's desires, she does not mention whom Hilarie loves. Instead, she treats this love as a secret [Geheimnis]. It remains unclear why the Baroness does not share the content of the secret with the Major, but it is certainly precisely the secret which creates a room for the Major to imagine Hilarie as lover and, as a consequence, begins to desire her. This initially only imagined desire becomes reality for the Major even though it remains unclear whether Hilarie, in turn, has any "authentic" feeling for the uncle. Instead, the narrator suggests that Hilarie's feelings are similarly nourished by the Baroness, Hilarie's mother:

> Die Baronin hatte ihren Bruder von Jugend auf dergestalt geliebt, daß sie ihn allen Männern vorzog, und vielleicht war selbst die Neigung Hilariens aus dieser Vorliebe der Mutter, wo *nicht entsprungen*, so doch gewiß *genährt* worden.[226]

> [The Baroness had loved her brother since childhood so much that she had thought him superior to all other men, and perhaps even Hilarie's inclination had, *if not sprung* directly from this preference of her mother's, surely been *fostered* by it.][227]

It is fascinating to see to what extent Hilarie's inclination is presented here as emptied of all genuine desire. Instead, it is presented by the narrator as an imitation of the mother's preferences. Even that origin of desire is not marked precisely, since it is kept in the realm of suggestion with vocabulary such as "vielleicht" [perhaps] and formulations such as "wo nicht entsprungen" [if not sprung]. By drawing on a juxtaposition of nature [*entsprungen*] and nurture [*genährt*], the narrator employs an opposition which increasingly gained currency in the assessment of selfhood (see chapter 3) and suggests that the object of desire, if not the desire's origin, is not located in the realm of nature but in that of nurture.

At first glance, the fact that the daughter's inclinations are an imitation of the mother's desires suggests an emergence of desire out of biological kinship situations. Such reading, however, is undermined not only by the narrator's hesitation to embrace the Major's naturalization of the situation and by the vocabulary of nurture, but also by the explanation of Hillarie's desires at the end of the novella. When, at the end of the novella, Hilarie cannot decide between the two men, the Major and his son Flavio; when she has scruples to reject the old man and to marry the younger lover, the mother tries to demonstrate to the daughter the legitimacy of a marriage with Flavio. The Baroness not only praises the younger lover's qualities but she tries to make clear that Hilarie's inclination to the older man was not true love, but the understandable, even necessary, confusion of a young girl's mind:

> Die Baronin . . . begann sogleich ein verständiges Darlegen der frühern Zeit, der Persönlichkeit ihres Bruders [der Major] und seiner Verdienste; sie gab den Eindruck zu, den der einzige Mann von Wert, der einem jungen Mädchen so nahe bekannt geworden, auf ein freies Herz notwendig machen müsse, und wie sich daraus, statt kindlicher Ehrfurcht und Vertrauen, gar wohl eine Neigung, die als Liebe, als Leidenschaft sich zeige, entwickeln könne.[228]

> [The Baroness . . . at once began a circumspect review of the past, and of her brother's personality and merits. She conceded that the one superior man with whom she was closely acquainted was bound to make a strong impression on a young girl's unattached heart, and that the girl might come to feel not childlike respect and trust but inclination, love, and finally passion.][229]

While in the beginning of the novella, the Baroness convinced the brother of the depth of Hilarie's feelings, now, in the end, she reads Hilarie's inclination as mere, albeit understandable and even "necessary" inclination to the "one superior man" in a girl's life—thereby putting into question whether her "ernstliches Gefühl" [pretty serious feeling] was not replacing quite consciously already in the first place the stronger adjective "ernst" [serious].

What the Baroness formulates here is a theory of desire in which desire emerges out of quasi incestuous contexts. However, if the daughter loves the father, then this inclination is not bound to natural conditions within the biological family, but to contingency: that male figure which the girl first gets to know in a more intimate setting, will become the object of her desires ("der einzige Mann von Wert, der einem jungen Mädchen so nahe bekannt geworden"). In most cases this will be the father (copying the mother's object of desire), but since Hilarie's father is already dead, the Major functions here as a father replacement.[230] Desire emerges out of incestuous contexts, even though not all of these contexts are incestuous from a biological point of view. The incestuous attraction between Major and Hilarie does not follow a biological necessity and refutes eighteenth-century incest etiologies which claimed that incestuous love follows a "voice of the blood."

Such stress on the contingency of the emergence of desire and its simultaneous incestuous quality must interest us in the context of the biological challenges of incest that were depicted in the *Lehrjahre*. Goethe's distinction between original nature (*wo nicht entsprungen*) and nurture (*so doch gewiß genährt*) suggests that Goethe assumes a biological origin of desire, which is however contingent in the choice of its object. If the girl's first love is directed to the father, then the structure of this desire might be naturally determined, but the object of this desire is contingent. The Baroness alludes to this fact. When the Major first declares Hilarie's inclination to himself as "something so unnatural"[231] the Baroness answers with reference to her own youth: "It is not at all that unnatural. . . . From my own girlhood I can recall a passion for a man older than you."[232] The fact that first love is devoted to the father (or his replacement) and has incestuous qualities is not an anomaly but the normal case in the Baroness' theory of love. The object of desire, however, is not biologically determined, but contingent. And it is this contingency, which makes a solution to the incest problem possible.[233] Since love is not conceptualized as dependent on nature and a "voice of the blood," a space is opened for a possible solution to the incest problem. At the same time that biology obtains determining power, contingency is discovered as the complementary and liberating force.

Soon after the Baroness's disclosure of the "secret," Major and Hilarie become a couple. Eventually, however, Hilarie is more attracted to the Major's son, Flavio. Nevertheless, she still hesitates to marry Flavio. A dominant line of argument has been that Hilarie's recognition of the interchangeability of the object of love is such deep shock for her that she can only overcome it in a process of renunciation. If we follow Erich Trunz, whose "Anmerkungen" to the Hamburger Ausgabe of the *Wanderjahre* is a standard reference, Hilarie's reaction is "heitere Entsagung," a term which Trunz defines in reference to Goethe's essay "Wohlgemeinte Erwiderung" (1832) as a renunciation which has become life control, spirit, and freedom.[234] According to Trunz, Hilarie can eventually overcome her scruples and marry Flavio, because she sees that the Major is not isolated any longer (he marries another woman fitting better to his age) and only after time has healed the wounds. Trunz concludes "Liebe will Absolutheit und glaubt an sie. . . . Welche Erfahrung mit dem Ich für einen Menschen wie Hilarie! Sofern es überhaupt einen Weg gibt, diese Erfahrung innerlich zu verarbeiten, bedarf es einer geraumen Zeit der Entsagung und Entwicklung"[235] [Love wants the absolute and believes in it. . . . What an experience with selfhood for a person such as Hilarie! If at all a way exists to work through this experience, it takes a significant time interval of renunciation and development]. Much as in the assessment of Mignon, the solution of the problem of desire and erotic attraction is seen in liberation via spiritualization.

However, in reading Hilarie's marriage to her cousin Flavio—of which we learn only outside the novella at a later point in the *Wanderjahre*—as the result of a process of renunciation, we ignore important moments not only in the development of Hilarie's desires, which suggest that these desires are far from being "absolute," but also in the novella's account of the transfer of love from father to son. The solution of the biological incest problem ultimately depends on how such transfer is possible. The narrator reports the precondition of Hilarie becoming Flavio's lover:

> [Die Baronin fand] aber auch unter andern des Bruders Miniatur-Portrait und mußte über die Ähnlichkeit mit dem Sohne lächelnd seufzen. Hilarie überraschte sie in diesem Augenblick, bemächtigte sich des Bildes und auch sie ward von jener Ähnlichkeit wundersam betroffen.[236]

> [(The Baroness) also came upon, among other things, a miniature of her brother, and sighed with a smile at the resemblance to his son. Hilarie surprised her at this moment, took possession of the portrait, and she, too, was strangely moved by the resemblance.][237]

The miniature image functions as a medium that produces similarity and, therefore, makes it possible for Hilarie to recognize in the son the father. This recognition via a medium that produces similarity is repeated in a scene soon afterwards. In this second scene, Flavio appears in his father's clothes, making it possible again for Hilarie to see Flavio as the Major.[238] Once Hilarie has recognized the father in the son by means of a medium, she can redirect her emotions to the younger lover. This redirection and transfer becomes possible precisely because love is not true, absolute, and natural love, but has emerged out of contingent situations. While the Major attempts to account for his feelings by casting them in the framework of a natural cycle, the novella does not only present the failure of these attempts but stresses moments of contingency and mediality to explain the development and transfer of desires. The recognition of this contingency is then not a moment of regret but has a liberating quality.[239] It is liberating not despite but because it abandons the idea of an absolute love. If it is renunciation, then renunciation is not defined as the giving up of desire, but as the transfer to a new object. Such desire offers a solution to a biological problem, precisely because it is not biologically determined.

Love and desire in *Der Mann von funfzig Jahren* are defined by contingency and exchangeability. If, at the beginning of the novella, Hilarie's desire is juxtaposed to the family's economic interests, it is not because it is the opposite of economy, but because it follows a different economy. While the Major had incorporated love and marriage in an economy of the household that follows "natural" cycles, Hilarie's love belongs to a new economy of exchangeability. Clearly marking his voice as auctorial, the narrator comments:

> Wir machen bei dieser Gelegenheit folgende Bemerkung. Den Enthusiasmus für irgend eine Frau muß man einer andern niemals anvertrauen;

> sie kennen sich unter einander zu gut um sich einer solchen *ausschließlichen* Verehrung würdig zu halten. Die Männer kommen ihnen vor wie Käufer im Laden, wo der Handelsmann mit seinen Waren die er kennt im Vorteil steht, auch sie in dem bestem Lichte vorzuzeigen die Gelegenheit wahrnehmen kann; dahingegen der Käufer immer mit einer Art Unschuld hereintritt, er bedarf der Ware, will und wünscht sie und versteht gar selten sie mit Kenner-Auge zu betrachten. Jener weiß recht gut, was er gibt, dieser nicht immer was er empfängt. Aber es ist einmal im menschlichen Leben und Umgang nicht zu ändern, ja so löblich als notwendig, denn *alles Begehren und Freien, alles Kaufen und Tauschen* beruht darauf.[240]

> [On this occasion we make the following observation: A man should never confide his enthusiasm for any woman to another one. They know each other too well to feel themselves worthy of such exclusive adoration. In their view, men are like customers in a shop, where the shopkeeper has the advantage of knowing the wares, and can take the opportunity to present them in the best light. The customer, by contrast, always enters the shop with a kind of innocence. He needs the wares, wants them, and thinks them desirable, but seldom knows how to examine them with expert eyes. The former knows full well what he is offering, but the latter not always what he is receiving. But this is one of those elements in human life and society that cannot be changed; indeed, it is as praiseworthy as it is necessary, since *all our wishing and wooing, all our buying and trading*, depends upon it.][241]

Love and desire, as the narrator concludes here, does not follow natural cycles, but economical principles of interchangeability. The recognition of this quality is not hailed or celebrated by the novella. Rather, it is introduced as necessary. It is, as the narrator suggests, "so löblich als notwendig" [as praiseworthy as it is necessary]. If Niklas Luhmann has argued that modernity has brought forward a new concept of love which wants to be grounded in a passion beyond class and parental restriction, Goethe shows that it also has to be understood within economic parameters.[242] When Hilarie and Flavio end up marrying each other after all, this is not to be read as a confirmation of the family's initial wish, but as a fundamental reinterpretation of this wish. While the father's rule followed a cosmologically understood natural order in which the individual subsumes individual desires to serve the whole and to maintain a land-based feudal estate, the children's marriage follows a passion which circulates as much as the money in the bourgeois society to which they belong. Married, Hilarie lives under this paradigm of exchangeability:

> Hilarie kam mit ihrem Gatten, der nun als Hauptmann und entschieden reicher Gutsbesitzer auftrat. Sie in ihrer großen Anmut und Liebenswürdigkeit gewann sich hier wie überall gar gern Verzeihung einer allzu großen Leichtigkeit, von Interesse zu Interesse übergehend zu wechseln, deren wir sie im Laufe der Erzählung schuldig gefunden. Besonders die Männer rechneten es ihr nicht hoch an. Einen dergleichen Fehler, wenn er einer ist, finden sie nicht anstößig, weil ein jeder wünschen und hoffen mag, auch an die Reihe zu kommen.[243]

> [Hilarie came with her husband, now a captain and decidedly wealthy landowner. She, with her great charm and loveliness, was gladly forgiven here, as everywhere, for the excessive ease in moving from one interest to another of which we found her guilty in the course of our narrative. The men especially did not hold this greatly against her. A fault of that sort, if it be one, does not strike them as very objectionable, because each of them could wish and hope to have his turn.][244]

In contrast to Trunz's assessment in which Hilarie appears as a figure motivated by high ethical standards, Hilarie appears here with a "Fehler" [blemish].[245] As regrettable as Hilarie's lack of devotion to one partner might seem at first sight, the liberation from absolute love is the precondition for the solution of the incest problem with which we have started this chapter. It is, therefore, not astonishing that we meet Hilarie, when she appears the second time in the novel, exactly at the place where the incest occurred of which Mignon is the offspring. At the Lago Maggiore, Hilarie and the Beautiful Widow (the woman that the Major ended up marrying) meet Wilhelm and the painter. In this scene, which Andrew Piper has called the "central turning point of the novel," not only the location but also the painter's artistic production refers to the Mignon episode. This, in Piper's words, "overwhelming accumulation of inter- and intratextual references" in the scene,[246] climaxes in the moment in which the painter decides to sing Mignon's song:

> Da ermannte, da entschloß sich der Sänger, auf seinem Instrumente kräftig präludierend, uneingedenk jener früheren wohlbedachten Scho-

> nung. Ihm schwebte Mignons Bild mit dem ersten Zartgesang des holden Kindes vor. Leidenschaftlich über die Grenze gerissen, mit sehnsüchtigem Griff die wohlklingenden Saiten aufgeregt, begann er anzustimmen:
> Kennst du das Land, wo die Zitronen blüh'n,
> Im dunklen Laub ——
> Hilarie stand erschüttert auf und entfernte sich, die Stirne verschleiernd; unsere schöne Witwe bewegte ablehnend, eine Hand gegen den Sänger, indem sie mit der anderen Wilhelm's Arm ergriff.[247]
>
> [Then the singer made bold to strike up his instrument, casting aside the discretion he had earlier observed. The image of Mignon rose before him with the first bars of the lovely child's delicate song. Forgetting all restraint, plucking the resonant strings with a force born of yearning, he began to sing:
> Know you the land where the lemon blossom blow,
> And though dark leaves ——
> Hilarie rose in great agitation and hurried away, covering her brow. Our fair widow gestured with one hand in warning to the singer, while with the other she grasped Wilhelm's arm.][248]

In reenacting Mignon's song, the painter repeats moments of transgression which have shaped the story of Mignon's parents. His total identification with the story of Mignon's fate, leads to a transgression of limits ("[l]eidenschaftlich über die Grenze gerissen"). Hilarie's shock becomes understandable both as a reaction to the painter's inability to distance himself from the story and the awareness for how close she herself came to a fate similar to Mignon's parents.

In *Der Mann von funfzig Jahren*, Goethe suggests that incest and its biological consequences, which limit the possibility of selfhood for the offspring, can be avoided by the replacement of partners. The relationship with the new partner might have psychologically the same incestuous quality as the old one, but that is irrelevant, since it does not pose a biological problem any longer.[249] If biological laws request an insistence on the incest taboo, then it becomes also clear that the emotions realized in the marriage are anticipated within the family's own, incestuous, boundaries.

In the *Lehrjahre*, Goethe formulated the incest problem in the context of contemporary experiments in livestock breeding, which proposed that incestuous reproduction might lead to sexual degeneration. Goethe's analysis of desire which he presents in *Der Mann von funfzig Jahren* is surprising considering this framework. It contradicts not only a standard explanation of incest such as the assumption that incestuous desire originates in a "voice of the blood," but also contemporary conceptualizations of erotic desire within studies on animal husbandry. Albrecht Thaer, the agrarian and physician to King George III who published extensively on English agriculture and its potential benefits for German farmers (see chapter 1) and on whose work Goethe drew when working on the *Wanderjahre*, discussed the question of erotic attraction in his treatise on sheep breeding and on Bakewell's inbreeding method.[250] He wrote:

> Die Bemerkung aber, daß die Böcke kaltsinniger gegen die Schaafe werden, womit sie täglich ausgehen, und daher bey der Begattung weniger Feuer zeigen, hat sich bestätiget, und man hält daher die Böcke abgesondert.[251]
>
> [However, the observation has been confirmed that rams become cold against those sheep with which they are going out on a daily basis and, therefore, show less fire during mating; and, therefore, the rams are kept separately.]

In contrast to the description of the development of desire in *Der Mann von funfzig Jahren*, Thaer concludes that the ram's desire is conditioned by *not* having contact with the sheep before mating. This difference between the theory promoted in the *Der Mann von funzig Jahren* and Thaer's observation is crucial. The fact that Goethe framed the Mignon episode in close proximity to contemporary beliefs in animal husbandry, but chose *not* to follow the same literature when theorizing desire suggests that Goethe did not intend to promote a "biologization" of human behavior (as Charles Darwin would propose half a century later). Instead, Goethe searched for a more differentiated approach. While accepting a certain degree of determinism, he also insisted on its limited significance. Instead of deciding for "biologization," Goethe pursued the option to distinguish between biological and what one could call psychological factors and attempted to define more precisely each of these factors. Around 1800, this is a highly innovative attempt and Goethe is at its forefront.

The wish for a distinction between nature and culture is articulated in the *Lehrjahre* in close connection to the Mignon episode. At the end of the *Lehrjahre* when the Abbé—speaking after Mignon's death and in the context of the Marchese's visit which discloses her incestuous origins—regrets the tendency to judge artworks according to the same principles as works of nature and stresses the necessity to perceive and judge everything according to its own rules:

> [D]as Kunstwerk soll ihm [dem Liebhaber] ungefähr wie ein Naturwerk behagen, und die Menschen glauben, die Organe, ein Kunstwerk zu genießen, bildeten sich ebenso von selbst aus wie die Zunge und der Gaum, man urteile über ein Kunstwerk wie über eine Speise. . . . Wie schwer ist es, was so natürlich scheint, eine gute Natur, ein treffliches Gemälde an und für sich zu beschauen, den Gesang um des Gesanges willen zu vernehmen, den Schauspieler im Schauspieler zu bewundern . . . ![252]

> [(T)he work of art is to appeal him just like a natural object. People tend to believe that the faculty of appreciating art develops as naturally as the tongue or the palate, and they judge a work of art as they do food. . . . How difficult it is to do what may seem so natural, to consider a good nature or a superb painting in and for itself, music as music, acting as acting . . . !][253]

In the *Hamburger* und *Münchner Ausgabe*, we still find the use of "eine gute Statue"[254] [a fine statue] instead of "eine gute Natur" [a good nature], which we find in the above quoted version of the *Klassiker* edition (Blackall's translation also assumes "statue"). In the *Hamburger Ausgabe*, Trunz comments: "Das Wort 'Statue' ist hier Konjektur. . . . Alle Drucke zu Goethes Zeit haben statt dessen *Natur*, was aber vermutlich ein Hörfehler des Schreibers oder ein Lesefehler des Setzers war"[255] [The word "statue" is here conjecture. . . . All prints during Goethe's lifetime have instead *nature*, but that was probably a hearing mistake of the scribe or a read mistake of the typesetter]. However, when reading the Abbé's deliberations in the context of the Mignon episode, the passage becomes a comment not so much on different art forms than on the necessity of a distinction between art and nature, a reading made possible by the *Klassiker* edition's return to the version used during Goethe's lifetime.[256]

The end of the *Lehrjahre* highlights the necessity of a distinction between different value spheres and thereby questions monistic frameworks.[257] If the Tower Society's "gedenke zu leben" [Remember to live] is a reference to one of Spinoza's propositions, as Oskar Walzel has first noted in the 1920s,[258] we have to be careful to read it at the same time as evidence of Goethe's unlimited endorsement of Spinozisism. As we have seen,

Spinoza's concepts of natura naturata and natura naturans are crucial for an understanding of the Mignon episode. At the same time, however, Goethe also limits their significance: they cannot answer questions of selfhood. Art, morals, and nature belong to different spheres.

In 1817, when looking back at his work on the metamorphosis of the plants in his essay "Schicksal der Handschrift" [Fate of the Manuscript], Goethe reported that already in the late 1780s he was thinking in terms of three "Weltgegenden" [world areas] of knowledge. He names them as: nature, art, and morals. According to his own account, he devoted to each of these areas an essay: "Einfache Nachahmung der Natur, Manier, Stil" (Simple Imitation of Nature; Manner; Style) to the arts, "the Metamorphose der Pflanzen" (Metamorphosis of the Plants) to nature, and to the "Dritte" (the Third) of morals and customs the essay "Das römische Karneval" (The Roman Carneval). Goethe comments on the last area:

> Das dritte was mich beschäftigte waren die Sitten der Völker. An ihnen zu lernen, wie aus dem Zusammentreffen von *Notwendigkeit und Willkür*, von Antrieb und Wollen, von Bewegung und Widerstand ein drittes hervorgeht, was *weder Kunst noch Natur*, sondern beides zugleich ist, *notwendig und zufällig*, absichtlich und blind. Ich verstehe die menschliche Gesellschaft.[259]

> [The third subject on which I was working involved national customs. My purpose in this study was to discover through the clash of *inevitability* and *accident*, compulsion and desire, impetus and inertia, the manner in which a third thing emerges that is *neither art nor nature*, but both at the same time—*inevitable and accidental*, premeditated and fortuitous. Human society is something I understand well.][260]

In creating a close network of antagonistic pairs such as inevitability [*Notwendigkeit*] and accident [*Willkür*], compulsion [*Antrieb*] and desire [*Wollen*], art [*Kunst*] and nature [*Natur*], Goethe suggests that necessity and contingency, drive and will relate to each other as nature to art. Keeping nature and art apart from each other, he proposes that morals emerge as a third area in which the antagonistic pairs are combined while still remaining distinct. In thinking about the incest taboo and in search for a possible solution to the threat of degeneration, Goethe engages with the question of morals as an area that is defined by the interplay of necessity and contingency. In *Der Mann von funfzig [sic] Jahren*, Goethe offers a possible solution to the incest problems which emerges out of an acknowledgment of biological necessity and the belief in the power of the social as a contingent force.[261] Goethe's novella can be read as taking part in a differentiation of value spheres in each of which incest is understood differently and poses a different problem. The ethical and the biological are closely connected in Goethe's thinking, but they are not the same.[262] While Mignon might suffer under a degenerative body, neither she nor her parents are morally condemned, and while biology poses challenges that ethics has to answer, answers become possible precisely because there are areas of human life that are not determined. In the Mignon episode Goethe introduced a distinction between biology and morality and by doing so articulates a differentiation of value spheres which, according to Max Weber, marks the advent of modernity.

EPIGRAPH SOURCE

Johann Wolfgang Goethe in *Rasse, Blut und Gene: Geschichte der Eugenik und der Rassenhygiene in Deutschland*, by Kurt Bayertz, Jürgen Kroll, and Peter Weingart (Frankfurt am Main: Suhrkamp, 1988), 44.

———. *Conversations of Goethe with Eckermann.* Translated by John Oxenford. Edited by J.K. Moorhead, pseud (London: Dent, 1930), 254.

NOTES

1. Dietmar Hipp, "Dangerous Love: German High Court Takes a Look at Incest," *Der Spiegel*, March 11, 2008, http://www.spiegel.de/international/germany/dangerous-love-german-high-court-takes-a-look-at-incest-a-540831.html (accessed May 29, 2013).

2. Johann Wolfgang Goethe, *Sämtliche Werke. Briefe, Tagebücher und Gespräche*, ed. Dieter Borchmeyer et al. (Frankfurt am Main: Deutscher Klassiker Verlag, 1985—), series 1, 9:881.

3. Johann Wolfgang Goethe, *Wilhelm Meister's Apprenticeship*, ed. and trans. Eric A. Blackall in cooperation with Victor Lange (New York: Suhrkamp, 1989), 307.

4. Goethe, *Sämtliche Werke*, ed. Dieter Borchmeyer et al., series 1, 9:908.

5. Goethe, *Apprenticeship*, 323.

6. Lynn Hunt, *The Family Romance of the French Revolution* (Berkeley: University of California Press, 1992).

7. Cf. Elisabeth Krimmer, "Mama's Baby, Papa's Maybe: Paternity and Bildung in Goethe's *Wilhelm Meisters Lehrjahre*," *German Quarterly* 77, no. 3 (Summer 2004): 257–77; and Heidi M. Schlipphacke, "'Vaterschaft beruht nur überhaupt auf der Überzeugung': The Displaced Family in *Wilhelm Meisters Lehrjahre*," *Journal of English and Germanic Philology* (July 2003): 390–412.

8. Krimmer, "Mama's Baby," 269; and 263.

9. Goethe, *Sämtliche Werke*, ed. Dieter Borchmeyer et al., series 1, 9:968.

10. Krimmer, "Mama's Baby," 258. Examples for such uncertainty are the narrator's stress on feeling (instead of fact): "mit dem Gefühl des Vaters" (Goethe, *Sämtliche Werke*,

ed. Dieter Borchmeyer et al., series 1, 9:881) [with the feeling of a father (Goethe, *Apprenticeship*, 307)] and Friedrich's statement: "Vaterschaft beruht überhaupt nur auf der Überzeugung, ich bin überzeugt, und also bin ich Vater" (Goethe, *Sämtliche Werke*, ed. Dieter Borchmeyer et al., series 1, 9:940) [Fatherhood rests only on conviction; I am convinced, therefore I am the father (Goethe, *Apprenticeship*, 342)]. Schlipphacke comments on Friedrich's assessment: "This highly modern concept of identity and truth must ultimately suffice for Wilhelm in his relationship to Felix: identity is not determined by biology and fate but by the choice of the individual" ("Vaterschaft," 396). Schlipphacke is right in pointing out that biology does not define the individual in the *Meister*. However, the focus on adoption overlooks that biology is a necessary albeit not sufficient condition of selfhood (see the discussion on Hippel below).

11. Theodor Gottlieb von Hippel, *Über die Ehe*, in *G. v. Hippel's Sämmtliche Werke* (Berlin: G. Reimer, 1828–1839), 5:9.

12. Theodor Gottlieb Hippel, *On Marriage*, trans. and introd. by Timothy F. Sellner (Detroit: Wayne State UP, 1994), 70.

13. Hippel, Über die Ehe, 152. However, Hippel's text displays a great interest in what we would call today biological questions and the biological transfer between the generations. For example, he discusses the question of whether or not late marriages—as they were common for male citizens of his time—had an impact on the offspring. Arguing against the convention of late marriage, he statees: "'Die Israeliten sind klein gewesen, sagt man, weil sie so zeitig geheiratet haben'; allein ich sehe keinen Nachttheil von kleinen Bürgern ein: wenigstens sind sie besser, als gar keine" (ibid., 11) [People say the Israelites were small in stature because they married so early; nevertheless, I see no disadvantage in smaller citizens; they are at least better than none at all (Hippel, *On Marriage*, 75)]. "Ein Licht zündet das andere sehr leicht an, und man könnte annehmen, daß Genies von einem noch ungeschwächten Vater gezeugt werden müßten; wenigstens sind die ersten Kinder von jeher immer die besten gewesen" (Hippel, *Über die Ehe*, 12) [One candle easily ignites another, and it could be assumed that geniuses would only be produced by a father in a still unweakened condition; the first children, at any rate, have since the beginning of time always been the best (Hippel, *On Marriage*, 75]. "Die Rechte der Erstgeburt und die Opfer der Erstlinge würden also auf gute physische und moralische Grundsätze berechnet seyn"(Hippel, *Über die Ehe*, 12) [Thus the rights of the firstborn and the sacrifice of firstborn children were based on good physical and moral principles (Hippel, *On Marriage*, 76)].

14. Hippel, *Über die Ehe*, 10–11.

15. Hippel, *On Marriage*, 74–75.

16. Hippel's recognition of a biological preconditions for citizenship does *not* imply in any way a condemmnation of those who might not fulfill these preconditions: "Es war von jeher gebräuchlich, den Menschen zum Besten des Bürgers leiden zu lassen; und es ist und bleibt barbarisch, daß man schwächliche Kinder, unter dem Vorwande, die könnten mit genauer Noth Menschen, nicht füglich aber Bürger werden, tödtete und, als thäte man dem Moloch des Staates einen Dienst daran, diese Kindermorde gesetzlich beförderte" (Hippel, *Über die Ehe*, 14) [From time immemorial it has been the custom to allow the human being to suffer for the good of the citizen, but it will forever remain barbaric to kill feeble and sickly children under the pretext that they might perhaps with difficulty become human beings, but never proper citizens of the state—and what is more, then

to sanction this action legally, as if one were doing a service to the Moloch of the state thereby (Hippel, *On Marriage*, 76).]

17. Eunuchs and hermaphrodites were both considered individuals whose civil rights were in question on the basis of their sexual "degeneration." When Johann Feiler published in 1820 a new book on hermaphroditism, he legitimized the publication of his insights with reference to their importance for the bourgeois society: "indem es keineswegs gleichgültig sein kann, ob ein neugeborenes Kind für einen Knaben oder ein Mädchen gehalten, als Knab oder als Mädchen erzogen werde, ob einem solchen Erwachsnen gewisse bürgerliche Rechte zu, oder abgesprochen werden" [since it cannot be a trivial matter at all whether a newborn child is considered a boy or a girl, whether it is educated as boy or as girl, whether such an adult is granted certain civil rights or not] (*Über angeborene menschliche Missbildungen im Allgemeinen und Hermaphroditen insbesondere. Ein Beitrag zur Physiologie, pathologische Anatomie, und gerichtlichen Arzneiwissenschaft* [Landshut: Philipp Krüll, Universitätsbuchhändler, 1820], 72). For a historical account of eunuchs and other sexual "degenerations," such as hermaphroditism, see Ann Rosalind Jones and Peter Stallybrass, "Fetishizing Gender: Constructing the Hermaphrodite in Renaissance Europe" in *Body Guards: The Cultural Politics of Gender Ambiguity*, ed. Julia Epstein and Kristina Straub (New York and London: Routledge, 1991), 90–91; and James Y. Simpson, "Hermaphroditism," in *The Cyclopaedia of Anatomy and Physiology*, ed. Robert B. Todd (London: Sherwood, 1839), 2:704.

18. See Thomas P. Saine "Was Wilhelm Meisters Lehrjahre Really Supposed to Be a Bildungsroman?" in *Reflection and Action: Essays on the Bildungsroman*, ed. James N. Hardin (Columbia: University of South Carolina Press, 1991), 118–41; and Krimmer, "Mama's Baby."

19. For an account of Goethe's interest in polarity in general see Andreas B. Wachsmuth, *Geeinte Zwienatur: Aufsätze zu Goethes naturwissenschaflichem Denken* (Berlin: Aufbau-Verlag, 1966); and Grete Schaeder, *Gott und die Welt: Drei Kapitel Goethescher Weltanschauung* (Hameln: Fritz Seifert, 1947).

20. Thomas W. Kniesche, "Die psychoanalytische Rezeption von Mignon," in *Goethes Mignon und ihre Schwestern*, ed. Gerhart Hoffmeister (New York: Peter Lang, 1993), 72.

21. Friedrich A. Kittler, "Über die Sozialisation Wilhelm Meisters," in *Dichtung als Sozialisationsspiel*, ed. Friedrich A. Kittler and Gerhard Kaiser (Göttingen: Vandenhoeck & Ruprecht, 1978), 36–43.

22. Konstanze Bäumer, "Wiederholte Spiegelungen-Goethes 'Mignon' und die 'Neue Melusine,'" in *Goethes Mignon und ihre Schwestern*, ed. Gerhart Hoffmeister (New York: Peter Lang, 1993), 113, emphasis mine.

23. Sabine Groß, "Diskursregelung und Weiblichkeit: Mignon und ihre Schwestern," in *Goethes Mignon und ihre Schwestern*, ed. Gerhart Hoffmeister (New York: Peter Lang, 1993), 83, emphasis mine.

24. Ulrike Rainer, "A Question of Silence: Goethe's Speechless Women," in *Goethes Mignon und ihre Schwestern*, ed. Gerhart Hoffmeister (New York: Peter Lang, 1993), 101, emphasis mine

25. Catriona MacLeod, "Pedagogy and Androgyny in *Wilhelm Meisters Lehrjahre*," *Modern Languages Notes* 108, no. 3 (April 1993): 408–11, italics mine.

26. An exception is Ronald Douglas Gray who reads Mignon in the alchemist tradition as a mixture of male and female principles (*Goethe the Alchemist: A Study of Alchemical*

Symbolism in Goethe's Literary and Scientific Works [Cambridge: Cambridge University Press, 1952; repr. 2010], see particularly the pages 221–49).

27. Robert Tobin, "The Medicinalization of Mignon," in *Goethes Mignon und ihre Schwestern*, ed. Gerhart Hoffmeister (New York: Peter Lang, 1993), 43–60. The connection between the figure of the "mignon" and the hermaphrodite is made already in Thomas Artus' *Description des l'Isle des Hermaphrodites* (ca. 1560), in which Artus satirizes the "mignon du roi" at the French court of Henri III by presenting them as hermaphrodites, quoted in Jones and Stallybrass, "Fetishizing Gender," 92. Wolfgang Riedel, in his commentary "Zur Textgestalt" in the Münchner Ausgabe, refers to the French origin of the word *mignon* as synonym for "Liebling" [Beloved] or "Günstling" [Favorite], a companion of the king who sometimes even shared his bed (Riedel, "Zur Textgestalt," in Johann Wolfgang Goethe, *Sämtliche Werke nach Epochen seines Schaffens*, ed. Karl Richter et al. [München: Carl Hanser, 1988–1998], 5:705, see also 5:736).

28. Eugen Wolff, *Mignon. Ein Beitrag zur Geschichte des Wilhelm Meister* (München: Beck, 1909), 110.

29. Goethe, *Sämthiche Werke*, ed. Dieter Borch Meyer et al., series 1, 9:924.

30. Tobin, "Mignon," 47.

31. Kurt Robert Eissler, *Goethe: eine psychoanalytische Studie 1775–1786*, trans. Peter Fischer and Rüdiger Scholz, ed. Rüdiger Scholz with Wolfram Mauser and Johannes Cremerius (Basel and Frankfurt am Main: Stroemfeld/Roter Stern, 1983–1985), 2:870.

32. Tobin, "Mignon," 49, italics mine.

33. Goethe, *Sämtliche Werke*, ed. Dieter Borchmeyer et al., series 1, 9:968, italics mine.

34. Goethe, *Apprenticeship*, 359, italics mine.

35. Goethe, *Sämtliche Werke*, ed. Dieter Borchmeyer et al. series 1, 9:966.

36. Goethe, *Apprenticeship*, 358.

37. Goethe, *Sämtliche Werke*, ed. Dieter Borchmeyer et al., series 1, 9:964, italics mine.

38. Goethe, *Apprenticeship*, 357, italics mine.

39. Hartmut Nonnenmacher, *Natur und Fatum: Inzest als Motiv und Thema in der französischen und deutschen Literatur des 18. Jahrhunderts* (Frankfurt am Main and New York: Peter Lang, 2002), 195.

40. Cf. Daniel W. Wilson, "Science, Natural Law, and Unwitting Incest in Eighteenth-Century Literature," in *Studies in Eighteenth-Century Culture*, ed. O M Brack (Madison, WI, University of Wisconsin Press, 1984), 252.

41. Lord Henry Home Kames declares incest a necessity for the maintaining of nobility in indigenous populations: "The tradition, that the Inca family were children of the sun, introduced incest among them; for it was a matter of religion to preserve their divine blood pure and unmixed," in *Sketches of History of Man* (Dublin: Printed by James Williams, 1775), 3:86. Eighteenth Century Collections Online: http://find.galegroup.com.myaccess.library.utoronto.ca/ecco/infomark.do?&source=gale&prodId=ECCO&userGroupName=utoronto_main&tabID=T001&docId=CB128044381&type=multipage&contentSet=ECCOArticles&version=1.0&docLevel=FASCIMILE (accessed February 9, 2014 from University of Toronto Libraries).

42. See Nonnenmacher, *Natur und Fatum*.

43. Goethe, Sämtliche Werke, ed. Dieter Borchmeyer et al., series 1, 9:965.

44. Goethe, *Apprenticeship*, 357.

45. Goethe, *Sämtliche Werke*, ed. Dieter Borchmeyer et al., series 1, 9:965.

46. Goethe, *Apprenticeship*, 357.

47. New American Standard Bible (NASB), http://www.biblegateway.com (accessed March 16, 2012), translation modified by author.

48. The motive of the lily reappears in the *Wanderjahre* in the context of the St. Joseph episode. There, it is a sign that innocence is possible exactly against nature, in the name of adoption.

49. The only exception that Nonnenmacher quotes is Morelly's *Le naufrage des Isles flottantes ou Basiliade du célèbre Pilpai* (1753), in which a botanical metaphoric is used to describe a father-daughter relationship. (*Natur und Fatum*, 198).

50. Ibid., 199.

51. For a different account of Goethe's ethics in the context of Lucretius and Spinoza see Julie D. Prandi, *Dare to Be Happy!: A Study of Goethe's Ethics* (Lanham, MD: University Press of America, 1993).

52. Goethe, *Sämtliche Werke*, ed. Dieter Borchmeyer et al., series 1, 9:965. Mignon's song, which the reader gets to know long before knowing Mignon's incestuous origin, alludes to this scene when it asks: "Kennst du das Land? wo die Zitronen blühn, / Im dunklen Laub die Gold-Orangen glühn . . .?" (9:503) [Know you the land where the lemon blossoms blow, / And through dark leaves the golden oranges glow . . .? (Goethe, *Apprenticeship*, 83)].

53. Goethe, *Apprenticeship*, 357.

54. Goethe, *Sämtliche Werke*, ed. Dieter Borchmeyer et al., series 1, 9:965.

55. Goethe, *Apprenticeship*, 357.

56. Goethe, Sämtliche Werke, ed. Dieter Borchmeyer et al., series 1, 9:968.

57. Goethe, *Apprenticeship*, 359.

58. Christian Heinrich Ernst Bischoff, *Darstellung der Gallschen Gehirn- und Schädel-Lehre. Nebst Bemerkungen über diese Lehre von Dr. C. W. Hufeland*. Zweite, vermehrte und verbesserte Auflage (Berlin: In Commission bei L. W. Wittich, 1805), 104.

59. Markwart Michler, "Lobstein, Johann Friedrich der Ältere," *Neue Deutsche Biographie* 14 (1985): 737–38.

60. Cf. Sigrid Oehler-Klein, *Die Schädellehre Franz Joseph Galls in Literatur und Kritik des 19. Jahrhunderts: zur Rezeptionsgeschichte einer medizinisch-biologisch begründeten Theorie der Physiognomik und Psychologie* (Stuttgart: Gustav Fischer Verlag, 1990).

61. Hellmut Ammerlahn, "Wilhelm Meisters Mignon-ein offenbares Rätsel. Name, Gestalt, Symbol, Wesen und Werden," *DVJS* 42, no. 1 (1968): 89–116.

62. Cf. Nonnenmacher, *Natur und Fatum*, 132. See also Eckhard Meyer-Krentler, "Erdichtete Verwandtschaft. Inzestmotiv, Aufklärungsmoral, Strafrecht in J. W. Goethes *Die Geschwister*," *Literatur für Leser* 4 (1982): 242n4.

63. *Allgemeines Landrecht für die Preußischen Staaten*, 1794, http://www.smixx.de/ra/Links_F-R/PrALR/PrALR_II_1.pdf (accessed March 9, 2012).

64. Gottlieb Carl Svarez, *Vorträge über Recht und Staat*, ed. Hermann Conrad and Gerd Kleinheyer (Köln: Westdeutscher Verlag, 1960), 317–18.

65. The question of whether or not experiments made in animal breeding in general and inbreeding in particular could become relevant for issues related to human reproduction was not yet settled in the eighteenth century. In 1667, one hundred years before

Robert Bakewell started his experiments with inbreeding in sheep, Lord William Cavendish (1592–1676) expressed the opinion in that "you cannot breed better than to breed to your own mares that you have bred; and let their fathers cover them; for there is no inceste [sic] in horses; and thus they are nearer, by a degree to the purity, since a fine horse got them, and the same fine horse covers them again" (quoted in Roger J. Wood and Vítězslav Orel, *Genetic Prehistory in Selective Breeding: A Prelude to Mendel* [New York and Oxford: Oxford University Press, 2001], 53). While inbreeding is here an accepted and desired way of stock improvement, at the same time the distinction between man and animal is emphasized by the rejection of the incest taboo for the animal and an implied acceptance of the taboo for the human being. In Germany, Johann David Michaelis (1717–1791), prominent German Lutheran theologian and professor for Oriental languages and philosophy in Göttingen, discussed the animal-human comparison in his *Abhandlungen von den Ehegesetzen Mosis welche die Heyrathen in die nahe Freundschaft untersagen* [Treatise on the Marriage Laws of Moses which prohibit Marriages among Close Friends], Diss., n.p., 1755; repr. (Zweite und vermehrte Auflage. Göttingen: Verleg Abram Vandenhoecks seel. Witwe, 1768). In referring to experiences in animal breeding, Michaelis aimed to clarify whether biological reasons provided a sufficient ground to accept or reject the incest taboo. However, he also wondered whether drawing conclusions from animal to human is a legitimate method in the first place. Eventually, he endorsed this comparison because he found link between human and animal: much like Kant, for whom some non-caucasian races were closer to animality (see chap. 1), Michaelis approached the question of the significance of breeding experiments by discussing non-caucasian races. By referring to the North American Indian, Michaelis eventually dared to make the suggestion that incestuous reproduction is not only anthropologically and morally sanctioned, but has also no biologically impact on the offspring in humans: "die Nordamerikaner, bey denen die Heyrath der Geschwister erlaubt ist, weder der Leibesstärke und Größe nach, noch auch wegen ihrer Gemüthskräfte, abgeartete Menschen zu sein scheinen" [the North Americans, which allow marriage among siblings, seem to be neither in bodily strength nor in size, nor in the powers of their mind degenerated] (Michaelis, quoted in Nonnenmacher, *Natur und Fatum*, 147). Michaelis's allusion to the man-animal connection—made possible via detour to the American Indian—is still not common at the time, but anticipated biological approaches to the incest problem.

66. Trends to free criminal law from non-legal arguments (e.g. moral or religious) increased around 1800. Albrecht Koschorke refers to Karl Grolman's *Grundsätze der Criminalrechtswissenschaft* [Principles of Criminal Law] (1797), and comes to the conclusion "Damit wird die Grundlage für die Verfolgung von Inzest brüchig, weil dieses Delikt dritte Personen nicht schädigt und insofern kriminalistisch nicht relevant ist" [Therefore, the basis for the prosecution of incest becomes fragile, because this delict does not harm third parties and, therefore, is not relevant from the perspective of criminal law] (Koschorke, *Körperströme und Schriftverkehr: Mediologie des 18. Jahrhunderts* [München: Fink, 1999], 5n14).

67. Daniel W. Wilson, "Science, Natural Law, and Unwitting Incest in Eighteenth-Century Literature," in *Studies in Eighteenth-Century Culture*, ed. O M Brack (Madison, WI, University of Wisconsin Press, 1984), 13:251.

68. Johann Georg Stumpf, *Versuch einer pragmatischen Geschichte der Schäfereien in Spanien und der Spanischen in Sachsen, Anhalt-Dessau &c.* (Leipzig: In der J.G. Müllerischen Buchhandlung, 1785), 59.

69. Wood and Orel, *Genetic Prehistory*, 156–57.

70. By the 1830s, scholars transferred results regarding hermaphroditism which were obtained in animal breeding to human beings without hesitations. See Johannes Müller, *Bildungsgeschichte der Genitalien* (Düsseldorf: Arn, 1830), 121–31; and Simpson, "Hermaphroditism."

71. Cf. John Saunders Sebright *The Art of Improving the Breeds of Domestic Animals* (London: John Harding, 1809); and John Sinclair, *Grundsätze des Ackerbaues nebst Bemerkungen über Gartenbau, Obstbaumzucht, Forst-Cultur und Holzpflanzungen* (Vienna: Bei Heubner, 1823). The occurrence of hermaphroditism in domestic animals was noted frequently. John Hunter remarks that he has never observed hermaphroditism in the human species, "but in the horse, ass, sheep, and black cattle, it is very frequent" ("Account of the Free Martin," in *Observations on Certain Parts of the Animal Oeconomy* [London: Sold at No. 13 Castle-Street, Leicester-Square, 1786], 46).

72. Parry was a well-known physician who had contact with famous contemporaries like statesman and philosopher Edmund Burke (1729–1797), Edward Jenner (1749–1823), pioneer of the smallpox vaccine, the naturalist Joseph Banks (1743–1820), and the astronomer Wilhelm Herschel (1738–1822). Jane Austen mentions Parry in letters to her sister Cassandra. For a short biography of Parry see Gillian Hull, "Caleb Hillier Parry 1755–1822: A notable provincial physician," *Journal of the Royal Society of Medicine* 91, no. 6 (June 1998): 335–38.

73. Caleb Hillier Parry, "Part II. History of the Merino-Ryeland Breed of the Author," in *Communications to the Board of Agriculture; on Subjects Relative to the Husbandry, and Internal Improvement of the Country* (London: Bulmer and Co., 1806), 5:471, italics mine). Available online at Eighteenth Century Collections Online: http://quod.lib.umich.edu/e/ecco (accessed November 5, 2011).

74. Goethe, *Sämtliche Werke*, ed. Dieter Borchmeyer et al., series 1, 9:443.

75. Goethe, *Apprenticeship*, 50.

76. Goethe, *Sämtliche Werke*, ed. Dieter Borchmeyer et al., series 1, 9:444.

77. Goethe, *Apprenticeship*, 50.

78. Goethe, *Sämtliche Werke*, ed. Dieter Borchmeyer et al., series 1, 9:568–69.

79. Goethe, *Apprenticeship*, 122.

80. Goethe, *Sämtliche Werke*, ed. Dieter Borchmeyer et al., series 1, 9:956, italics mine.

81. The switch between male and female pronouns is so puzzling that it is ignored in Blackall's translation: "Sadly we brought *her* here, here shall *she* stay . . . Her head we crowned with roses, sweet and friendly was her gaze" (Goethe, *Apprenticeship*, 352, italics mine).

82. Goethe, *Sämtliche Werke*, ed. Dieter Borchmeyer et al., series 1, 9:895. Mignon's quote refers to Matthew 22:30: "For in the resurrection they neither marry nor are given in marriage, but are like angels in heaven" (New American Standard Bible). Similar Luke 20: 34–36: "Jesus replied: 'The sons of this age marry and are given in marriage, but those who are considered worthy to attain to that age and the resurrection from the dead, neither marry nor are given in marriage; for they cannot die anymore, because they are like angels, and are sons of God, being sons of the resurrection'" (New American Standard Bible).

83. Goethe, *Apprenticeship*, 316.

84. Riedel, "Zur Textgestalt," in Goethe, *Sämtliche Werke nach Epochen seines Schaffens*, ed. Karl Richter et al., 5:705.

85. Goethe *Sämtliche Werke*, ed. Dieter Borchmeyer et al., series 1, 9:553.

86. Blackall's translation: "a silly androgynous creature" (Goethe, *Apprenticeship*, 113).

87. It is also Jarno who has a hermaphrodite aide once he appears as Montan in the *Wanderjahre*. Like Mignon, this aide speaks in a way which is difficult to understand.

88. See MacLeod, "Pedagogy." See also Krimmer who argues that the body is the primary site of gender for Goethe: "The stark contrast between the stereotypically female character traits of Goethe's fictitious cross-dressers and their incongruous male attire illustrates that, in *Wilhelm Meisters Lehrjahre*, the body is defined as the primary site of a person's gender identity" ("Mama's Baby," 260). In *Der androgyne Mensch*, Helmut Fuhrmann presents the thesis that Goethe defines the human being as an androgynous being (*Der androgyne Mensch: "Bild" und "Gestalt" der Frau und des Mannes im Werk Goethes* [Würzburg: Königshausen & Neumann, 1995]).

89. Hence, my reading does not confirm earlier claims that Therese's and Natalie's androgyny is a sign of a natural wholeness or human perfection, as in the following example: "Die Anziehungskraft, die Figuren wie . . . Therese (die 'wahre Amazone') und Natalie (die 'schöne Amazone') im Roman auf Wilhelm ausüben, verweist auf das Streben nach 'harmonischer Ausbildung' seiner menschlichen Natur. Die Verbindung von männlicher und weiblicher Geschlechtseigentümlichkeiten . . . deutet in der Tradition des platonischen Mythos auf das Ideal der Vollkommenheit der menschlichen Gattung" [The attraction, which figures . . . such as Therese (the "true amazone") and Natalie (the "beautiful amazone") have for Wilhelm in the novel, alludes to the striving for "harmonious formation" of his human nature. The combination of male and female qualities . . . points to the tradition of the platonic myth of the ideal of the perfection of the human race] (Wilhelm Voßkamp, "Stellenkommentar," in Goethe, *Sämtliche Werke*, ed. Dieter Borchmeyer et al., series 1, 9:1388).

90. For a discussion of the distinction between androgyny and hermaphroditism see also Ruth Gilbert, *Early Modern Hermaphrodites: Sex and Other Stories* (New York: Palgrave, 2002), 9–12. In his article on "Androgyny Seen Through Shakespearean Works" (1982), Robert Kimbrough comes to the conclusion that "[m]uch of the recent literature on androgyny has been confused and muddled because of the simple failure to observe that androgyny is a mythic concept which represents an inner, psychic state of experience, whereas hermaphroditism is an objective, physical state of being" (quoted in Gilbert, *Early Modern Hermaphrodites*, 11). Gilbert's counter-argument that "representations of early modern hermaphroditism incorporated the enactment as well as the embodiment of sexual ambiguity" (11) is convincing. However, Kimbrough's distinction can be productively used if we consider eighteenth-century discussions on the topic. I argue that the distinction is historically bound to the time around 1800 when biology emerged as an independent discipline.

91. Because Mignon remains bound to her natural, biological status—which makes it impossible for her to conceive and to speak—she remains also outside the cultural system of kinship. Mignon's inability to transcend her nature into culture finds its symbolic expression in the fact that she is unable to verbalize her experience and write her songs. It is always somebody else who has to do it for her (e.g. Goethe *Sämtliche Werke*, ed. Dieter Borchmeyer et al., series 1, 9:504).

92. Grace Tiffany, *Erotic Beasts and Social Monsters: Shakespeare, Jonson, and Comic Androgyny* (London and Cranbury, NJ: Associated University Press, 1995), 11.

93. Gilbert, *Early Modern Hermaphrodites*, 9.

94. Jones and Stallybrass, "Fetishizing Gender," 81.

95. Augustine, *City of God*, trans. Marcus Dods (Edinburgh: T & T Clark, 1913), 97.

96. St. Augustine's assessment of the hermaphrodite in the context of monstrosities remained influential well into the eighteenth century. For some, the hermaphrodite became the prime example for monstrosities, as the following publication from the late eighteenth century suggests D. Georg Heuermann, ed., "Bemerkungen von einigen in Seeland befindlichen Missgeburten, welche man gemeiniglich Zwitter (Hermaphroditi und Androgynos) benennet" [Notes on some monstrosities in Seeland which one usually calls hermaphrodite (hermaphroditi and androgynos)], in *Vermischte Bemerkungen und Untersuchungen der ausübenden Arzeneiwissenschaft* (Koppenhagen and Leipzig: Gabriel Christian Rothens Wittwe, 1765), vol. 2.

97. Gilbert, *Early Modern Hermaphrodites*, 22.

98. Jacob Rueff, *Hebammen Buch* (Gedruckt zu Franckfort am Mayn: Sigmund Feyerabendt, 1580), 122.

99. Kathleen P. Long discusses in particular medical treatises in late sixteenth-century France and Switzerland to support her claim of a shift from reliance on classical sources to more empirical methods of inquiry (*Hermaphrodites in Renaissance Europe* [Aldershot, Hampshire and Burlington, VT: Ashgate, 2006], 2–3 and 29).

100. Maximilian Schochow, *Die Ordnung der Hermaphroditen-Geschlechter: Eine Genealogie des Geschlechtsbegriffs* (Berlin: Akademie Verlag, 2009), 14.

101. Ibid., 17.

102. Ibid., 91–101.

103. Christian Ludwig Mursinna (1744–1823) was professor for surgery in Berlin since 1789. Goethe was familiar with his work and when Mursinna died in 1823, Goethe commented the death with the following words "Mögen junge heranstrebende Männer mit gleichem Ernst und reiner kräftiger Tätigkeit sich in gleichem Falle an so edlen Beispielen ermutigen" [May young and striving men with equal seriousness and pure forceful activity be encouraged in the same situation by such noble example (June 8, 1823)] (Goethe, *Sämtliche Werke*, ed. Dieter Borchmeyer et al., series 1, 24:653).

104. Christoph Wilhelm Hufeland, "Beschreibung und Abbildung eines zu Berlin beobachteten weiblichen Hermaphroditen," in *Journal der practischen Arzneykunde und Wundarzneykunst*, (Berlin: Ungers Journalhandlung, 1803), 170, quoted in Schochow, *Die Ordnung*, 220.

105. Hufeland, "Beschreibung," 171–72, quoted in Schochow, *Die Ordnung*, 220.

106. D. Johann Christian Stark, ed., "Nro. VI. Kurze Beschreibung eines sogenannten Hermaphroditen oder Zwitters, . . ." in *Neues Archiv für die Geburtshülfe* . . . (Jena: Wolfgang Strahl, 1801), 551, quoted in Schochow, *Die Ordnung*, 223. Already in the often discussed case of the hermaphrodite Marie/Marin, which had occupied French physicians like Jacques Duval in the seventeenth century, ejaculation had played a crucial role in determining the hermaphrodite's sex. Referring to Duval's ability to arouse Marie/Marin's "penis" to ejaculate, Stephen Greenblatt has argued that "medical authority had masturbated Marin's identity into existence" ("Fiction and Friction," in *Reconstructing Individualism: Autonomy, Individuality, and the Self in Western Thought*, ed. Thomas C. Heller, Christine Brooke-Rose et al. (Stanford, CA: Stanford University Press, 1986), 32.

107. Stark, "Nro. VI. . . .," quoted in Schochow, *Die Ordnung*, 223.

108. Thomas Walter Laqueur, *Making Sex: Body and Gender from the Greeks to Freud* (Cambridge, MA: Harvard University Press, 1990), 20.

109. Aristotle, *Generation of Animals*, trans. A. L. Peck (Cambridge, MA: Heinemann and Harvard University Press, 2000), 401.

110. Rueff, *Hebammen Buch*, 122.

111. Looking at the history of hermaphroditism, Johann Feiler concluded that all hermaphrodites are men: "Aus allem bisher Gesagten folgt nun ohne weiteres, daß es immer nur Männer waren, die man für Hermaphroditen ansah" (From all what has been said until now, it follows without hesitation that it were always men which were considered hermaphrodites) (Feiler, *Über angeborene*, 96). Similarly, James Parsons denied altogether that hermaphrodites exist. Instead, he claimed that the hermaphrodite is a woman with enlarged clitoris (*A Mechanical and Critical Enquiry into the Nature of Hermaphrodites* [London: J. Walthoe, 1741]).

112. C. L. Schweickhard. "Geschichte eines lange Zeit hindurch für einen Hermaphroditen gehaltenen wahren Mannes" [History of a real man who was long considered a hermaphrodite] (in Schochow, *Die Ordnung*, 236).

113. Feiler, *Über angeborene*, 116. The first surgeon who did a surgical intervention and described it is Georg Arnaud. See his *Anatomisch-Chirurgische Abhandlung über die Hermaphroditen, aus dem Französischen übersetzt, nebst Kupfertafeln* (Straßburg: Amand König, 1777).

114. Goethe, *Sämtliche Werke*, ed. Dieter Borchmeyer et al., series 1, 9:599. In reports on hermaphrodites, the discovery of a person's sexual condition is often connected to a doctor's visit for another reason. See for example Marguerite Malaure, "Bittschrift der Margarethe Malaure an den König," in *Eine Frauensperson, die fälschlich für einen Zwitter gehalten wird. Erzählung sonderbarer Rechtshändel, sammt deren gerichtlichen Entscheidung*, ed. François Gayot von Pitaval (Vierter Theil. Leipzig: Gottfried Kiesewetter, 1748), 404–15. See also Maria Dorothea Derrier's case of which Hufeland reports in "Beschreibung," quoted in Schochow, *Die Ordnung*, 185 and 220.

115. Goethe, *Apprenticeship*, 140.

116. Goethe, *Sämtliche Werke*, ed. Dieter Borchmeyer et al., series 1, 9:498.

117. Blackall translates: "[Mignon's] rigid limbs unfroze, her whole inner self poured itself out" (Goethe, *Apprenticeship*, 82).

118. Despite the hermaphrodite's ambiguous sex and infertility, the hermaphrodite was seen as a figure with strong sexual desires: "[M]en with testicles *inside* their body were . . . considered . . . notoriously lustful" (Jones and Stallybrass, "Fetishizing Gender," 88). James Y. Simpson, in his article on hermaphroditism in *The Cyclopaedia of anatomy and physiology* (1836–1847), refers to a case, related by Jean Chroker, in which a nun who "changed . . . into a male, seven years after taking the vows" ("Hermaphroditism," 692). After his transformation, this hermaphrodite's "sexual desires became extremely strong, and he is said to have been ultimately condemned, whether justly or not, under an accusation of rape" (693).

119. The combination of erotic desire and infertility was considered a defining feature of the hermaphrodite already in the seventeenth century. Joshua Poole writes in the English Parnassus (1657): "*Hermaphrodite.* Ambiguous, promiscuous, mixed, sex-confused, mongrell, neuter, effeminate" (Poole, quoted in Gilbert, *Early Modern Hermaphrodites*, 9).

120. Stark, "Nro. VI. . . .," 540, quoted in Schochow, *Die Ordnung*, 224.

121. Kniesche, "Die psychoanalytische Rezeption von Mignon," 73.

122. Goethe, *Sämtliche Werke*, ed. Dieter Borchmeyer et al., series 1, 9:926, italics mine.

123. Goethe, *Apprenticeship*, 334, italics mine.

124. See Moritz Baßler, "Goethe und die Bodysnatcher," in *Von der Natur zur Kunst zurück*, ed. Moritz Baßler et al. (Tübingen: Niemeyer, 1997), 181–97. See also Irmgard Egge,

"'Verbinden mehr als Trennen': Goethe und die plastische Anatomie," *Germanisch-romanische Monatsschrift* 51, no. 1 (2001): 45–53.

125. The anatom says to Wilhelm: "'Ihnen gesteh' ich's, mein Freund, in dieser Stadt hat man gemordet, um den dringenden, gut bezahlenden Anatomen einen Gegenstand zu verschaffen'" (Goethe, *Sämtliche Werke*, ed. Dieter Borchmeyer et al., series 1, 10:611). ["I will confess to you, my friend: murder has been committed in this city in order to supply the insistent anatomist who pays well for the object he needs" (Johann Wolfgang Goethe, *Wilhelm Meister's Journeyman Years*, ed. Jane K. Brown and trans. Krishna Winston [New York: Suhrkamp, 1989, 329].)

126. Goethe was bothered by the necessity of dissection for medical training. In fact, as Gerhard Müller has pointed out, in *Wilhelm Meisters Wanderjahre* this criticism is included in the third book when Wilhelm starts his training as a surgeon (Müller 211–23). Here, the narrator explains Wilhelm's hesitation to disect the arm of a drowned person: "Der Widerwille dieses herrlichen Naturzeugnis noch weiter zu entstellen stritt mit der Anforderung, welche der wissensbegierige Mensch an sich zu machen hat" [His reluctance to mutilate this magnificent product of nature any further struggled with the demands which any man striving for knowledge must place on himself (Goethe, *Journeyman Years*, 323).] Also mentioned is the lack of corpses for this study "Unangenehm hindernd war beim Studium die immer wiederholte Klage vom Mangel der Gegenstände, über die nicht hinreichende Anzahl der verblichenen Körper. . . .Solche, wo nicht hinreichend, doch in möglichster Zahl zu verschaffen, hatte man harte Gesetze ergehen lassen, nicht allein Verbrecher . . ., sondern auch andere körperlich geistig Verwahrloste wurden in Anspruch genommen. Mit dem Bedürfnis wuchs die Strenge und mit dieser der Widerwille des Volkes, das in sittlicher und religioser Ansicht seine Persönlichkeit und die Persönlichkeit geliebter Personen nicht aufgeben kann" (series 1, 10:601–2) [An annoying hindrance to these studies was the constant complaint at the lack of specimens. . . . To provide as many as possible, though not enough, harsh laws had been enacted, so that not only criminals . . . , but also other poor wretches, victims of physical or mental ills, were claimed for this use. As the need grew, so did the stringency, and with it the opposition of the people, who, on moral and religious grounds, cannot surrender their own person and that of their loved ones (Goethe, *Journeyman Years*, 322)]. The problem of mere body as scientific object and the neglect of the feelings of the bereaved because of strong interest in the deceased body occupied Goethe up to the very end of his life. In the text on essay "Plastische Anatomie" (completed in February 1832), one month before his death, as in the *Wanderjahre* the suggestion is to use wax models instead (Goethe, Sämtliche Werke nach Epochen seines Schaffens, ed. Karl Richter et al., 18.2:539–41).

127. C. F. Neickelio, (a.k.a. Caspar Friedrich Jenckel), *Museographia oder Anleitung zum rechten Begriff und nützlicher Anregung der Museorum oder Raritätenkammern* (Leizpig and Breslau: Michael Hubert, 1727), italics mine. For the popularization of anatomical studies at the end of the eighteenth century, see Irmela Marei Krüger-Fürhoff, *Der versehrte Körper: Revisionen des klassistischen Schönheitsideals* (Göttingen: Wallstein, 2001).

128. Ibid., 5, italics mine.

129. After Mignon's death, the Abbé reports that Mignon had a tattoo which contained a crucifix and other signs: "Kruzifix, von verschiedenen Buchstaben und Zeichen begleitet" (Goethe, *Sämtliche Werke*, ed. Dieter Borchmeyer et al., series 1, 9:959) [crucifix, together with various letters and signs (Goethe, *Apprenticeship*, 353)] and that Mignon kissed this image passionately shortly before her death. The fact that Mignon's tattoos have

a certain resemblance with the signs on the body of Rueff's "Monster of Ravenna" further increase the close relationship between Mignon and the iconography of the hermaphrodite monster (see fig. 2.1).

130. Goethe, *Sämtliche Werke*, ed. Dieter Borchmeyer et al., series 1, 9:958, italics mine.

131. Goethe, *Apprenticeship*, 353, italics mine.

132. Barbara Maria Stafford, *Body Criticism: Imaging the Unseen in Enlightenment Art and Medicine* (Cambridge, MA: MIT Press, 1991), 52.

133. Johann Jacob Scheuchzer, *Kupfer-Bibel, in welcher die Physica Sacra oder geheiligte Natur-Wissenschafft derer in Heiliger Schrifft vorkommenden natürlichen Sachen deutlich erklärt und bewährt* (Augspurg [sic] und Ulm: Christian Ulrich Wagner, 1733–1735), 2:150.

134. The section on the technique of injection is indebted to Rüdiger Schultka and Luminita Göbbel, "Präparationstechniken und Präparate im 18. und frühen 19. Jahrhundert, dargestellt an Beispielen aus den anatomischen Sammlungen zu Halle (Saale)," in *Anatomie: Sektionen einer medizinischen Wissenschaft im 18. Jahrhundert*, ed. Jürgen Helm and Karin Stukenbrock (Stuttgart: Franz Steiner Verlag, 2003), 59–60.

135. Johann Leonhard Fischer, *Anweisungen zur praktischen Zergliederungskunst. Nach Anleitung des Thomas Pole* (Leipzig: Weygandsche Buchhandlung, 1791), 172–77, quoted in Schultka and Göbbel, "Präparationstechniken," 65.

136. Goethe, *Sämtliche Werke*, ed. Dieter Borchmeyer et al., series 1, 9:926.

137. Goethe, *Apprenticeship*, 334.

138. Goethe, *Sämtliche Werke*, ed. Dieter Borchmeyer et al., series 1, 9:951.

139. Goethe, *Apprenticeship*, 349.

140. Goethe mentions Frederik Ruysch's daughter Rachel, an aid to Ruysch and a painter, in his essay "Einfache Nachahmung der Natur, Manier, Stil" [Simple Imitation of Nature; Manner; Style] (Goethe, *Sämtliche Werke*, ed. Dieter Borchmeyer, series1, 12:33). Rachel Ruysch's paintings of flowers were extremely popular and she, as well as Jan van Huysum (the second painter mentioned in Goethe's essay), is known as having replaced the Frankfurt born Abraham Mignon (sic) (1640–1679) as leading still life artist of the time. Excursus: The worldwide greatest collection of Mignon's paintings can be found in the Dresden gallery, a collection that Goethe visited in 1768. Abraham Mignon is also mentioned as a teacher of Maria Sibylla Merian (1647–1717) who became famous through her paintings of Surinam flora and fauna.

141. Cf. Harold J. Cook, "Time's Bodies: Crafting the Preparation and Preservation of Naturalia," in *Merchants and Marvels: Commerce, Science, and Art in Early Modern Europe*, ed. Pamela H. Smith and Paula Findlen (London: Routledge, 2002), 241. See also Julie V. Hansen, "Resurrecting Death: Anatomical Art in the Cabinet of Dr. Frederik Ruysch," *Art Bulletin* 78, no. 4 (December 1996): 673.

142. Cf. Markus Schnöpf, *Die Kunstkammer Peters des Großen. Geschichte der Sammlung*, 2000, http://www.schnoepf.de/m/kukammer/geschichte.html (November 20, 2011).

143. Dániel Margócsy, "Advertising Cadavers in the Republic of Letters: Anatomical Publications in the Early Modern Netherlands," *British Journal of The History of Science* 42, no. 2 (June 2009): 201.

144. Valentini, *Musaeum Musaeorum, oder vollständige Schaubühne aller Materialien und Specereyen, nebst deren natürlichen Beschreibung . . .* (Frankfurt: Heirs of Johann David Zunner and Johann Adam Jungen, 1714), 59–61, quoted in Margócsy, "Advertising Cadavers," 201.

145. Margócsy, "Advertising Cadavers," 199.

146. Among the institutions for which Goethe, together with C. G. Voigt, became responsible after returning from Italy was an anatomical collection. Goethe was responsible for the maintenance of existing collections (in particular the one in the Jenaer castle) and he was also involved in creating a new collection of anatomical preparations for teaching purposes. See Imtraut Schmid, "Amtliche Tätigkeit," in *Goethe Handbuch in vier Bänden,* ed. Hans-Dietrich Dahnke and Regine Otto (Stuttgart and Weimar: Metzler, 1998), 4:39–42.

147. Ruysch, quoted in Margócsy, "Advertising Cadavers," 199.

148. Goethe, *Sämtliche Werke*, ed. Dieter Borchmeyer et al., series 1, 9:958, italics mine.

149. Blackall translates: "Draw near, my friends, and observe the wonders of art, the sum of solicitude!" (Goethe, *Apprenticeship*, 353).

150. For a comprehensive discussion of the significance of references in Goethe's text and beyond, see Andrew Piper, *Dreaming in Books: The Making of the Bibliographic Imagination in the Romantic Age* (Chicago: University of Chicago Press, 2009).

151. Cf. Esther Schor, *Bearing the Dead: The British Culture of Mourning from the Enlightenment to Victoria.* (Princeton, NJ: Princeton University Press, 1994). The disruptive quality of Mignon's body has historical precedents. In her book on the hermaphrodite in Classical Antiquity, Marie Delcourt has described how the birth of a hermaphrodite child was considered a threat to the entire community already in ancient Greece (*Hermaphrodite: Myths and Rites of the Bisexual Figure in Classical Antiquity*, trans. Jennifer Nicholson. [London: Studio Books, 1961], 43).

152. Goethe, *Sämtliche Werke*, ed. Dieter Borchmeyer et al., series 1, 9:920.

153. Goethe, *Apprenticeship*, 331.

154. Hans-Jürgen Schings, "Einführung," in Goethe, Sämtliche Werke nach Epochen seines Schaffens, ed. Karl Richter et al., 5:640.

155. Fredrich Schiller, in Goethe, *Sämtliche Werke*, ed. Dieter Borchmeyer et al., series 1, 9:1491.

156. Marion Maria Ruisinger, "Schön oder Hässlich? Die Ambivalenz medizinischer Museumsdinge," in *Schönheit: Traum–Kunst–Bildung*, ed. Eckart Liebau and Jörg Zirfas (Bielefeld: transcript Verlag, 2007), 131–32.

157. Cf. ibid. See also Ragnhild Münch, "Theater des Todes—Museum des Lebens," in *Theater der Natur und Kunst: Theatrum naturae et artis; Wunderkammern des Wissens*, ed. Horst Bredekamp, Jochen Brüning, and Cornelia Weber (Berlin: Henschel, 2000–2001), 135–42.

158. Goethe, *Sämtliche Werke*, ed. Dieter Borchmeyer et al., series 1, 9:896.

159. Goethe, *Apprenticeship*, 316.

160. Goethe, *Sämtliche Werke*, ed. Dieter Borchmeyer et al., series 1, 9:957.

161. Goethe, *Apprenticeship*, 353.

162. Neickelio [Jenckel], *Museographia*, 2.

163. Ibid., 410.

164. The representation of Mignon's corpse in the Hall of the Past poses not only a moral, but also an aesthetical challenge. As Julie Hansen has shown, the confrontation with the dissected body and its representation in art and literature posed a significant aesthetic problem. In her work on Frederic Ruysch, Hansen describes how in "seventeenth-century Holland, a cultural paradigm was constructed whereby the narrative of the anatomy lesson was fictionalized and aestheticized. Representation of anatomical practice adopted the prevailing themes

of emblematic morality, and were often presented in the guise of modern science." Much as in Ruysch's work, the Tower Society anesthetizes Mignon's corpse and the scandal of her monstrous body. However, such anesthetization is only possible as a replacement ("an der Stelle des Bluts" [*in place* of real blood], see n122 above). Mignon's monstrous body is ultimately not grasped by the Tower Society's ritualistic representation.

165. Ruisinger, "Schön oder Hässlich? 131. "In der Kunst- und Wunderkammer der Renaissance fanden Kunstwerke von Menschenhand und die Wunder der Kreatur zusammen. . . . In dem ganzen Durcheinander herrscht dennoch eine Ordnung, allerdings eine Ordnung, die nicht den Gesetzen der Naturgeschichte, sondern den Gesetzen der Ästhetik gehorcht" [In the curiosity cabinet of the Renaissance, artworks from human hand and the miracles of creation came together. . . . In the entire chaos, there was nevertheless an order; an order, however, that did not follow the laws of natural history, but the laws of aesthetics] (ibid., 131–32). Despite their aesthetic and moral features—as Ilse Jahn has pointed out —premodern collections of the kind that we see in the uncle's house were decisive for the emergence of the natural sciences, and anatomical collections and the cabinet of curiosities have been placed in the genealogy of the laboratory. See Ilse Jahn, "Die Museologie als Lehr- und Forschungsdisziplin mit spezieller Berücksichtigung ihrer Funktion in naturhistorischen Museen, Geschichte, gegenwärtiger Stand und theoretische Grundlagen," *Neue Museumskunde*, (1979).

166. Stafford, *Body Criticism*, 47.

167. Ibid., 34.

168. Ibid., 74; cf. Manfred Engel, *Der Roman der Goethezeit. Anfänge in Klassik und Frühromantik: Transzendentale Geschichten* (Stuttgart: J. B. Metzler, 1993), 1:260; see also Monika Fick, *Das Scheitern des Genius: Mignon und die Symbolik der Liebesgeschichten in "Wilhelm Meisters Lehrjahre"* (Würzburg: Königshausen & Neumann, 1987).

169. Goethe, *Apprenticeship*, 316.

170. New American Standard Bible, Matthew 22:30.

171. Blackall translates: "let me seem" (Goethe, *Apprenticeship*, 316).

172. Hannelore Schlaffer, *Wilhelm Meister: Das Ende der Kunst und die Wiederkehr des Mythos* (Stuttgart: Metzler, 1980), 72–78, here 78.

173. See Aaron J. Atsma, *Theoi Project*. 2000, http://www.theoi.com/Protogenos/Phanes.html (accessed March 29, 2012). See also Thomas Taylor who, in his 1787 translation of the orphic hymns, connects the Phanes figure with "the duad, or number two" (*The Mystical Initiations; Or, Hymns of Orpheus* [London: T. Payne and Son, 1787], 94).

174. Goethe, *Sämtliche Werke*, ed. Dieter Borchmeyer et al., series 1, 9:895.

175. Benjamin Hederich, *Gründliches mythologisches Lexicon. Durchgesehen, ansehnlich vermehrete und verbessert von Johann Joachim Schwaben* (Leipzig: In Gleditschens Handlung, 1770), 1968, italics mine.

176. Ibid., 1968.

177. Drawing on Genesis 1:27 ("male and female created he them" in New American Standard Bible), early Christian theologians such as Philo and Origen assumed "that the first and original man was an androgynous; that the division into male and female belonged to a later and lower state of creation" (Wind in Tiffany, *Erotic Beasts*, 24). Similar beliefs existed in alchemistic thought (cf. R. D. Gray, *Goethe the Alchemist*).

178. Cf. Stafford, *Body Criticism*, 47.

179. Walter Müller-Seidel, "Naturforschung und Deutsche Klassik," in: W. M.-S. *Die Geschichtlichkeit der deutschen Klassik. Literatur und Denkformen um 1800* (Stuttgart: Metzler, 1983), 105–18.

180. Friedrich Schiller, in Müller-Seidel, "Naturforschung und Deutsche Klassik," 109.

181. Ibid.

182. Friedrich Schiller, "Kallias, oder über die Schönheit," in *Werke und Briefe*, ed. Rolf-Peter Janz (Frankfurt am Main: Deutscher Klassiker, 1992), 8:303.

183. Friedrich Schiller, "Kallias or Concerning Beauty: Letters to Gottfried Körner (1793)," in *Classic and Romantic German Aesthetics*, ed. J.M. Bernstein (Cambridge: Cambridge University Press), 2003, 164.

184. Schiller, "Kallias, oder über die Schönheit," 303.

185. Schiller, "Kallias or Concerning Beauty," 164.

186. Schiller, "Kallias, oder über die Schönheit," 303–4.

187. Schiller, "Kallias or Concerning Beauty," 164.

188. Schiller, "Kallias, oder über die Schönheit," 304.

189. Schiller, "Kallias or Concerning Beauty," 164.

190. Schlaffer, *Wilhelm Meister*, 76.

191. Hederich, *Gründliches*, 1968.

192. Horst Lange, "Goethe and Spinoza: A Reconsideration," *Goethe Yearbook* 18 (2001): 17.

193. Goethe, *Sämtliche Werke*, ed. Dieter Borchmeyer, series 1, 5:731.

194. Johann Wolfgang Goethe, *From My Life. Poetry and Truth*, Part Four, ed. Thomas P. Saine and Jeffrey L. Sammons, trans. Robert R. Heitner (New York: Suhrkamp, 1987), 524.

195. Schiller, "Kallias, oder über die Schönheit," 304.

196. Schiller, *Kallias or Concerning Beauty*, 164.

197. Lange, "Goethe and Spinoza," 28.

198. Stefan Blessin, *Goethes Romane: Aufbruch in die Moderne* (Paderborn: Schöningh, 1996), 184.

199. Schlipphacke, "Vaterschaft," 396.

200. Claude Lévi-Strauss, *Structural Anthropology*, trans. Claire Jacobson and Brooke Grundfest Schoepf (New York: Basic Books, 1963), 50.

201. René Girard, *Violence and the Sacred* (London and New York: Athlone Press, 2005), 225.

202. Ibid., 227.

203. For an interpretation of the incest-story in *Die Geschwister* see Meyer-Krentler, "Erdichtete."

204. We assume that the story takes place in post-revolutionary Europe (cf. Terrence James Reed, "Revolution and Rücknahme: *Wilhelm Meisters Lehrjahre* im Kontext der Französischen Revolution," *Goethe Jahrbuch* 107 [1990]: 27–43). See the *Allgemeine Landrecht* in the passages quoted earlier in this chapter. The French *Code Civile*, which was adopted in the "linksrheinische" [left bank of the Rhine] German regions in the beginning of the nineteenth century, formulated similar principles. Articles 161 and 162 state: "In direct lineage, marriage is prohibited between all ascendents and descendents and the relatives by marriage in the same lineage." "Marriage is further prohibited between uncle and niece, aunt and nephew" (quoted according to the version of 1803).

205. Cf. Elisabeth Joris, "Kinship and Gender: Property, Enterprise, and Politics," in *Kinship in Europe: Approaches to Long-Term Development (1300–1900)*, ed. David Warren Sabean, Simon Teuscher, and Jon Mathieu (New York: Berghahn, 2007), 231–57.

206. Only when practitioners and scientists began to discuss whether reproduction via inbreeding might cause "deafness, blindness, insanity, and infertility" (Kuper 83), the enthusiasm for intermarriage came to an end. As Adam Kuper reports in *Incest and Influence: the Private Life of Bourgeois England* (2009), books like Alexander Walker's *Intermmarriage: Or the Mode in Which, and the Causes Why, Beauty, Health, and Intellect Result from Certain Unions, and Deformity, Disease and Insanity from Others* (1838) became popular among bourgeois families in the nineteenth century. Charles Darwin consulted the new book shortly before his own marriage (83). Much like other nineteenth-century bourgeois families, the Darwins and the Wedgwoods had strengthened the ties between the families repeatedly by cousin marriages.

For the question of the age of married partners in the eighteenth century see Imhof (1983).

207. Goethe, *Sämtliche Werke*, ed. Dieter Borchmeyer, series 1, 10:446.

208. Goethe, *Journeyman Years*, 221.

209. For an account of how inheritance could be passed down or distributed see also Jack Goody, "Inheritance, Property and Women: some Comparative Considerations," in *Family and Inheritance: Rural Society in Western Europe, 1200–1800*, ed. Jack Goody, Joan Thirsk, and Edward P. Thompson (Cambridge: Cambridge University Press, 1976), 10–36.

210. Volker Bauer, "Die gedruckte Ahnentafel als Ahnensformular: zur Interferenz von Herrschafts-, Wissens- und Medienordnung in der Universalgenealogie des 17. Jahrhunderts," in *Die Ahnenprobe in der Vormoderne*, ed. Elizabeth Harding and Michael Hecht (Münster: Rhema, 2011), 125–56.

211. Hans-Jörg Rheinberger and Staffan Müller-Wille, *Vererbung. Geschichte und Kultur eines biologischen Konzepts* (Frankfurt am Main: Fischer- Taschenbuch-Verl., 2009), 156.

212. Cf. Richard Friedenthal, *Goethe: His Life and His Times* (New York: Transaction Publishers, 2010), 110–11.

213. Goethe, *Sämtliche Werke*, ed. Dieter Borchmeyer et al., series 1, 10:446–47, italics mine.

214. Goethe, *Journeyman Years*, 221, italics mine.

215. Ibid., 221.

216. Ohad Parnes, "Generationen," in *Geneaologie des Konzepts—Konzepte von Genealogie*, ed. O. Parnes et al. (München: Fink, 2005), 205.

217. Rheinberger and Müller-Wille, *Vererbung*, 157.

218. Goethe, *Sämtliche Werke*, ed. Dieter Borchmeyer et al., series 1, 10:434, italics mine.

219. Goethe, *Journeyman Years*, 213.

220. For this new definition of "love" see also Niklas Luhmann, *Love as Passion: The Codification of Intimacy* (Cambridge, MA: Harvard University Press, 1986).

221. Cf. Otto Brunner, ed, "Das 'ganze' Haus und die alteuropäische 'Ökonomik,'" in *Neue Wege der Sozialgeschichte* (Göttingen: Vandenhoeck & Ruprecht, 1956), 33–61.

222. Goethe, *Sämtliche Werke*, ed. Dieter Borchmeyer et al., series 1, 10:435. For the English version see Goethe, *Journeyman Years*, 214.

223. Goethe, *Sämtliche Werke*, Dieter Borchmeyer et al., series 1, 10:436.

224. Goethe, *Journeyman Years*, 214.

225. As is well-known, the *Wahlverwandtschaften* were first thought of as a novella within the *Wanderjahre*. The close relationship between the two texts also surfaces in the failure of metaphors that are taken from nature or the natural sciences to describe human relationships. Like the Major's reference to the cyclical structure of nature in the *Wanderjahre*, the chemical formula of "elective affinities" fails to grasp the human relationships in the *Wahlverwandtschaften*.

226. Goethe, *Sämtliche Werke*, ed. Dieter Borchmeyer et al., series 1, 10:447, italics mine.

227. Goethe, *Journeyman Years*, 222, italics mine.

228. Goethe, *Sämtliche Werke*, ed. Dieter Borchmeyer et al., series 1, 10:488–89.

229. Goethe, *Journeyman Years*, 249.

230. The possibility that the Major is also Hilarie's biological father is not excluded (see ibid., series 1, 10:436).

231. "Etwas so Unnatürliches," in ibid., series 1, 10:436. For the English version see Goethe, *Journeyman Years*, 214.

232. Goethe, *Sämtliche Werke*, ed. Dieter Borchmeyer et al., series 1, 10:436. For the English version see Goethe, *Journeyman Years*, 214.

233. The difference between determinism, here articulated in the framework of fate [*Schicksal*], and contingency [*Zufall*] and the decise activity of guidance (here "lenken") is already brought up in the first book of the *Lehrjahre*. When Wilhelm, in a conversation with the *Fremde*, declares his belief in fate, his conversation partner disapproves: "Das Gewebe dieser Welt ist aus Notwendigkeit und Zufall gebildet, die Vernunft des Menschen stellt sich zwischen beide, und weiß sie zu beherrschen, sie behandelt das Notwendige als den Grund ihres Daseins, das Zufällige weiß sie zu lenken, zu leiten und zu nutzen" [The texture of this world is made up of necessity and chance. Human reason holds the balance between them, treating necessity as the basis of existence, but manipulating and directing chance, and using it.] (Goethe, *Sämtliche Werke*, ed. Dieter Borchmeyer et al., series 1, 9:423). For the English version see Goethe, *Apprenticeship*, 38.

234. Erich Trunz, ed., "Anmerkungen des Herausgebers. *Wilhelm Meisters Wanderjahre oder Die Entsagenden*," in *Johann Wolfgang Goethe Werke* (Hamburg: Wegner, 1950), 8:607.

235. Ibid., 8:665.

236. Goethe, *Sämtliche Werke*, ed. Dieter Borchmeyer et al., series 1, 10:476.

237. Goethe, *Journeyman Years*, 241.

238. Goethe, *Sämtliche Werke. Briefe, Tagebücher und Gespräche*, ed. Neumann and Dewitz, series 1, 10:476.

239. It is noteworthy that the narrator, at the very moment in which a solution is suggested, provides a reading of "Heitere." In reference to Flavio's appearance in the father's clothes, he comments: "Wie aber oft in bedeutenden, ja schrecklichen Momenten etwas Heiteres, ja Lächerliches sich zu ereignen pflegt, so glückte es auch hier" (ibid.) [But often in significant, even terrible moments, something cheerful, even comic will occur, and this happened now (Goethe, *Journeyman Years*, 241).] If renunciation accompanies the transfer from father to son, then we have to read this renunciation not as "heitere," but as "lächerliche Entsagung." A renunciation, that is to say, which distances itself from total identification.

240. Ibid., series 1, 10:461, italics mine.

241. Goethe, *Journeyman Years*, 230–231, italics mine.

242. Niklas Luhmann, *Liebe als Passion: Zur Codierung von Intimität* (Frankfurt am Main: Suhrkamp, 1983), in particular 163–66.

243. Goethe, *Sämtliche Werke*, ed. Dieter Borchmeyer et al., series 1, 10:432. Hilarie who resembles in so many ways Ottilie from the *Wahlverwandtschaften*, from the eyes to her silent retreat, acts in the end—albeit outside the novella—quite differently. If Ottilie "ent-sagt" [renounces] as Jochen Hörisch has pointed out, Hilarie eventually accepts the world of exchangeable signs and, therefore, can live (*Gott, Geld und Glück: Zur Logik der Liebe in den Bildungsromanen Goethes, Kellers und Thomas Manns* [Frankfurt am Main: Suhrkamp, 1983], 156).

244. Goethe, *Journeyman Years*, 401–402.

245. Although the couple has taken over the parents' estate, the narrator also marks that their nobility is more performed than lived ("als . . . Gutsbesitzer auftrat" [presenting himself as . . . wealthy landowner]).

246. Piper, *Dreaming in Books*, 36; 37.

247. Goethe, *Sämtliche Werke*, ed. Dieter Borchmeyer et al., series 1, 10:510–11.

248. Goethe, *Journeyman Years*, 263.

249. Nonnenmacher claims that "Inzest wird im *Wilhelm Meister* also zum Signum des Anderen, Fremden, Zivilisation und Bildung Entgegengesetzten. . . ." [In *Wihelm Meister*, incest turns into the sign of the Other, the Foreign, that which is opposite to civilization and education. . . .] (*Natur und Fatum*, 198). However, our reading has shown that incest is not representing the *Other*, but an experience which belongs quintessentially to every individual's development. See also Michel Foucault's description of the nuclear family as an erotically charged place (1990).

250. In their commentary on "Das nussbraune Mädchen" and the discussion of agriculture in this novella of the *Wanderjahre*, Fink, Bauman, and John note that among the works on agriculture from which Goethe drew his information was Thaer's *Grundsätze der rationale Landwirtschaft* (1812) (Gonthier-Louis Fink, Gerhart Baumann, and Johannes John, "Kommentar," in Goethe, *Sämtliche Werke nach Epochen seines Schaffens*, ed. Karl Richter et al., 17:1131).

251. Albrecht Dainel Thaer, *Einleitung zur Kenntnis der englischen Landwirthschaft* (Hannover: Bey den Gebrüdern Hahn, 1798), 680.

252. Goethe, *Sämtliche Werke*, ed. Dieter Borchmeyer et al., series 1, 9:954, italics mine.

253. The translation follows Eric A. Blackall (in cooperation with Victor Lange); however, Blackall (Lange) translates "consider a fine statue or a superb painting" (Goethe, *Apprenticeship*, 351).

254. Johann Wolfgang Goethe, *Werke*, ed. Erich Trunz, 14 vols. (Hamburg: C. Wegner,1950), 7:573; and Goethe, *Sämtliche Werke nach Epochen seines Schaffens*, ed. Karl Richter et al., series 1, 5:574.

255. Trunz, "Anmerkungen," 8:646–47.

256. See also Goethe's letter to J. H. Meyer from 20 May 1796 in which Goethe speaks of "[d]ie ewige Lüge von Verbindung der Natur und Kunst" [the eternal lie of the unity of nature and art]. Hans-Jürgen Schings, "Kommentar," in Goethe, *Sämtliche Werke nach Epochen seines Schaffens*, ed. Karl Richter et al., 5:850.

257. As Waltraud Naumann-Beyer has pointed out, Goethe's definition of nature is nuanced and encompasses a number of different notions, which makes it impossible to speak

about "Nature" in Goethe. Naumann-Beyer attributes these different notions of nature to different moments in Goethe's life. While nature in the work of the young Goethe is an infinitely productive force in the sense of Spinoza's *deus sive natura*, which does not establish a value hierarchy between essence and appearance, the later Goethe turns to platonic-neoplatonic notions of nature that distinguish between essence and appearance (assuming, however, that essence too is immanent). Naumann-Beyer concludes that in the language of the "classical" Goethe, there are "at least two notions of nature: the raw material of the sensual appearance or the 'lower real nature' (vgl. Eckermann, 20.10.1828) and the true or ideal nature which is thought as unity of essence and appearance" (Waltraud Naumann-Beyer, "Ästhetik," in *Goethe Handbuch in vier Bänden*, ed. Hans-Dietrich Dahnke and Regine Otto [Stuttgart and Weimar: Metzler, 1998], 4:10–11).

258. Walzel, quoted in Goethe, *Sämtliche Werke*, ed. Dieter Borchmeyer et al., series 1, 9:1490.

259. Goethe, *Sämtliche Werke*, ed. Dieter Borchmeyer et al., series 1, 24: 415, italics mine.

260. Johann Wolfgang Goethe, *Goethe's Botanical Writings*, trans. Bertha Mueller, with an introduction by Charles J. Engard (Woodbridge, Connecticut: Ox Bow Press, 1989), 168, italics mine.

261. For an account that stresses the significance of Goethe's scientific projects for his understanding of aesthetics see Elizabeth Wilkinson, "The Poet as Thinker: On the Varying Modes of Goethe's Thought," in *Goethe: Poet and Thinker*, by Elizabeth M. Wilkinson and Leonard A. Willoughby (London: Edward Arnold, 1962), 167–84.

262. Trunz argues that a biolocial law is the basis for Goethe's ethics: "Das Ethische hängt hier zusammen mit dem Biologischen. Goethe hat als Naturforscher oftmals das Naturgesetz ausgesprochen, das der Entsagungslehre zugrunde liegt, das Gesetz der Bedingtheit, des niemals symmetrischen, aber dennoch ausbalancierten biologischen Gleichgewichts" [The ethical is here connected to the biological. Goethe as scientist has often expressed the natural law that is the basis of the theory of renunciation, the law of limitation, (the law) of the never symmetrical, but nevertheless balanced biological harmony] ("Anmerkungen," 8:590).

CHAPTER THREE

Welcher Staat hat noch Preise auf Einliefern von monstris gesetzt, geschweige auf Erzeugung derselben, wie doch bei Blumen geschehen? . . . Wer kann denn aber eine Mißgeburt, die sich so wenig als ein Genie fortpflanzt—denn sie ist selber ein körperliches, eine Einzigperle—nicht einmal ein Sonntagkind, sondern ein Schalttagkind—, ersetzen, ich bitte jeden? —Jean Paul Richter, Dr. Katzenbergers Badereise

[*Is there yet a state which has given prices for the delivery of monsters, not to speak of their production, as it happened for flowers after all? . . . But who can replace a monster, which, much like a genius, does not reproduce—because it is itself a corporeal (genius), a unique pearl—not even a Sunday child, but a leap day child—I ask everybody?*]

Cross-breeding and Hybridization

BASTARD AND SYSTEM

SINCE ALBERT EINSTEIN'S formulation of relativity theory, no scientific discovery has been as significant for the understanding of human and nature as the discovery of the double-helix structure of the DNA, which James D. Watson and Francis H. C. Crick published in the science journal *Nature* in 1953.[1] This discovery has handed to us not only the knowledge of the building blocks of life, but also the power of creation, a fact Watson never hesitated to advertise and endorse. In a recent documentary on the discovery of the DNA double-helix structure, *DNA: The Secret of Life. Pandora's Box*, Watson visits the bio-technology giant Monsanto and raises two issues that demonstrate the persistence of questions first formulated around 1800. First, the question of the transfer of genetic material from animal to plant comes up. One of the leading figures of the company, Rob Horsch, says, "[W]e won't put animal genes in plants. It obviates the problem of: is it still vegetarian or not?," to which Watson replies "Oh Jesus," and thus suggests that such taxonomy is anachronistic.[2] Later, Watson closes his encouragement of genetic engineering with the words: "You know, there are people who say, well, we're playing God. And, you know, I have a straightforward answer: If we don't play God, who will?" (laughter and clapping).[3] Both the blurred line between life forms previously considered distinct and the biological engineer displacing the divine creator are at the heart of eighteenth-century breeding experiments.

This chapter delineates how the eighteenth century engaged with the issues raised by Watson. It illustrates the challenges to the definition of the human in the context of new breeding efforts and the methods eighteenth-century authors such as Jean Paul Richter employed to deal with them. While in the Mignon episode, Goethe thought through the problem of incest and incestuous reproduction—the problem of inbreeding—Jean Paul Richter works through questions raised by new efforts in cross-breeding between species—the problem of out-breeding. Richter was fascinated with the work of contemporary botanists who produced new forms by cross-breeding. This fascination, however, also confronted Richter with a dilemma. How, he asked himself, can we account for the creativity of nature seen in these new forms without giving up a stable order of nature and a divine creator? Or, if the creativity is not in nature, but in the human being who manipulates nature, how can we account for human creativity without undermining the creativity of God?

The question of whether human creativity could threaten the singularity of God's creativity had been asked before, but such questions had been answered with reference to nature. Nature was the one entity which man could *not* create. If humans could now create nature, as it seemed, would a reassessment of human creativity be necessary? Richter tried to answer these questions by defining more precisely the limits of nature's stability and by drawing a strict line between nature and nurture.

The first part of this chapter focuses on Richter's direct discussion of contemporary cross-breeding efforts and their implications for the conceptualization of nature; the second part asks what kind of literal, metaphorical, and conceptual possibilities are opened by these breeding experiments,[4] especially regarding the impact on character formation and the conceptualization of selfhood. If selfhood was understood in the last third of the eighteenth century through an organic totality that emerges from its own origin, how, we will ask, did the conceptualization of selfhood need to be changed for Jean Paul Richter in light of new research on cross-breeding?

MONSTROUS FLOWERS

In the epigraph to this chapter, Dr. Katzenberger evokes the eighteenth-century fascination with plant breeding experiments that followed Carl Linnaeus' (1707–1778) musings about whether or not bastardization of plants is possible and whether such bastardization occurred in nature. That is, whether it is possible that a new form, called hybrid or bastard, is created when two species are crossed.[5] Even more, the question was whether this new form, if it existed, could be fertile. Such questions were explosive in the eighteenth century, because the answer could undermine the belief in the stable and god-created system of species. If a bastard flower could breed with another bastard flower of the same type, there was potential for a new productive form. An evolutionary development could take place by redefining the relationship between individual and species. Dr. Katzenberger's wish for the production of "monstrosities" betrays both the eighteenth-century fascination with new forms of nature and the extent to which these new forms were, at the same time, still considered "deformities."[6]

In writing "Plantae hybridae" (1751), Linnaeus had established the parameters which would shape hybridist activities for the rest of the century and beyond. Although he had little evidence to support his statement, Linnaeus claimed in this paper that the formation of hybrids was a natural occurrence.[7] This was outrageous, not only because it assumed the creation of a new form (even if infertile), but also because the novelty would have to be attributed to cross-pollination and not to climatic conditions. Linnaeus addressed this crucial question in the context of the difference and similarity of contemporary plants and plants known from antiquity:

> It was said, namely, that these varieties are produced by *soil*, *locality*, *climate*, so that they are by all means accidental. In the meanwhile, some [plants] have been subjected to this reform, which have troubled botanists. . . . These were similar to two others in such a way, that no single character of the plant in question, even if most distinct and singular, could be found, which not also occurred in one of the two others. These hybrids (*hybrida*) also grow in the same soil with both others, and yet were not observed before. This has first led us to the question, whether two different plants may have produced a third, or if they should be considered hybrids.[8]

Linnaeus' observation implied three questions: first, should novelty be attributed to accidental (and, therefore, reversible) conditions such as soil or climate? Or, second, was the change the result of a crossing of two different plants ("whether two different plants may have produced a third")? And, third, was the new plant fertile or was it a hybrid? The answers to all three questions were crucial in establishing the relationship between species and variety and to assess the creativity of nature. The conviction was that those plants which exposed the same form through several generations when planted under the same conditions, for example in the highly regulated conditions of the botanical garden, belonged to one species. Changes in accidental conditions would produce varieties. As Staffan Müller-Wille and Vitezslav Orel observe, "[s]ince Linnaeus, at least, species and varieties [were] distinguished along a nature-nurture divide,"[9] with nurture being understood as that force which modifies nature by means of contingent forces such as climate and soil. One crucial question implied by Linnaeus' observation was: if the new form was to be attributed to cross-breeding and not to climatic conditions, could cross-breeding itself be considered a contingent force? This was necessary, if one wanted to obtain the stability of nature—an option which the German botanist Joseph Gottlieb Koelreuter preferred, as we will see below.

The strict division between species and varieties was, however, challenged already in Linnaeus's time by what Linnaeus called "constant varieties."[10] These were plants which exposed a mixture of characteristics from two plants even in the stable climate of the botanical garden. These forms challenged eighteenth-century assumptions regarding species/variety divisions and regarding the stability of taxonomies. As James Larson had observed in *Interpreting Nature* (1994), significant energy was spent on developing and documenting these taxonomies. "Systems of nature and natural methods," Larson states, "are among the most characteristic and enduring achievements of the eighteenth century. All of these compendia, from the most circumspect regional flora to the most ambitious inventory of the terraqueous globe, are inspired by the belief that nature's own plan is not only fixed but known, at least in outline."[11] Consequently, the most dominant conviction in the eighteenth century was that if change was observed as a result of crossing, the new forms would either be hybrids or revert to their original forms once human intervention stopped. Constant varieties challenged such convictions.

And yet, it was only in the mid-nineteenth century that the significance of man-made conditions for the production of hybrids was fundamentally questioned, and with it the division between species and variety. In a now famous but then ignored paper published in 1866, the Austrian monk Gregor Mendel (1822–1884) stated that "[i]t has so far been found to be just as impossible to draw a sharp line between the hybrids of species and varieties as between species and varieties themselves."[12] The novelty of Mendel's work derived not so much from his style of experimentation, which was in the tradition of experiments with hybrids from the eighteenth-century onward, but by the bold move to accept that a "sharp line" between varieties and species cannot be drawn. By negating the possibility of distinguishing clearly between varieties and species, Mendel helped to prepare what one could call a genetic turn in the understanding of nature and its history. However, eighteenth-century botanists had paved the way.[13]

One eighteenth-century botanist in particular, Joseph Gottlieb Koelreuter (1733–1806), did more than any other before Mendel to further the research on cross-breeding and hybridization. Having studied in Tübingen and then worked at the Academy of Science in St. Petersburg, he conducted thousands of hybridization experiments in his South German birthplace Sulz and in nearby Calw after returning from Petersburg and before he became professor for natural history in Karlsruhe.[14] Koelreuter published the results of his groundbreaking experiments in plant breeding in the years from 1761 to 1766 in his multi-volume work with the humble title *Vorläufige Nachricht von einigen das Geschlecht der Pflanzen betreffenden Versuchen und Beobachtungen* [Preliminary Report on some Experiments and Observations Regarding the Sex of Plants]. In this work, he traces the motivation for his experiments back to Linnaeus, whose provocative claim that plants had sexual organs which would allow them to cross-pollinate was still highly contested in the mid-eighteenth century (see fig. 3.1).[15]

Koelreuter set out to establish the significance of plant sexuality in hybridization by means of artificial pollination. The importance of pollen for fertilization had been discovered already by Rudolf Camerarius (1665–1721) Professor of medicine in Tübingen who helped to further the rich botanical knowledge among Tübingen scholars on which Koelreuter could draw when being a student there some decades later.[16] Koelreuter's attention was focused on the demonstration of two facts: first, the fact that a hybrid presents a mixture of the features of its parents such that the new being contains a "middle proportion" of its parents' qualities. Koelreuter assumed that by back-crossing the mixture could be diluted and finally returned to the original parental forms. And, second, that all hybrids were unable to reproduce.[17]

Koelreuter's entire creative energy was spent on producing new forms, but almost as much energy was spent on declaring these new forms the result of contingent forces and thus not stable. Two methods were used to

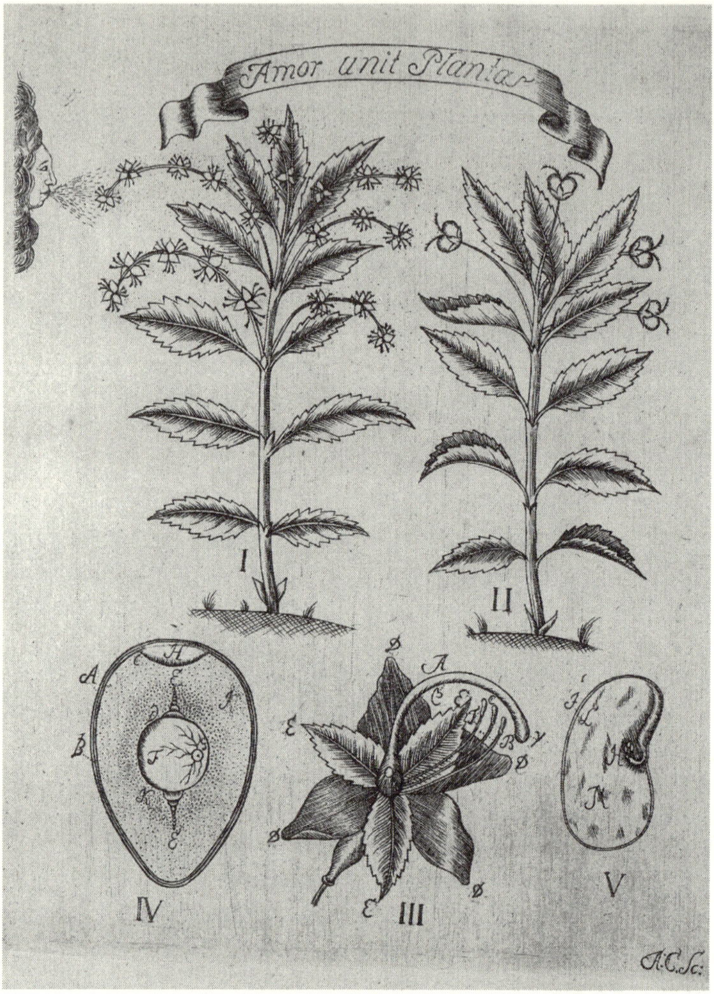

fig. 3.1. Sexuality of Plants. Herzog August Bibliothek Wolfenbüttel: Ng 330. Carl von Linné, Johan Gustaf Wahlbom. Sponsalia plantarum. *Stockholm: Salvius, 1746.*

maintain the belief in a stable nature despite the acknowledgment of new forms: insistence on the necessity of human intervention and insistence that the cross-bred forms were infertile. Although both helped to maintain the nature/nurture divide, they also redefined it. Climatic conditions had been considered as having contingent influence on species, they somehow modified the plant but not (irreversibly) the seed; they were imagined as a cloth which modifies a body superficially.[18] Cross-breeding went a step further by initiating changes within the plant and its seed. The assumed artificiality and reversibility of this operation made it possible to think this action still as nurture, but the nature/nurture divide had just shifted.

Scholars in the late-eighteenth century sensed the looming danger of an increasingly blurred line between nature and nurture. In an attempt to maintain it, Koelreuter aimed for an absolutely stable ground for his theoretical framework. Like Descartes (1596–1650) and Kant in philosophy, Koelreuter searched in botany for one principle, or rather one test, from which he could build the rest of his theory. The "Verbindungsversuch" (joining experiment), that is the experiment whether or not two individuals can reproduce, became for Koelreuter that indicator that guarantees god-given stability even if change in nature is acknowledged. He stated "so werde ich den Verbindungsversuch mit allem Grunde für den einigen wahren, sichern und untrüglichen Probierstein aller besondern Gattungen und Varietäten halten können"[19] [therefore, I can consider with every reason the joining experiment as the one true, sure and undoubtable touchstone of all particular species and varieties]. Koelreuter's use of the word "Probierstein" [touchstone] reveals that he sought an indicator outside personal judgment and which was not accidental. Kant, too, used the term in the *Anthropology*, first conceptualized roughly in the same years as Koelreuter's experiments, where he defined it as a "'*criterium veritatis externum*' a 'means for assuring the truth of our judgment' that is situated *outside* of the solipsistic individual."[20] Through the use of the term "Probierstein," Koelreuter stressed the centrality of the reproduction criterium for eighteenth-century understandings of nature as a proof for a stable system which is guaranteed by a creator God and not merely imagined by the observing human.

Much as in the context of animal breeding, in botany the intense wish to maintain a true and original nature to which all changes were accidental led botanists such as Koelreuter to formulate the oxymoronic concept of a second nature produced artificially ("durch Kunst").[21] The hybrid was considered to be part of the latter (see fig. 3.2).

In an implicit attack on Linnaeus, who had claimed the natural occurrence of hybridization, Koelreuter stated:

> Die von einigen der neuern Kräuterlehrer mit vieler Dreistigkeit für Bastarte ausgegebene Pflanzen mögen in dieser Absicht wohl nichts anders als unzeitige Geburten einer übertriebenen Einbildungskraft seyn. . . . Wie kann man sie mit Gewissheit dafür ausgeben, ehe man sie durch die Kunst, und zwar durch die zuverläßigsten Versuche, hervorgebracht

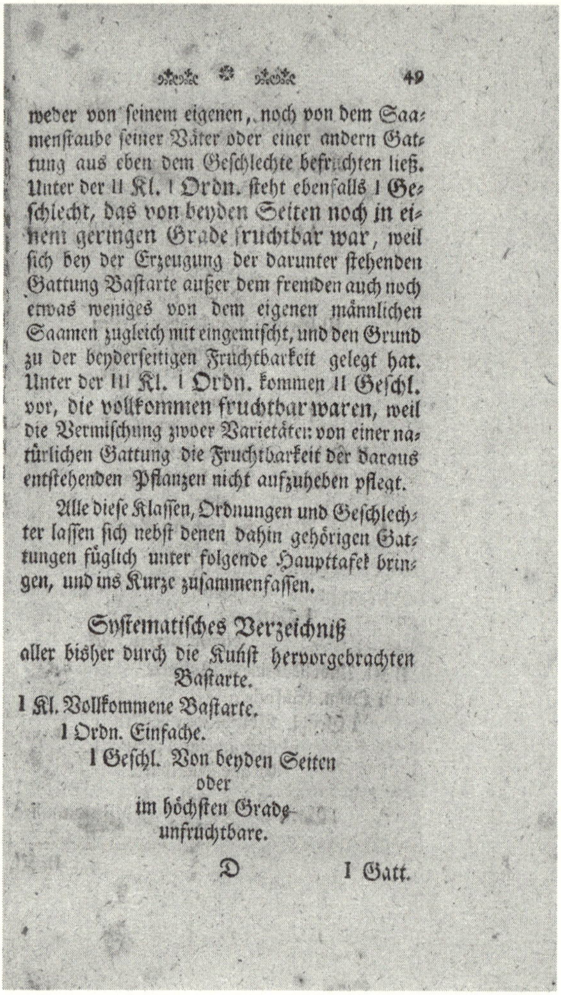
49

weder von seinem eigenen, noch von dem Saamenstaube seiner Väter oder einer andern Gattung aus eben dem Geschlechte befruchten ließ. Unter der II Kl. I Ordn. steht ebenfalls I Geschlecht, das von beyden Seiten noch in einem geringen Grade fruchtbar war, weil sich bey der Erzeugung der darunter stehenden Gattung Bastarte außer dem fremden auch noch etwas weniges von dem eigenen männlichen Saamen zugleich mit eingemischt, und den Grund zu der beyderseitigen Fruchtbarkeit gelegt hat. Unter der III Kl. I Ordn. kommen II Geschl. vor, die vollkommen fruchtbar waren, weil die Vermischung zwoer Varietäten von einer natürlichen Gattung die Fruchtbarkeit der daraus entstehenden Pflanzen nicht aufzuheben pflegt.

Alle diese Klassen, Ordnungen und Geschlechter lassen sich nebst denen dahin gehörigen Gattungen füglich unter folgende Haupttafel bringen, und ins Kurze zusammenfassen.

Systematisches Verzeichniß
aller bisher durch die Kunst hervorgebrachten Bastarte.

I Kl. Vollkommene Bastarte.
I Ordn. Einfache.
I Geschl. Von beyden Seiten
oder
im höchsten Grade
unfruchtbare.

D I Gatt.

fig. 3.2. Hybridization. Courtesy of the Smithsonian Institution Libraries, Washington, D.C. Joseph Gottlieb Koelreuter. Vorläufige Nachricht *(Preliminary Report). Leipzig: in der Gleditischischen Handlung, 1761–1766.*

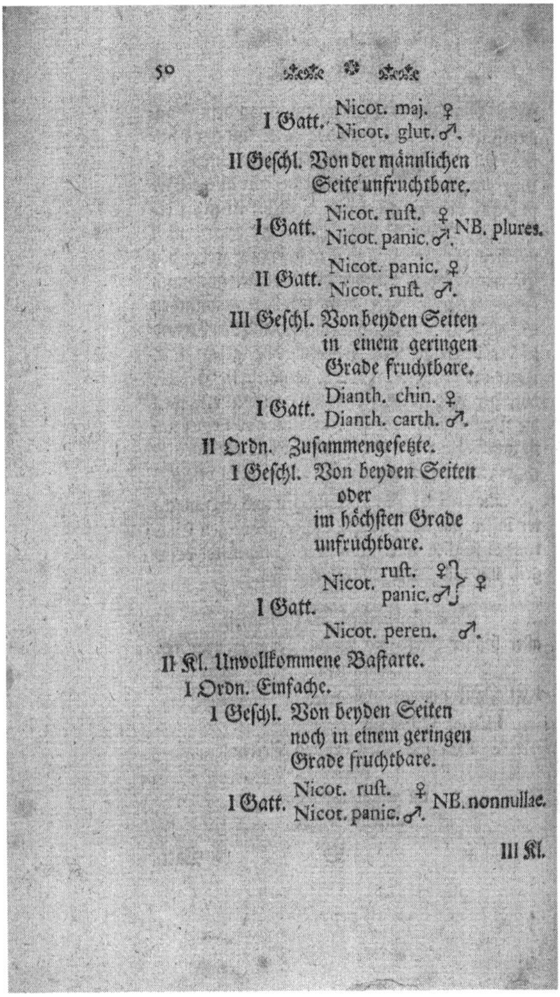

I Gatt. Nicot. maj. ♀
Nicot. glut. ♂.

II Geschl. Von der männlichen Seite unfruchtbare.

I Gatt. Nicot. rust. ♀
Nicot. panic. ♂. NB. plures.

II Gatt. Nicot. panic. ♀
Nicot. rust. ♂.

III Geschl. Von beyden Seiten in einem geringen Grade fruchtbare.

I Gatt. Dianth. chin. ♀
Dianth. carth. ♂.

II Ordn. Zusammengesetzte.

I Geschl. Von beyden Seiten oder im höchsten Grade unfruchtbare.

I Gatt. Nicot. rust. ♀ / panic. ♂ } ♀
Nicot. peren. ♂.

II Kl. Unvollkommene Bastarte.

I Ordn. Einfache.

I Geschl. Von beyden Seiten noch in einem geringen Grade fruchtbare.

I Gatt. Nicot. rust. ♀
Nicot. panic. ♂. NB. nonnullae.

III Kl.

fig. 3.2. (continued)

> hat? So unwahrscheinlich es ist, daß von zwoen verschiedenen Gattungen von Thieren, die in ihrer natürlichen Freiheit gelebt, jemals ein Bastart erzeugt worden, so unwahrscheinlich ist es auch, daß bey der ordentlichen Einrichtung, die die Natur by dem Pflanzenreiche gemacht hat, eine Bastartpflanze entstanden sey.[22]
>
> [The plants, which have been passed off as bastards by some recent botanists with great boldness, are in this respect probably nothing else than untimely deliveries of an exaggerated imagination. . . . How can one pass them off as such with certainty before having them produced artificially, namely by means of the most reliable experiments? As unlikely as it is that from two different species of animals, which live in their natural freedom, a bastard is ever produced, so unlikely is it that in the right order, which nature has made in the plant realm, a bastard plant has developed.]

In rejecting the idea that bastards could occur naturally, Koelreuter not only highlighted the necessity of human intervention to create new and contingent forms, but also stressed the need for a specific, empirical method to obtain results and to formulate laws in the sciences. In evoking imagination in the context of claims of naturally occurring bastards, Koelreuter suggested that such claims are themselves, to use a phrase that Friedrich Schiller employed, "monstrosities of the mind."[23] There is, however, a way in which imagination also opens a possibility for Koelreuter to accept the bastardization of nature. "Einbildungskraft" much as "Kunst" (and when thought as part of it) can be accepted as tool for modifying nature because—this is the assumption—it does not permanently alter its original. The products which art and imagination deliver are not permanent. Koelreuter stated: "Eben die Gründe, die mir die Erzeugung derselben by natürlichem Zustande verdächtig machen, bewegen mich, sie unter diesem widernatürlichen zuzugeben"[24] [Precisely those reasons which arouse my suspicion regarding the production (of hybrids) under natural conditions, persuade me to admit them under those artificial (unnatural) conditions]. Precisely because he created a clear split between nature and art, Koelreuter was able to observe and accept change in nature within his framework of stability. By decoupling imagination from original nature, Koelreuter opened a space for an emphatic notion of imagination, which Jean Paul Richter, as we will see, embraced. At the time of the first volume of *Vorläufige Nachricht* in 1761, Koelreuter had not yet succeeded in transferring one species into another, but he had already established a framework in which such transfers had to be thought either impossible or contingent.

Koelreuter's belief in the possibilities of human intervention in natural development and rejection of contingency as a force able to permanently alter nature contradicts the dominant historiography on the period, which speaks of a temporalization [*Verzeitlichung*] of human experience. If one of the key paradigms in the understanding of that period is indeed temporalization, as Reinhart Kosellek and others have argued,[25] Koelreuter's experiments and his interpretations of them suggest that this temporalization of human self-understanding and nature was only possible against (and conditioned by) the backdrop of an absolutely stable system. Whether new forms were produced in contemporary breeding experiments or differences were observed in the comparison of fossilized skeletons and still existing life forms, conceptualizing change meant thinking variation and modification, not change.

The urgency with which contingency was kept away from nature proper can be observed in Koelreuter's report on his discovery of plant pollination by insects, a discovery still attributed to him:

> Ich erstaunte, als ich diese Entdeckung an einer von diesen Pflanzen zum erstenmal gemacht hatte, und sahe, daß die Natur eine so wichtige Sache, als die Fortpflanzung ist, einem solchen *Ungefähr*, einem *glücklichen Zufalle*, überlassen hat. Mein Erstaunen verwandelte sich aber bey fortgesetzten Beobachtungen nach und nach in eine Bewunderung eines, *dem ersten Ansehen nach zufälligen*, aber *in der That allersichersten* Mittels, dessen sich hier der weise Schöpfer bey der Fortpflanzung bedienet.[26]
>
> [I was astonished when I made this discovery on one of these plants for the first time, and saw that nature left such an important thing as reproduction to such randomness—a happy coincidence. My astonishment, however, transformed in the course of continued observation into admiration of a device, which first looks random but is in fact the most certain, that is used here by the wise creator in the case of reproduction.]

This passage is shaped by a shift from a moment of contingency [*Zufalle*] to a moment of absolute certainty [*allersichersten*]. It demonstrates the extent to which eighteenth-century scholars attempted to distinguish nature from contingent forces. Even where, as in the case of insect pollination, they observed contingency, such contingency needed to be declared part of the divine plan in order to be considered part of nature. Nature was that which is not contingent and not temporalized.

Perhaps nothing helped Koelreuter to understand his cross-breeding experiments as contingent as much as his perception that the plant on which he experimented was without agency. Both its location and its pollination could be manipulated by the human. Even without human intervention, pollination depended on forces beyond the plant, such as the visits of insects or the occurrence of wind. While it was relatively easy for Koelreuter to show that cross-breeding depends on human activity (either in form of planting arrangements or in form of direct fertilization), the proof of the necessity of human intervention became more difficult in the case of animals, given that animals can move around freely. Since his system depended on the artificiality of hybridization, Koelreuter tried hard to explain why hybridization does not occur in nature even in the case of the animal. In order to maintain his claim, Koelreuter introduced an instinct [*Trieb*] which prevents animals of different species to mate with each other: "Die Natur, die jederzeit, auch bey der grössesten scheinbaren Unordnung, die schönste Ordnung beobachtet, hat dieser Verwirrung bey den wandelnden Thieren außer andern Mitteln hauptsächlich durch die natürlichen Triebe vorgebeuget"[27] [Nature, which observes at all times, even in case of the greatest apparent disorder, the most beautiful order, has prevented this confusion in case of moving animals mainly—among other means—by natural instincts]. We observe here the same strategy as the one used in the passage on insect pollination: even what seems accidental is part of a divine plan.

In the eighteenth-century, the concept of "Trieb" [drive] tends to be invoked when no rational motivation can be found and when—at the same time—the direct reference to God or miracle seems no longer appropriate. However, "Trieb" remains closely linked to the divine, since God instituted it. In Johann Heinrich Zedler's *Universal Lexicon*, "Natur-Triebe" and "Naturae Instinctus" are used interchangeable.[28] Here, instinct is defined as something which has been introduced by God to guarantee the joyful self-preservation of the human being. As examples for such instincts the following (among others) are given: instincts for eating, for sexuality, for love of children, and for truth. Hence, instinct guarantees here, much as in Koelreuter, the stability of a God-given order without being itself a spiritual force.[29]

ALCHEMICAL TRANSFORMATION

We have learned to tell the story of the long eighteenth century in terms of an abandoning of non-empirical explanations of nature. As Lorraine

Daston and Peter Gallison have recently demonstrated, *objectivity* not only becomes a core value in the sciences but it is also redefined as an empirical category, and thus leaves behind the Kantian model of thinking objectivity as opposite to the empirical and sensual.[30] In this narrative of a differentiation of value spheres in which the subjective becomes distinguished from the objective, science from aesthetic accounts of the world, the overcoming of alchemical thought and the turn to what are considered rational approaches to the world have played an important role. What has attracted less attention is the degree to which alchemical models continue to inform the sciences in the Enlightenment and beyond and the degree to which these models were important for facilitating concepts which still shape our current science culture. Joseph Gottlieb Koelreuter's work is an important case study in this reconsideration of the continued significance of alchemical thinking.

Koelreuter's thousands of experiments attest to the fact that his work is located within a new, enlightened and empirical paradigm in which experimentation is central. However, we also encounter an author who drew on what seem to be anachronistic models to explain what he observed in these experiments. Of the concepts which Koelreuter used to understand his experiments, meticulously described in the four volumes of the *Vorläufige Nachricht*, none is perhaps so surprising for modern readers as the usage of the concept of alchemical transformation of metals. Koelreuter employed alchemical language in order to boast about his most important achievement, the transformation of the tobacco plant Nicotiana rustica into Nicotiana paniculata (see fig. 3.3).

Having relegated his work to contingent forces and having declared his results as hybrids, in the third volume of *Vorläufige Nachricht* (1764), Koelreuter announced rather surprisingly:

> Ich glaube durch diese Entdeckung in den Augen derjenigen, die eine Sache nach ihrem innern Werthe zu schätzen wissen, wo nicht mehr, doch zum wenigestens eben so viel geleistet zu haben, als wenn ich Bley in Gold, oder Gold in Bley verwandelt hätte. Man hat die Verwandlung der Metalle schon von uralten Zeiten her für möglich gehalten; es ist aber, meines Wissens, noch niemand eingefallen, das es möglich wäre, eine Pflanze in die andere, oder ein Thier in das andere zu verwandeln; vermuthlich, weil man sich die Schwierigkeiten, die der Verwandlung organischer Wesen im Wege stehen möchten, unendlich größer, als bey den unorganischen, metallischen Körper vorgestellt hat. Und doch ist

fig. 3.3. Tobacco plants (Nicotiana rustica and Nicotiana paniculata). Engraving by James Caldwall, c.1805, after Peter Henderson, Wellcome Library, London.

dieses durch so viele Jahrhunderte hindurch von so vielen vergeblich unternommen, jenes hingegen in wenigen Jahren und zwar vom ersten, der es gesucht, größtentheils glücklich zu Stande gebracht worden.[31]

[I believe to have achieved through this discovery in the eyes of those who know to judge things according to their inner value at least as much

> as if I had transformed lead into gold, or gold into lead, if not more. One has considered the transformation of metals as possible already from ancient times onwards; however, as far as I know, nobody has had yet the idea that it could be possible to transform one plant into another, or one animal into another; probably, because one imagined the difficulties, which may stand in the way of the transformation of organic beings, infinitely greater than in the case of inorganic, metallic bodies. And yet, the latter has been tried in vain by many through so many centuries, the former, on the other hand, has been achieved for the most part successfully within a few years by the first who tried.]

With unconcealed pride, Koelreuter elevated both the idea and result of his breeding experiment by comparing it with having achieved the highest goal in the context of alchemical activities. However, it is too easy to relegate it to the realm of metaphoric speech. In fact, it was for him literally to describe the developments leading to his breeding results.

It was difficult for eighteenth-century breeders to understand how best to obtain new plants. Linnaeus's work on the sexuality of plants made it possible to conceive artificial pollination. It remained difficult, however, to understand the function of what Linnaeus called the male and female plant organs for reproduction, since both ovist and animaculist theories were still hotly debated in the middle of the eighteenth century and in both of them the importance of either male or female contribution to the process of generation was ignored. In fact, at the end of the eighteenth century, Erasmus Darwin could still claim in *Zoonomia* (1794–1796) that reproduction requires "no mother."[32] Therefore, it was no small thing to explain plant reproduction as a process in which male and female are equally involved. By drawing on alchemy, Koelreuter found a model to explain the confluence of male and female contributions to the process:

> Vielleicht erweckt es bey einigen meiner Leser Vergnügen, wenn ich ihnen zeige, daß die Theorie der Alchymisten von dem Wachsthum und der Veredlung der Metalle mit derjenigen, die ich von der Erzeugung der Pflanzen und von der Verwandlung einer Pflanze in die andere gegeben, sehr viel übereinkömmt. Die Alchymisten nehmen zweyerley Saamen an, vermittelst deren die Vermehrung und Verwandlung der Metalle geschehen soll. . . . Die Erzeugung und Verwandlung der Pflanzen geschieht ebenfalls durch einen männlichen und weiblichen Saamen. . . . Bey der Verwandlung einer Pflanze in

> die andere geschieht nach und nach eben das, was nach der Theorie der Alchymisten bey der Verwandlung eines Metalls in das ander auf einmal geschehen soll; es wird nämlich bey einem Bastarte im aufsteigenden Grade die eigenthümliche Natur von der fremden nach eben dem Maaße verdrungen, nach welchem die letztere von einer Zeugung zur andern über die erstere das Uebergewicht bekömmt.[33]
>
> [Perhaps my readers will enjoy when I show them that the alchemists' theory from the growth and refinement of metals coincides in many ways with the one I have given of the generation of plants and of the transformation of one plant into the other. The alchemists assume two kinds of seeds by which the propagation and transformation of metals should take place. . . . The generation and transformation of plants also happens by means of a male and female seed. . . . In case of the transformation of one plant into another, that which, according to the theory of the alchemists, happens in the transformation of one metal into another, takes place step by step; namely, in the case of a hybrid in ascending degree, the particular nature is superseded by the foreign one in that measure in which the latter obtains predominance over the former from one generation to the next.]

While acknowledging that his understanding of heredity is, much as the alchemists' understanding of transmutation, a theory, Koelreuter embraced this theory as one that allows him to think the form of the new being in terms of mixture. Although he was eventually proven wrong regarding his idea that both female and male seminal fluids exists, he was right in stressing the necessity of female and male contributions to reproduction at a moment in the history of science in which animaculist and ovist accounts of generation were still not completely rejected. The model of mixture of seminal fluids, which Koelreuter could conceive by relying on alchemical models, allowed him to describe hereditary patterns. Koelreuter observed in the first generation a "middle proportion" of the parents' qualities. Most of the time though, Koelreuter expressed the conviction in his work that these "Bastarte" could be diluted and finally returned to the original father or mother form by back-crossing.[34]

Two options for the understanding of alchemical experiments became clear early on and they are repeated structurally in later texts on species transformation. These two patterns were transmutation as variation or transmutation as creation. Alchemists tended to understand their findings in terms of variation, while their opponents understood it in terms of creation not achieved by alchemical processes. In his treatise in support of alchemy, *The Necessity of the Art of Alchemy*, the ninth-century Arab writer al-Farabi (ca. 870–950) for example, presented the argument that all metals "belong to a single Aristotelian genus, and are distinguished from one another only by their accidents: hence they should be able to undergo mutual transmutation."[35] The eleventh-century Persian thinker Avicenna (Ibn Sina, ca. 980–1037), on the other hand, argued against alchemy by insisting that alchemists cannot produce genuine products of nature. He drew on the Aristotelian "opposition of species (*nau*) and genus (*jins*), as al-Farabi did. Unlike al-Farabi, however, Avicenna believed that the mere fact of belonging to a single genus (metallic substance) did not mean that the individual species (the different metals) can be transmuted among themselves."[36] If changes were made to the metals, these could only refer to the accidental qualities of a metal, such as color or density. Ultimately, Avicenna stated: "Art is weaker than nature and does not overtake it, however much it labors. Therefore let the artificers of alchemy know that the species of metals cannot be transmuted" (*Quare sciant artifices alkimie species metallorum transmutari non posse*).[37] This *Sciant artifices* statement gained a particular religious interpretation in the Christian Middle Ages, where it was closely related to the idea that alchemical attempts were blasphemous because they challenged God's creation.

The same distinction between accidental modification and transformation appears in discussions of crossbreeding. Drawing on a long line of alchemical thought, William R. Newman demonstrates how in Charles Darwin's notebooks from the period between 1836 and 1844 the alchemical expression of a "transmutation of species" is repeatedly used in the biological context.[38] As is well known, in contrast to medieval and early modern thinkers, the innovation, and scandal, of Darwin's work was precisely that he believed such species transformation could indeed occur. While Darwin's suggestions went well-beyond what most alchemists would have been willing to claim; nevertheless, his idea of species transformation is closely linked to the alchemical tradition. It is precisely the work of Joseph Gottlieb Koelreuter which serves as one of Darwin's sources when developing his theory of evolution.[39]

Koelreuter's own use of alchemical images can be read in at least three different ways. The first option is that Koelreuter still considered alchemy a real option and that he considered botanic patterns as essentially chemical patterns. In other words, what is active in the chemical pattern is active in the crossing of plants (Koelreuter's focus on seminal fluids provides possibilities for such reading). The second option is that Koelreuter used alchemy as an analogy to better understand the biological process. The third option is that Koelreuter was not so much interested in the actual process, but wanted to stress the significance of his achievement. By comparing it to the production of gold, he elevated his discovery to the level of the highest achievement in the alchemical tradition.

Indeed, Newman's central thesis is "that alchemy provided a uniquely powerful focus for discussing the boundary between art and nature."[40] As he reports, the greatest concern of medieval and early modern thinkers was that their work would place them in a god-like position and transgress the limits imposed by the Creator.[41] Individuals who questioned the alchemical enterprise most profoundly were artists. They accused alchemists of claiming divine qualities, while they, the artists, themselves had the proper relation to nature by imitating and representing, not creating it. Newman reads the rivalry between artists and alchemists as an anticipation of today's rivalry between the arts and the sciences. Whether or not one agrees on this last point with Newman (the understanding of art has changed dramatically from the here mentioned mimetic definition), he does remind us of the fact that a changing understanding of nature had also important implications for the definition of aesthetics.

In the second half of the eighteenth century, the ground shifts in the life sciences. Experiments are conducted and observations are made which threaten to destabilize nature as it was known and conceived until then. In light of such insecurity, the insistence on clear limits between nature and nurture might be understood not as ignorance of the shift, but reaction to it. Evolutionary thought was well on its way.

FILAMENT OF FLESH—WEAVING OF CREATION

As it is well documented, throughout almost his entire life Jean Paul Richter produced excerpts on a vast number of topics, beginning in 1778 when he was sixteen.[42] Starting in 1778, Richter took notes on biological questions of his day, such as the debate between animaculists and ovists, and that between preformationists and epigenesists. As Helmut Pfotenhauer has observed the motive of conception recurs again and again in Jean Paul Richter's work.[43] It is often noted that Richter, despite his comprehensive knowledge of the natural sciences, seems to have held on to antiquated models of nature, such as preformation, in which each species has its predetermined place in the order of nature.[44] However, the way in which Richter revisited these models and the reasons for which he insisted on a finite concept of nature, as well as the conclusions that he drew from these assumptions for notions such as selfhood and creativity are underexplored. Hans Esselborn has argued that the use of outdated scientific models should not be understood as an expression of Richter's anachronistic position, but in terms of the metaphorical and ironic use of these models.[45] While Esselborn has highlighted the strategic significance of the anachronism, Götz Müller has stressed that in fact both progressive and outdated scientific models are employed by Richter and that the author's opinion concerning the question of whether nature has to be imagined as preformed or as a productive force changed often between these two extremes.[46] Such assessment is in line with a general trend in Richter research which considers the author's thinking as defined by an eclectic, albeit vast, knowledge base which cannot be brought in a coherent system.

The failure and critique of available systems of nature gain significance around 1800. In 1802, Gottfried Reinhold Treviranus published the first volume of his *Biologie, oder Philosophie der lebenden Natur* [Biology or Philosophy of the Living Nature]. In this work, Treviranus, who had studied under Johann Friedrich Blumenbach in Göttingen, introduced the term *Biologie* as a superordinate concept for what we might call today the life sciences. By establishing the term, Treviranus had not only inaugurated a new discipline, but had also announced the need to search for a unifying principle instead of amassing further data within "artificial systems"[47] like the Linnaean taxonomy. This principle was nothing less than the "Triebfedern, wodurch jener große Organismus, den wir Natur nennen, in ewig reger Thätigkeit erhalten wird"[48] [driving forces, by which that great organism which we call nature is kept in eternal lively activity]. The new field of biology signaled then, both the turn to a specific set of objects and replacing collection with the search for principles, or life's governing force itself. Dismissing an entire body of work, Treviranus exclaimed: "Der Verfasser kennt kein traurigeres und geisttödtenderes Geschäft, als das Schreiben und Lesen bloßer Compilationen"[49] [The author does not know any sadder and more spirit killing activity than the writing and reading of mere compilations]. Having compiled excerpts of scientific works for twenty years at that point and aware that these excerpts and their content could no longer be brought into a coherent system, one of Treviranus's most avid readers, Jean Paul Richter, must have understood the comment as a direct attack on his own work. The question of how to account for life's governing principles, how to find that which organizes the world, in light of a limited grasp of empirical reality became one of the most burning questions for Richter in the first decades of the nineteenth century.

Richter's most direct and comprehensive discussion of the biology of his time took place in his essay *Frage über das Entstehen der ersten Pflanzen, Thiere und Menschen* [Question about the emergence of the first plants, animals, and human beings] (1812), a text often understood as a continuation of the immortality question raised in Richter's narration *Kampaner Tal* [Valley of Kampan] (1797). In *Frage über das Entstehen*, Richter summarized important claims by contemporary scientists regarding the emergence of life forms, albeit often only in order to reject their evolutionary approach. The immensity of Richter's knowledge base is documented in the sheer number of authors cited, which includes Blumenbach, Buffon, Erasmus Darwin, Dupont, Lamarck, Kant, Koelreuter, Herder, Humboldt, Linné, Meiners, Cuvier, Oken, and Treviranus.

Richter opened his essay with a summary of what he calls "Organische Maschinenlehre"[50] [theory of the organic machine]. By framing evolutionary thought in the context of the eighteenth-century conceptualization of nature as machine, Richter not only created what many would consider an oxymoron (organic machine), but he also set the tone for his attack on the materialism of evolution, especially in Erasmus Darwin, who had suggested that an abundance of life forms had developed from a simple living filament. Confronted with Erasmus Darwin's theory, Richter articulated an existential fear:

> Schon die bloße *Angst*, die jeden bei Darwins . . . Satze . . . befällt und ihm das *Herz einkerkert*, daß aus einem Lebensfädchen sich der ganze Weltknäuel aufzwirnt zur Webe der Schöpfung, treibt zu weitern, sogar kühnen Forschungen und—Annahmen. Woher aber der *angeborne*, kaum der Theoriensucht weichende *Abscheu* vor einem geistigen Entstehen aus Körper-Mächten, vor jedem Uhr- und Räderwerk, das den Uhrmacher macht? Ich frage woher; aber ich antworte: daher, weil wir selber ein viel höheres Bilden und Schaffen nicht nur kennen, sondern auch treiben, ja jedem niedrigern, um es nur einigermaßen zu begreifen, unseres *unterlegen müssen*.[51]

> [The fear alone, which grips everybody when considering Darwin's statement . . . and which incarcerates one's heart—(namely the statement) that the entire world tangle develops from a living filament to the web of creation—drives us to further, even daring questions and

> assumptions. But where from the innate repulsion, which barely gives way to the addiction for theory, for a spiritual development from corporeal forces, (where from the repulsion) for every watch- and clockwork, which generates the watchmaker? I ask where from; but I answer: (it derives from the fact) that we do not only know, but also pursue a much higher forming and creating; yes, even more, we have to assume our (way of forming and creating) for all lower forms in order to grasp them at least approximately.]

Richter's rejection highlights what is at stake for him. It was directed in particular against the idea that spiritual life could have emerged from matter ("repulsion . . . for a spiritual development from corporeal forces").[52] The a-spiritual, material quality of the filament is stressed in another section of the text where Richter translated Darwin's "living filament" not only with "Lebensfädchen," as in the given quote, but with "Fleischfädchen"[53] [filament of flesh]. An inductive development from matter to spirit, which he saw implied in the evolutionary model provided by scholars such as Darwin, seems unimaginable to him. Richter's rejection of evolutionary thought is directed less against the occurrence of new forms, which became possible in the new framework, than against the loss of a spiritual, divine, justification of these forms.

While Erasmus Darwin had thought of the evolutionary development of the living filament as taking place through a mixture of outside stimuli and hereditary transmission of acquired traits (habit), he had focused on embryology and generation and, therefore, not provided concrete examples to suggest the possible transformation of species. This step was taken by Lamarck (1744–1829) and Treviranus who had, much like Erasmus Darwin, suggested the ongoing evolution of nature and thereby of the human race by means of habit transfer.[54] Rejecting Treviranus as much as Darwin, Jean Paul Richter commented: "Noch kühner ist seine [Treviranus's . . .] Behauptung, daß sogar der Mensch sich in ein höheres Erdgeschöpf hinaufbilden und verlieren könne" (Even more daring is his [Treviranus's . . .] claim that even the human being could develop into a higher earth creature).[55] Thus, Jean Paul Richter rejected both the idea of a development from simple to more complex life forms and the idea that the human race could lift itself to higher states of being. Referring to the theory that new forms emerge after being triggered by natural catastrophes, he stated ironically: "Zu wünschen wäre der Menschheit ein solcher Untergang zum Uebergange [sic], und zumal jetzo wären ein Paar Hochmenschen, gegen welche wir nur Untermenschen und Affen wären, eine Erlösung durch ein messianisches Paar"[56] [One would wish humanity such decline for transformation, and a couple of supermen, compared to which we were only subhumans and apes, would be—particularly now—a salvation by a messianic couple]. In the notes to the *Entstehen* essay in the Richter edition by the Akademie der Wissenschaften, Eduard Berend commented: "Den folgenden Nietzscheschen Gedanken wollte Jean Paul Richter, wie er in den Vorarbeiten bemerkt, eigentlich 'zu einer Satire vernützen'"[57] [Jean Paul Richter wanted to use the following Nietzschean thought (Treviranus's evolutionary idea) "for a satire," as he stated in his notes]. While Berend assumed that Jean Paul Richter never wrote this satire ("eigentlich"), I will suggest below that the *Komet*, written eight years after *Entstehen,* is the novel in which Richter thinks about the question of the "superman."

The degree to which evolutionary thought was perceived as putting pressure on models which assumed spiritual forces can be measured in the fact that Richter, in light of Darwin's and Treviranus's suggestions, revisited in his essay Blumenbach's idea of the *Bildungstrieb*. Blumenbach's epigenetic model, while considered a precondition for evolutionary thought, did not suggest or necessitate evolutionary models of nature. However, Richter detected in it a looming materialism which he tried to ban by providing a particular reading of the formative drive. Revisting Blumenbach's groundbreaking *Ueber den Bildungstrieb* [An Essay on Generation] (1781) published over thirty years earlier, Richter stated: "der Blumenbachische [sic] Bildungstrieb kann . . . nur im Einzelwesen, aber schon im Gebildeten wohnen, er kann Leben nur fortpflanzen, nicht pflanzen. Dabei setzt ja der Bildtrieb seinen eignen Bildner voraus und dazu sein Gebildetwerden zu einem *bestimmten* Ziele und Bilde"[58] [Blumenbach's formative drive can . . . only live in the individual being, but in that one which is already formed; it can only reproduce life, not produce. The formative drive is conditioned by having its own creator and also by its formation with respect to a specific goal and form]. By insisting on an agent and telos provided from outside the organism as condition for the organism's development, Richter reversed everything what eighteenth-century thinkers found fascinating about Blumenbach. As Müller-Sievers has demonstrated, epigenesis became a model to think through philosophical and literary problems precisely because it allowed one to conceive the organism as a self-generating system. By placing the origin of the *Bildungstrieb* outside the organism and by assuming that this drive acts in an already (pre)formed entity [*im Gebildeten*], Richter virtually reversed the entire idea of the formative drive—and questioned the agenda of Treviranus's *Biologie* which was to find such life-generating force within organisms. Richter's extreme position demonstrates the degree of irritation created by evolutionary theory—with its implication of matter-driven development.

The difference between Blumenbach and Richter or rather the difference that Richter opened up between himself and Blumenbach is nowhere clearer than in the author's assessments of Koelreuter's work. Both writers referred to Koelreuter to illustrate their claims, but their readings are diametrically opposite. Blumenbach had alluded to Koelreuter's experiments already in *Über den Bildungstrieb*. He did so in order to support his rejection of preformationism. He supposed, he writes, that any preformationist would be converted by "eine einzige Erfahrung wie die, da Hr. Kölreuter durch wiederholte Erzeugung fruchtbarer Bastardpflanzen, endlich eine Gattung von Taback (Nicotiana rustica) so vollkommen in eine andere (Nicotiana paniculata) verwandelt und umgeschaffen, daß sie nicht eine Spur von ihrer angestammten mütterlichen Bildung übrigbehalten hat" [59] [a single experiment, such as that of Mr. Koelreuter, who by repeatedly producing prolific bastard plants, transformed one species of tobacco (nicotiana rustica) into another (nicot. paniculata) so completely, that it had not the smallest resemblance to its maternal parent].[60] In referring to Koelreuter, Blumenbach felt that he had illustrated clearly why preformationism could not work, since it could not explain the transformation of one species into another. Richter, on the other side, used Koelreuter's hybridization experiment to prepare his defense of precisely the opposite, namely that indeed preformationism is the better model to explain life forms. As if arguing directly against Blumenbach's text, Richter observed:

> Zwar Köhlreuter verwandelte wirklich eine Gattung Tabak (nicotiana rustica) durch lange Bastard-Bestäubungen in eine andere (nicot. paniculata); aber hier bringe man außer menschlichem Scharf- und Vorsinn, und Vorrichtung, noch den Hauptpunkt in Rechnung, daß Tabak nur in Tabak verwandelt wurde.[61]
>
> [Although Köhlreuter indeed transformed one species of tobacco (nicotiana rustica) into another (nicot. paniculata) by means of lengthy hybrid pollination, one has to consider not only human intelligence and planning, and arrangement, but also the main point that tobacco was transformed only into tobacco.]

To stress his point, Richter added in a footnote: "Auch bemerkte Köhlreuter selber, daß fruchtbare Bastarde nach einigen Zeugungen wieder in der ganzen alten Natur ihrer Stammeltern erscheinen"[62] [Köhlreuter observed himself that fertile hybrids revert after a few crossings again in the original nature of their stem parents]. Richter highlighted three points. First, the transformation of species was artificially obtained and dependent on human intervention. Second, in an argument reminiscent of arguments made in the debate on alchemy, the change in form was not really considered a transformation but rather a variation. And, third, the changes were not maintained beyond a few generations and, therefore, had no lasting impact on nature.

In rejecting evolutionary thought, Richter did not exclude the possibility that new life forms could emerge. He too saw the abundance of evidence in the form of new breeding results and shared, as his figure Dr. Katzenberger attests, the fascination for them. However, Richter hoped still to account for these new forms in a preformationist framework guaranteed by a divine creator in which fertile new forms are explained as having been dormant until their appearance and in which infertile, monstrous forms, are explained as temporary modifications triggered by contingent forces. To illustrate this point, we might consider how Dr. Katzenberger plans to produce monstrous specimen which he hopes to add to his anatomical collection. As we recall, Katzenberger had alluded to monstrous flowers to express his wish that similar specimens might also be produced in the human form. That is to say, he placed monstrous forms in the context of hybridization experiments. When he records his own efforts to produce teratological specimens, he tells his daughter, Theoda:

> Ich für meine Person könnte für dergleichen [eine Mißgeburt] viel hingeben, ich könnte z.B. mit einer weiblichen Mißgeburt, wenn sie sonst durchaus nicht wohlfeiler zu haben wäre, in den Stand der Ehe treten; und ich will dirs [sic] nicht verstecken, Theoda—da die Sache aus reiner Wissenschaftsliebe geschah und ich gerade an der Epistel de monstris schrieb—, daß ich an deiner sel.[igen] Mutter während ihrer guten Hoffnung eben nicht sehr darauf dachte, aufrechte Tanzbären, Affen oder kleine Schnecken und meine Kabinetts-Pretiosen fern von ihr zu halten, weil sie doch im schlimmsten Falle bloß mit einem monströsen Ehesegen mein Kabinett um ein Stück bereichert *hätte*.[63]

> [I, for my person, could give for such (monstrosity) a lot, I could for example marry a female monstrosity, if I could not get her cheaper oth-

erwise; and I don't want to hide it from you (Theoda)—since this happened for the pure love of science and I was writing the Epistel de monstris at that moment—that I didn't take much care to keep away upright dancing bears, apes or small snails and my rare cabinet specimens from your late mother while she was pregnant, because in the worst case she would have enriched the collection with a monstrous marriage blessing.]

Katzenberger describes here monstrosities produced by means of breeding and by means of the traditional "technology" of imagination as happening on the same level. Thus he conceptualizes the production of monstrosities in terms of environmental forces and not as a potentially progressive development to new forms. Monstrosities are for Katzenberger marked by their contingent production and their exceptionality, namely that they stay outside the genealogical line. Katzenberger speaks for Richter here, I think, because his monstrous technologies fit with Richter's position expressed in the *Entstehen* essay. Confronted with evolutionary theories, Richter returned to preformationsm. In the *Entstehen* essay, he asked:

Warum aber will man nicht lieber *annehmen*, daß alle diese Organisazionen [sic] schon als Eier und Körner vorher in den Elementen vielleicht Jahrtausende lang umgeschwommen, ehe sich die entwickelnde äußere Mutterhülle für sie vorgefunden? . . . Vielleicht ist der Luftkreis und das Wasserreich das unendliche Eiweiß zahlloser kleiner Eidotterpünktchen, die nicht erst einen Vater brauchen, sondern nur eine warme Federbrust. . . . Es ist kühn, aber auch weiter nichts, zu vermuthen, daß vielleicht seit der Schöpfung lebendige Keime kalt-unentwickelt umherfliegen, welche nur im jetzigen Jahrhundert eine eben jetzo recht gemischte Feuchtigkeit ins Leben brütet, so wie nach den Sternkundigen manche Sonne oben leuchtet, die erst nach Jahrhunderten ihr Licht zu uns herunterbringt. Was gilt die Zeit denn der Natur? Der Ewige wird nicht mit Jahren kargen, der Unerschöpfliche nicht mit Geschöpfen. Die Ewigkeit hat zu allem Zeit und zu allem Kraft.[64]

[However, why does one not rather assume, that all these organizations have swum already as eggs or germs in the elements perhaps for thousands of years, before they found the developing womb? . . . Perhaps the air and the water are the infinite egg white of countless small egg yolk dots that do not need a father but a warm feather breast. . . . It is bold, but nothing more, to assume that perhaps since the creation living

> germs fly around cold-undeveloped which only in the current century are brought to live by an only now rightly mixed humidity, like many a sun—according to the astronomers—shines above that brings light to us only after centuries. What is time to nature? The eternal (God) is not stint on years, the infinite (God) not with creatures. Eternity has time for everything and force to everything.]

This recapitulation of preformationist understandings of the development of life forms is offered here in the mode of an endorsed assumption. Like preformationists, Richter accounted for potential new forms by referring to climate or other contingent changes which trigger the development of preformed life. Drawing on a conceptualization which can be traced back to Aristotle, Richter defined the force which guarantees the stable system as male [*Vater*], while the nurturing powers are defined female [*warme Federbrust*]. In this account, time and development of natural forms exist, but they are understood within an eternal system which has provided the potential for development.[65] Richter's exclamation "What is time to nature?" indicates that he thought of new life forms less in temporal than in a spatial parameters. The development of new forms is considered a progression in so far as different environments or geographical regions provide different developmental conditions.

In rejecting evolutionary thought and in his wish to explain new forms within the framework of stable species, Richter makes at one point an astonishingly progressive suggestion. He wonders: "wenn 24 Buchstaben tausend Quintillionen Male zu versetzen sind: wie oft nicht die Millionen Thiere selber wieder"[66] [if twenty-four letters can be combined a thousand Quintmillion times: how often not the millions of animals again]. Precisely because Richter wanted to insist on stable forms, he could suggest a model of what one could call *recombination* which comes astonishingly close to a modern understanding of the recombination of genes in genetics. While the two models differ in size—animal organism vs. gene—both rely on units which can be recombined and both models think this recombination in terms of letters and words.[67]

In his attempt to justify the world with reference to a spiritual force, Richter not only revisited Blumenbach's *Bildungstrieb*, but also the idea of the chain of being. The main proponent of the chain of being in the eighteenth century and the first who provided a visual depiction of it (see fig. 3.4) was a naturalist from Geneva, Charles Bonnet (1720–1793).[68] Drawing on Aristotle's hierarchical structure of life forms in his *Historia animalium*, Bonnet's chain of being was organized according to perceived degrees of

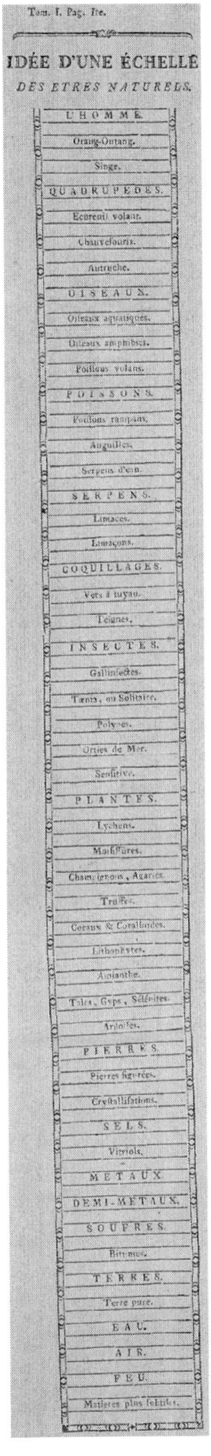

fig. 3.4. Chain of being. Herzog August Bibliothek Wolfenbüttel: Lm 500. Charles Bonnet. Oeuvres d'histoire naturelle et de philosophie *(Works of natural history and philosophy). Neuchatel: Fauche, 1779–1783.*

perfection. Bonnet's chain started with inorganic elements such as fire and water and continued to plants, animals, and the human being. Angels could also be part of the ascending ladder. God alone was perceived as standing outside the chain, in his function as the creator and guarantor of the beings presented in the chain.[69] Although the chain was considered a more or less stable structure, it allowed for transitional stages since it was conceptually connected to Aristotle's organization according to degree of perfections. In the reinterpretation of the chain in the medieval Christian context, connections to the idea of a gradual development to greater and greater spirituality became possible. One important advantage of the chain compared to a taxonomic organization was for many eighteenth-century thinkers the teleological progression within a stable system. As a model of thinking the organization of nature, it remained a powerful influence even for scholars who were interested in taxonomy such as Linnaeus.[70] Despite the chain's potential for conceptualizing development, it did not easily allow for new forms. Contemporary breeding experiments had a potentially destructive influence on the organization of the chain, since they created new forms and it was unclear where they could fit in the chain which was assumed to be preestablished by God.

The attraction of the chain of being for Richter lay in its teleological structure, which was ultimately guaranteed by a spiritual force. It is interesting though, that the chain of being, as static as it seemed to be, was also, among the two leading models of organizing nature available in the century before Charles Darwin—Bonnet's chain of being and Linnaeus's taxonomy—more likely to allow for a temporalization of species relations, since it assumed a gradual development of forces from simpler to more complex forms. Richter sensed this potential and in an attempt to keep it at bay, stressed the importance of a spiritual origin. He asked: "Begehen wir überhaupt nicht einen Fehler, daß wir die höheren Kräfte aus niedern entwickeln wollen und entstehen lassen, anstatt die Leiter umgekehrt an den Himmel anzusetzen, um auf ihr zur Erde herab zu steigen?"[71] [Do we not make a mistake in wanting to develop higher forces from lower, instead of putting the ladder up against the heavens in order to descend from it to earth?].

However, this last quotation complicates Richter's assessments. Strangely enough, the agent which descents the ladder is a "we," apparently humanity. It becomes possible to consider the human being itself as the spiritual origin emanating from above. This reading is lent further support in a second comment of Richter on the chain:

> Wenn wir unsre Begriffe von Geist und seinen Kräften auf andre Geister, die nicht Menschen sind, anwenden; so müssen wir notwendig irren. . . . All' unsre Begriffe vom Tier bis zum Engel, bis zum Schöpfer hinauf—ist der Begriff von einer Menschenreihe. Unter dem Tier stellen wir uns den Menschen auf seiner untersten Stufe vor—den Engel denken wir uns als einen veredelten, und Gott als den volkommensten Menschen.[72]
>
> [If we apply our notions of spirit and its powers to other spirits who are not humans, we necessarily have to err. . . . All our notions from animal to angel, and to the Creator—is the notion of a human chain. We think the animal as the human being on the lowest step—we think the angel as a refined, and God as the most perfect human being.]

In thinking about the chain of being, Richter highlights the epistemological difficulty of grasping any order outside human parameters. At first sight, Richter's reference to the chain of being seems simply to repeat Bonnet's idea of the ladder that bridges the gap between human and divine in incremental steps. However, a second look reveals that in revisiting the chain of being, Richter redefined it. Bonnet's structure was defined by a gradual progression of life forms which could be thought in terms of growing spirituality. The crucial point of this progressive structure was, however, that God, as the divine creator and guarantor of the system, remained outside of it. By including God as part of the chain, Richter highlighted the fact that the chain is not guaranteed any longer by a stable point outside the system; or, if this point is still posited, it is thought from within the system. Much as in the conceptualization of creation which Richter tried to understand by means of an analogy to human creation, the thought on the chain is marked by awareness for the difficulty, even inability to think outside the anthropomorphic—and anthropocentric—framework.[73] Richter's apparent anachronism in endorsing Bonnet's idea of the chain of being and the idea of a divine creation have to be considered within the context of Richter's sensibility for epistemological questions. In revisiting preformationism and the chain of being, Richter redefined them as fundamentally shaped by human concepts [*Begriffe*]. Being aware of such conceptual uncertainty, Richter searched for a ground outside the concepts.

While Richter seems to arrive at a rejection of what we would consider today more progressive notions, what is so remarkable about his rejection is the way he justified it. In highlighting the epistemological difficulties, Richter arrived at a theoretical impasse. However, he did not stop there.

If we recall Richter's statement on the rejection of evolutionary thought, we will remember that he did not claim to know whether or not there can be any empirical or rational proof or whether the spiritual force which he assumed has an ontological status, but he was convinced that the assumption of such a force is necessary. Jean Paul Richter provided two reasons to prove this necessity. First, as we have already described, he claimed a certain analogical necessity: since we know that human creation is governed by spiritual forces, we can conclude, by analogy, that this is also true for divine creation. Since we know that our own creative process works as a spiritual process, in fact, we have no other option than to think that all creation originates from spiritual activity.[74] The second reason is rather astonishing. Jean Paul Richter put the emphasis on an entity which remains outside full reflective clarity in what one could call an instinctive realm. We have to assume, he said, a spiritual force behind creation, because otherwise we feel a reaction in us which he described with reference to an emotional or corporeal manifestation. He spoke of a feeling of fear [*Angst*], a feeling as if the heart was incarcerated [*das Herz einkerkert*] and he named repulsion [*Abscheu*].[75] Richter also called this reaction "innate" [*angeborne*]. What he seems to circumscribe here is a kind of instinctive reaction to the idea that spirit has emerged from matter.

In pointing to this non-conceptual, instinctual realm, Richter's order is stabilized by a force which is not unlike Koelreuter's instinct [*Trieb*]. Reading Jean Paul Richter's notion of innate repulsion in the context of Koelreuter's "Trieb" and Zedler's definitions of the term, suggests that Richter's notion of instinct is to a very limited extent intellectually graspable. Götz Müller has stressed an important point when defining instinct in Richter as "Freiheit, die dem Menschen vorausläuft—es ist die Freiheit des Schöpfers, der den Instinkt in den Menschen gelegt hat, es ist die göttliche Freiheit"[76] [freedom which runs ahead of the human being—it is the freedom of the Creator who has put the instinct in the human being, it is divine freedom], and by highlighting that instinct in Richter's work is not driven by human agency because it originates outside the human.[77] If there is freedom in instinct, it has to be stressed that this freedom lies only in the divine, but not in the individual which has no choice in acting according to instinct. Much like the insect that confirms God's order by visiting flowers randomly, the human being has a proof of God's creative spirit by noticing an innate reaction to concepts which would undermine it. This instinct-like feeling is defined precisely by not coming to full intellectual clarity. If we understood this instinct completely, we would be God: "Wären wir unserer ganz bewußt," says Richter, "so wären wir unsre Schöpfer und schrankenlos. Ein unauslöschliches Gefühl stellet in uns etwas Dunkles, was nicht unser Geschöpf, sondern unser Schöpfer ist, über alle unsre Geschöpfe"[78] [If we had full consciousness of ourselves, we were our own creator and without limits. An indestructible feeling presents us something dark, which is not our creature (creation) but our creator, beyond all our creatures (creations)].

Placing Richter's notion of instinct in the context of the natural sciences and in particular of Koelreuter's instinct sheds new light on what has been called the author's skepticism. If Dorothee Sölle is right in noting the intellectual crisis that Richter experienced from the 1780s onwards and in noting that neither the system of Leibniz and Wolff nor French materialism was an answer for him, there is still more to be said about it then "radikaler Skeptizismus"[79] [radical scepticism]. A look at the *Entstehen* essay in the context of Koelreuter's work suggests that by the first decade of the nineteenth century Richter had found the vaguely defined but solid ground in a pre- (or post-) rational notion of instinct.

Jean Paul Richter's considerations can be read within a historical moment when epistemological questions attracted unprecedented attention. These questions were formulated to a significant extent in the context of questions around nature and related to the ones Richter addressed in *Entstehen*. Jean Paul Richter's argumentation is particularly interesting, if we consider how Kant dealt with a similar problem. In the *Kritik der Urteilskraft* (1790), Kant had been confronted with the question of whether purpose can be assigned to nature. As often discussed, Kant came to the conclusion that "[D]er Begriff eines Naturzwecks seiner objektiven Realität nach durch die Vernunft gar nicht erweislich ist (d. i. er ist nicht für die bestimmende Urteilskraft konstitutiv, sondern die reflektierende bloß *regulativ*)."[80] For Kant then, purpose in nature had only a heuristic function for the observing human.

How to assign meaning to nature once it was not any longer considered a medium of God's message was, of course, one of the most provocative questions of the day around 1800. Georg Christoph Lichtenberg (1742–1799) also considered the question in his *Sudelbücher*. With reference to the chain of being, he stated in entry 392:

> Sinn ist Ordnung und Ordnung ist doch am Ende Übereinstimmung mit unserer Natur. Um unserem Gedächtnisse etwas einzuverleiben suchen wir daher immer einen Sinn hineinzubringen oder eine andere Art von Ordnung. Daher Genera und Species bei Pflanzen und Thieren.

> Ähnlichkeiten bis auf den Reim hinaus. . . . Die Frage ist aber, ob alles für uns lesbar ist. Gewiß aber läßt sich durch vieles Probieren, und Nachsinnen auch eine Bedeutung in etwas bringen was nicht für uns oder gar nicht lesbar ist. . . . Auch die Stufenleiter in der Reihe der Geschöpfe, alles das ist nicht in den Dingen, sondern in uns.[81]
>
> [Sense is order and order is in the last resort conformity with our nature. When we speak rationally we are only speaking in accordance with the nature of our being. That is why to annex something to our memory we always seek to introduce sense or some other kind of order into it. That is why we devise genera and species in the case of plants and animals. . . . The question is, however, whether everything is legible to us. Certainly experiment and reflection enable us to introduce a significance into what is not legible, either to us or at all. . . . Likewise the gradation we establish in the order of creatures: all this is not in the things but in us.][82]

Lichtenberg brought the question of taxonomy into connection with the structure of memory, by acknowledging that species are not expressions of existing differences in nature, but an expression of necessity within the perceiving individual. While all three authors highlighted epistemological questions in light of the order and purpose of nature, Jean Paul Richter's grounding of the hesitation to endorse Erasmus Darwin's evolutionary model is different in that it refers to a ground that is not fully accessible to reason. While Kant considered the question of order in nature under a heuristic aspect and Lichtenberg stressed the structure of memory, Jean Paul Richter referred to an emotional-corporeal ground, which is not fully accessible to the intellect, but which is ultimately the only entity which can provide meaning and purpose.

What is so remarkable about Richter's assessment of the relationship between nature and time is not its anachronism but the fact that he depicts this relationship as dependent on the human observer. Just like Richter's account of the chain of being, his account of preformationism is intertwined with thinking about the conditions of the assumption of such conviction. Such reflectivity is expressed not only in using words like "annehmen" [to assume], "vielleicht" [perhaps], and "vermuthen" [to guess] but also in highlighting the human perception of natural phenomena. Richter's reference to, and comparison with, astronomy highlights how much our perception of the world is shaped and limited by a human perspective. Much as it might be wrong to assume that the light that we see today, has been emanated today ("manche Sonne oben leuchtet, die erst nach Jahrhunderten ihr Licht zu uns herunterbringt"[83] [many a sun shines above that brings light to us only after centuries]), it might be wrong to assume that apparently new life forms have emerged autonomously.

Against the backdrop of a growing awareness of the fact that human lives are fundamentally shaped by history, which is not justified by an eschatological framework, we can observe in Richter's work an increasing tendency to search for a stable origin, for a place outside of history. Richter's contribution to evolutionary thought as formulated in *Entstehen* is then less a contribution to the science of his time than it is a negotiation of the epistemological limits of such assumptions and the human possibilities to deal with it. Although Richter's work draws on premodern frames to understand nature and history, this very move is the expression of a modern state of mind. If Rousseau returned to the primitive because he knew civilization too well, Richter returned to preformationism precisely because he knew of evolutionary theories. Much as Rousseau in the *Discourse on the Origin of Inequality* (1755) labeled his account of primitive man conjectural, so did Richter highlight epistemological questions when returning to preformationism. By doing so, Richter's work directs our attention to the fact that in both historical and biological accounts of the time, even where those accounts referred to moments of novelty, the religious framework remained an important reference.

A COMET

We need to name here a third concept which Koelreuter employed to understand his results before we move on to a discussion of Richter's novel *Der Komet*. In examining Koelreuter's assessment of his experiments, we have seen that Koelreuter oscillated between considering his results as creative and in competition with God's creation and considering them merely contingent phenomena. The reference to alchemic thought and thereby to potentially autopoietic processes, however, remained an exception. For most part, Koelreuter relegated his experiments to contingency by stressing the necessity for human intervention and the hybrid's infertility. For the difficult balance between an absolute fascination with the exceptionality of the hybrid and the wish to maintain the stable system, Koelreuter found a fitting image. In an attempt to understand, and highlight, the importance of his work and the epistemological significance of the hybrid, Koelreuter referred to the comet:

> Ich bin vollkommen überzeugt, daß die Kräuterverständige . . . eine Menge Pflanzen, die in der Kräuterwissenschaft auf eine gewisse Art

eben das sind, was die Cometen vor Zeiten in der Sternkunde waren, ihre gehörige Stellen werden anweisen, und in wenigen Jahren dasjenige leisten können, was man schon so viele Jahre her vergeblich gewünscht hat.[84]

[I am completely convinced that the botanists . . . will be able to place a number of plants—which are in botany in a certain way that which the comets were some time ago in astronomy—in their right place and that they will be able to achieve in a few years that which one has hoped in vain for so many years.]

By understanding the hybrid in terms of the comet, Koelreuter was able to explain the hybrid's exceptionality which was so puzzling for contemporary botanists. Much as the comet's exceptionality was crucial for a better understanding of the universe, so, Koelreuter stated, botanists will be able to place the bastard within the system thereby confirming—but also possibly redefining it (although the latter is not an option which Koelreuter seemed eager to highlight).

Each of the frameworks encountered in Koelreuter's work provided, a conceptual challenge but also metaphorical and associative space for Jean Paul Richter to think through notions of selfhood and creativity. In naming his last novel *Der Komet* (1820–1821), Richter revisited a natural phenomenon which had fascinated the eighteenth century ever since Edmond Halley (1656–1742) had predicted with astonishing precision the appearance of the comet which later was named after him for the year 1759, and ever since its dramatic actual appearance on Christmas day of 1758. We can see that Richter drew on Koelreuter's hybridization experiments and their conceptualization in order to tell the story of the *Komet*'s protagonist Nikolaus Marggraf. He is the child of two very dissimilar parents—a bastard like Koelreuter's hybrids—who is heavily invested in alchemic projects. Much as for Koelreuter, the crucial question for Jean Paul Richter is how to guarantee an absolute belief in creativity and autonomy without losing the reference to the system guaranteed by God. What makes the image of the comet so attractive in this situation for both Koelreuter and Richter is the fact that it points toward the system without representing it. If we read the comet as an image for selfhood, it expresses the absolute uniqueness of the event of the appearance of the comet from the perspective of the individual human. However, based on eighteenth-century knowledge, once we remove ourselves from the position of the individual perspective and look from a vantage point somewhere outside the planetary system, we can see that the comet belongs to a larger, stable system. In the next section, we will look at how Richter defined selfhood in the context of hybridization experiments and evolutionary thought.

AUTONOMY—HEREDITY

Few concepts have been so influential and persistent in scholarship on German literature as the conviction that selfhood around 1800 is best described as an attempt to attain complete autonomy. This autonomy, in turn, has been defined in aesthetic terms. As Daniel Purdy has observed, literature around 1800 is shaped by the idea of what he has named a "neo-classical aesthetics of artistic autonomy."[85] Scholars such as Martha Woodmansee and most recently Elliott Schreiber have presented Karl Philipp Moritz (1756–1793) as the author who inaugurated the concept of the autonomy of the artwork.[86] As Schreiber has described brilliantly, it was Moritz's (1788) essay "Über die bildende Nachahmung des Schönen" [On the Artistic Imitation of the Beautiful], which initiated the eighteenth-century debate on aesthetic production as "intuiting absolute totality."[87] In Moritz's words, the artwork is "ein für sich bestehendes Ganze"[88] [a whole existing for itself alone],[89] which has its telos in itself. This autonomous totality via art has been understood as a response to, and corrective of, the modern experience of a fractured world in which the reference to a unifying principle is lost. It is the artwork alone, as Manfred Engel has stressed, in which the self can still experience the totality of its capacities.[90]

Ever since Helmut Müller-Sievers' *Self-Generation* (1997), it is understood that one of the most important models for thinking this autonomy was the organism as it was conceived in the life sciences of the time. This model, according to Müller-Sievers, is epigenesis, namely the idea that organisms are autopoietic systems which generate themselves driven by a formative drive (*Bildungstrieb*), as Blumenbach had suggested. "Epigenesis," Müller-Sievers states, "is thus the condition of the possibility of any claim to absoluteness, be this a philosophical or literary absolute."[91] Drawing on epigenesist models, authors could conceptualize a self which did not need an outside stimulus for development (as it was needed in preformationism), because it had its origin in itself.

Müller-Sievers's explanatory tool is powerful and has proven extremely productive for the reading of a number of texts, but telling the history of the conceptualization of organisms in a straight line from preformationism to epigenesis, and stretching the significance of this shift well into the nineteenth century risks overlooking both the multiplicity of models available at that time and the changes which occurred during the Romantic period, even if this narrative is viewed as one of progression. While the epigenetic model remained influential for conceptualizing selfhood well into the nineteenth century, Jean Paul Richter's work reminds us of competing and possibly more progressive models explaining life, models which became available in the late eighteenth century. As Erasmus Darwin and Joseph Koelreuter's studies demonstrate, the idea of the organism as a self-contained totality was undermined by the transmission of traits which came in the focus of conceptualization efforts in the natural sciences around 1800. There was no intrinsic need to consider such conceptual changes as significant for the understanding of selfhood. After all, authors had been well aware of the model character of epigenesis and as such it could remain the reference for notions of selfhood. Richter's work, however, demonstrates that research on heredity did indeed trigger some irritation regarding the question of whether or not existing concepts of selfhood and aesthetics, and their claim to totality and self-regulation, were still viable. A strongly held conviction—which has also shaped the historiography of the period—was at stake: it became more difficult to think selfhood and aesthetics in terms of autonomy. Or, differently put, if writers continued to consider models developed in biology as significant for conceptualizing selfhood, how could selfhood and aesthetics be thought as autonomous?

Richter's *Komet* can be read both as a novel which is interested in character formation and as a novel in which figures function as allegories with poetological significance. Here, I would like to focus first on the question of selfhood in the context of heredity and return in a second step to poetological questions.[92] The idea that the child was shaped by the parents and in turn could shape her own children, as it was assumed in the context of trait transmission, was not a new one. After all, this had been precisely one of the discoveries—and main playgrounds—of Enlightenment thinkers from Locke (1632–1704) to Rousseau (1712–1778), Moritz, Campe (1746–1818), and beyond. What was new around 1800 was the recognition that education did not start with a blank slate and that the optimism which shaped Rousseau's *Émile* (1762) might have been ill-conceived. However, what was also new was the "discovery" that the child *could* be shaped in such a way that it had impact on the future of the species. In both cases, this discovery was connected to an emerging belief that psychological qualities were as transmittable as bodily features. This was certainly a liberating thought since it handed agency to the human, but it also created fears that the individual freedom and potential for development was limited by heredity. Furthermore, it created the fear, as expressed in Richter's *Entstehen*, that nature could be forever changed because individual developments could lead to species transformation. It is in this context that we observe in Richter an abandonment of whatever monist thought might have existed in Germany in the wake of Spinoza[93] and a new interest in a division between nature and aesthetics. In employing the idea of a strict divide between nature and nurture, Richter took up a thought which was, as we have seen, pursued by several eighteenth-century plant and animal breeders. It is precisely by drawing on this differentiation between nature and nurture (nature "produced by art"[94]), that Richter distinguished the literary creation from an organic growth model.

NATURE—NURTURE—AUTOPOIESIS

Richter's last novel has often been read within its socio-historical context. Written in the first decades of the nineteenth century, it has attracted attention as either a parody or celebration of the achievements of the French Revolution in the context of a climate of growing dissatisfaction in the wake of the social and political restoration initiated by the *Wiener Kongress*.[95] It has also been noted that the novel stands in close connection to Richter's project of the *Selberlebensbeschreibung* and, in fact, can be considered its replacement once the author gave up on the autobiography in 1819 in order to further his work on the novel. The novel is read not only as an account of periods in Richter's life (e.g., his study years in Leipzig), but also as a coming to terms with his failed literary ambitions.[96] Whether socio-historical or biographical, readings of the *Komet* have tended to stress moments of failure of the dream of autonomy which is negotiated in the novel.[97] The protagonist's ambition to become duke is read as the unfulfilled dream of the bourgeois to come to aristocratic power, and his love for the wax image of Princess Amanda has been interpreted as a scathing self-critique of literary ambition.[98] Both the social and biographical reading of the *Komet* point in important directions, but, I would like to suggest, they fall short of grasping the novel's epistemological depth. Richter's approach to autonomy, I argue, cannot be sufficiently explained by reference to resignation, but needs to be read in light of the redefinition of the possibilities and limits of autonomy within the context of contemporary thought on heredity, hybridization, and evolution. In attending to the multifaceted depictions of Nikolaus Marggraf's search for father and wealth, I want to trace the ways in which contemporary hybridization research offers literal, metaphorical and conceptual possibilities for Richter's negotiation of the concept of autonomy.

Der Komet is a story which evolves around heredity and inheritance. It is a story about the search for the (biological) origin which ends with finding (economic) autonomy: the bastard Nikolaus Marggraf is on a search for his biological father with the hope to gain his aristocratic title and wealth. In delineating Nikolaus Marggraf's life, the novel is driven by constant references to heredity, a heredity whose significance is, however, continuously questioned and ultimately shown to be irrelevant as a determining factor. Indeed, from the start of the novel, the protagonist's biological origin is put into question. Although born as the son of the beautiful Italian singer Mara and the pharmacist Henoch Elias Marggraf, there is some doubt about the hero's conception. First, the marriage took place in great haste and Henoch wonders how such a beautiful woman has chosen precisely him. Then, Nikolaus Marggraf is born just "neun kurze . . . Februar- oder Hornung-Monate"[99] [nine short . . . February months] after his parents' wedding and he shows little resemblance to the old Marggraf. Finally, he is born with mysterious, extraordinary signs: twelve pockmarks on his nose and a halo. His father Henoch—like most other people, as the narrator concedes—has none of them. The uncertainty regarding Nikolaus's biological father shapes the novel's beginning and drives the entire plot.

If uncertainty about the father motivates the story, the mother's narration fills the gap. Following four not too happy years of marriage, Mara dies after giving birth to a third daughter. Shortly before her death, Mara confesses to a Franciscan monk. Old Henoch eavesdrops on the confession behind a secret door. He learns that Nikolaus is the son of a catholic duke—whose name Mara has sworn not to tell—and that "der eben seinen Heiligenschein und seine Nasen-Narben auf den Kleinen fortgepflanzt"[100] [precisely this one transmitted his halo and his nose scars to the little one]. Previously, when Henoch had wondered about the child's halo and scars, Mara had invoked the ancient theory of imagination in order to explain them: she had seen a person with just such halo and scars during pregnancy and this image imprinted itself on the unborn. Now, shortly before her death, Mara declares the marks the result of hereditary transmission ("fortgepflanzt"). In revisiting the marks, Mara implicitly claims that heredity has a greater truth value. She also discloses the place where she has hidden a diamond ring received from the child's biological father which should now support Nikolaus's education. After the mother's confession, it seems that a solid ground has been established on which Nikolaus's identity formation can take place.

However, this is not the case, for the claim's veracity is questioned as soon as it is expressed. As the narration unfolds, the significance of the *biological* origin for the emergence of selfhood vanishes. The signs that Nikolaus inherited, pockmarks and halo, are signs which are defined precisely by the fact that they are not hereditary. The halo, as an attribute given either to Saints or to Jesus—in fact there are several similarities between Nikolaus and the Jesus child—emanates the presence of the divine and is the opposite of contingency. The pockmarks, on the other hand, are a sign of a contingent event which leaves its traces on the body.[101] Therefore, if Mara assigns the scar to some aristocratic biological father, we can hardly trust her. In fact, the narrator points the reader's attention to the contradictions in her claim:

> [Nikolaus] hatte nämlich auf der Nase zwölf Blatternarben auf die Welt gebracht, als hätt' ihn die Natur schon ungeboren mit diesen Stigmen (Wundmalen) für das Leben gestempelt und tätowiert, was aber nicht gewesen sein kann, da er später die wahren Pocken bekam und also die Narben früher als die Wunden hatte.[102]
>
> [(Nikolaus) had brought to the world twelve pock marks on his nose, as if nature had stamped and tattooed him for life with these stigmas (scars) already before he was born, which, however, is not possible, since he got later the real pocks and, therefore, had the scars before he had the wounds.]

The uncertainty regarding Nikolaus's birth marks is carried through to the end of the novel, but indications that Mara's claim about the aristocratic father was just that, a claim, mount as the novel unfolds. At stake here is not so much the mother's faithfulness or the sincerity of her confession, but the destabilization of the connection between corporeal sign and the genealogical reference framework to which it is supposed to point. Heredity and origin are claimed, but the claim is immediately presented as ludicrous.

Yet despite the uncertainties about its truth, the mother's story becomes the dominant force in Nikolaus's selfhood formation. Nikolaus's (adoptive) father Henoch, instead of being shocked by the mother's confession, starts immediately to act as if it were true. He educates Nikolaus as the son of a duke, speculating that once the biological, aristocratic, father is found, he will be repaid abundantly. Nikolaus himself lives according to the mother's account and, when studying in Leipzig, searches for his biological father among each and every aristocrat that comes through the city. If this is a story about selfhood in the context of new concepts of heredity, it seems to choose nurture over nature as the determining force. In fact, in the course of the novel nurture replaces a nature which is not accessible to the protagonist.[103]

During Nikolaus's study years in Leipzig, however, the plot-line takes a decisive turn. As much as Nikolaus has looked for a father with the same birth signs, twelve pockmarks and a halo, he has been unable to find anybody that would resemble him:

> Nikolaus hatte dort [in Leipzig] unter allen Vätern von Geburt, welche zuweilen durchreiseten, nie seinen eignen angetroffen, sondern war mit seinen Heiligenstrahlen und zwölf Nasennarben ohne Vorbild ungekannt stehen geblieben. . . . Lange konnte er sich ohnehin . . . nicht auf der Universität aufhalten, da . . . der Diamant durch ihn und Peter [sein Freund] so glücklich verflüchtigt und geschmolzen war, als wäre der eine ein Brennspiegel, der andere Bockblut.[104]

> [There (in Leipzig), Nikolaus had never met his own (father) among all fathers of (aristocratic) birth who traveled through from time to time. Instead, he remained unrecognized with his halo and his twelve nose scars without model. . . . Anyway, he could not stay long . . . at the university, since . . . the diamond was so happily evaporated and melted by him and (his friend) Peter, as if the one were a burning glass and the other goat blood.]

Nikolaus cannot find his aristocratic father. Although his search has failed, there seems to be the assumption that this aristocratic father exists. If we look more closely at the quotation, the existence of the father is actually put into question. We learn that in the Leipzig years, Nikolaus Marggraf could not find his father and that during this time the diamond, which Mara had supplied as evidence for the existence of the aristocratic father, has been all "vaporized and melted" [*verflüchtigt und geschmolzen*]. The money realized from the diamond has run out. Even more, if the diamond was the guarantee for the truth of the mother's claim, then it is precisely during the years in Leipzig that this guarantee expires. It is in the Leipzig years, after both (adoptive) father and mother have died, that Nikolaus realizes that the mother's claim might not have a basis in reality. And it is precisely at this moment in which the referent for his birth marks is lost, that Nikolaus can start to create his own self. The narrator frames this autopoietic process with implicit reference to Joseph Gottlieb Koelreuter's hybridization experiments. He states:

> [G]erade eine Nase erhob des Reiseapothekers [Nikolaus Marggrafs] gemeines Leben zum Epos, zum Pik mit Nasenlöchern, in welche nicht nur Tabakpflanzungen, sondern ganze Tabakpflanzer gehen.[105]

> [Exactly a nose raised the traveling pharmacist's (Nikolaus Marggraf's) ordinary life to an epos, to a pic (picaresque novel) with nostrils in which not only tobacco plantations fit but entire tobacco cultivators.]

The passage, with its strange and apparently nonsensical reference to tobacco and tobacco cultivators becomes most relevant to our reading of the text when viewed in the light of Koelreuter's hybridization experiments. As we have seen, it was with tobacco plants that Koelreuter claimed his success in the transmutation of one species into another by means of hybrid creation. As we will recall, Koelreuter had understood this process as artificial [*widernatürlichen*] and had considered the new forms as obtained by art [*durch die Kunst*].[106] It is with reference to Koelreuter's art that Nikolaus's formation of the self can be understood as a creative process. The pockmarks become places for Nikolaus to become a cultivator of tobacco, a creator of new forms. What he is cultivating in these pockmarks, however, is not the tobacco plant, but the "Tabakpflanzer" [tobacco cultivator] himself. The oxymoronic constellation of hereditary pockmarks becomes for Nikolaus not a place of nature, not a place of nurture, but a place for autopoietic self-nurturing. Once Nikolaus has realized that the biological origin might not be accessible and that the mother's narration of it might be false, he can gain access to selfhood. Instead of relying on genealogy, he becomes a creator of himself.

FROM BLOOD TO MONEY

Koelreuter had relied on alchemical concepts to understand both the processes involved in his hybridization experiments and his own contribution to it and now Richter takes up this alchemical vocabulary to articulate Nikolaus's process of self-creation. In the moment in which Nikolaus realizes that the search for the father might be in vain and in the moment in which the mother's diamond has vanished, Nikolaus becomes not only his own "tobacco planter" but he starts an alchemical production:

> Nikolaus ist seit der Zurückkehr aus Leipzig teils um einige Jahre älter geworden, teils um manches Goldstück ärmer [der Diamant ist ohnehin längst fort]. . . . An Thronbesteigung denkt vor der Hand kein Mensch, und Gott dankt man schon in der verschuldeten Apotheke, wenn man nur etwas zu beißen anstatt zu beherrschen hat. —übrigens legte sich Nikolaus, noch bevor er Weisheitszähne hatte, etwas auf den Stein der Weisen . . .[107]
>
> [Since his return from Leipzig, Nikolaus has both grown some years older and become some gold pieces poorer (the diamond is anyway long gone). . . . Nobody thinks any longer openly about accession to the throne, and in the indebted apothecary one thanks God already if one has only something for eating instead of ruling. —By the way, Nikolaus occupied himself with the stone of the wise even before he had wisdom teeth . . .]

Only at the moment when Nikolaus is solely concerned with mere survival ("etwas zu beißen anstatt zu beherrschen" [to have something to bite instead of something to rule]), when he cares solely about satisfying his most basic needs and not his social status, can he begin to become a creator of himself. Richter introduces this moment of recognition which marks a turning point in Nikolaus's life and which becomes the precondition for self-creation in a remarkably non-organic way. The passage is shaped by a weird development which is driven by a kind of associative transmutations from word to word. The sentence moves from *denken* to *danken*, and from *beißen* to *Weisheitszähne*, and from *Weisheitszähne* to the *Stein der Weisen* in order to present Nikolaus's development. The associative play that is evoked here stresses a non-organic development which is driven much more by contingent combination than by any inner law.

In the course of the novel the alchemical transubstantiation turns out not to produce gold (or the diamond which Nikolaus wants to produce), but to bring money into circulation on the mere assumption that a diamond has been created.[108] Instead of a story of creation, the novel tells the story of speculation. Richter in fact plays here with the two possibilities that the alchemical metaphor offers: creation and manipulation. The conviction that the alchemical process is a creative act is placed on the level of the main character (Nikolaus believes—or pretends to believe—in the creative act of alchemical transubstantiation), while the recognition that this act is only an act of manipulation is placed on the level of the narrator and reader. When producing the diamond, Nikolaus's hopes are reported in the language of creation: "er wollte den allergrößten Schöpfungstag erleben, den sechsten, nämlich den Geburttag eines größten Diamanten"[109] [he wanted to experience the very greatest day of creation, the sixth, namely the birth of a biggest diamond]. Alluding to biblical language, the narrator highlights the degree to which Nikolaus wants to see his alchemical endeavor as the act of creation of a new Adam. The eventual success of Nikolaus's diamond production is, however, staged in such a way that the reader is immediately forced to doubt it. When Nikolaus sends his servant to his laboratory to check whether the diamond production was successful, Stößer, the servant, returns quickly with the message: "Die Kohlen im Heinze sind alle maustot und kohlschwarz, aber es scheußt etwas darin Strahlen über Strahlen, und muß etwa der Demant sein"[110] [the coals in the oven are all dead like a mouse and black like coal, but something among it emanantes rays over rays, and (it) has to be the diamond]. While a diamond has indeed been found, Stößer's words shed doubt whether it has been produced by means of transubstantiation. His words indicate that none of the coals have turned into a diamond and that the one diamond that he finds lies among (*darin*) the coals. The narrator comments dryly: "Sollte man nicht denken, ich [der Erzähler] hätte den Zufall selber ersonnen, daß gerade jetzo ein Polizeibedienter auf der Gasse klingelt, der das Wiederbringen eines weggekommenen Diamantringes gegen ansehnliche Erkenntlichkeit verlangt"[111] [Should one not think that I (the narrator) have invented the coincidence myself that a police servant, who asks for the return of a disappeared diamond ring for considerable appreciation, rings the bell on the street precisely now].

While the narrator casts significant doubts on the fact that Nikolaus's alchemy has been successful, it is remarkable how little the answer to the question of whether or not he was successful matters for Nikolaus. Despite all doubts, Nikolaus Marggraf is able to exchange this diamond for cash; a circulation of money can be started. Whether the stone is authentic or not remains in doubt in the same way as the mother's claim of Nikolaus's nobility. What matters is that Nikolaus is able to establish a new circulation system, which has the potential to make him a duke, which he believes his father is: "[A]ber gesetzt, ich würde Fürst, weil ich natürlicherweise, insofern ich so viel Gold machte (denn nähere Ansprüche verschweig' ich), daß ich eine und die andere verpfändete Markgrafschaft um das Doppelte auslöste und Spaßes halber z.B. wirklich Hohengeis zu regieren bekäme: so weiß ich kaum, was ich täte vor Freude"[112] [(B)ut assumed I would become a duke, because of course—insofar as I would produce so much gold (because I withhold more precise claims) that I would redeem the one or the other indebted dukedom for double the prize just for fun, I would really get to rule Hohengeis: I would hardly know what to do because of joy].[113] For Nikolaus, money has replaced blood. Fiction has created reality.[114]

The success of Nikolaus's money production is presented to us as originating in a fictitious, possibly self-delusionary, moment. Parallel to the diamond production, Nikolaus continues his searches in the hope to find his father. This father search is presented as similarly self-delusional. Standing in front of a mirror together with Stoß, his servant, just around the time when the diamond production "succeeds," Nikolaus exclaims:

> "O, du teuerster, liebster Vater, du willst deinen Sohn aufnehmen?—Ach sieh, ach sieh! Er machts [sic] so mit den Armen nach mir!" rief Marggraf und streckte die seinigen aus. Er sah nämlich im fernen Spiegel seine eigne Gestalt, die er der Ähnlichkeit wegen für die väterliche hielt. . . . "Ich sehe vom alten Herrn gar nichts," —versetzte Stoß, welcher glaubte, der alte Apotheker Henoch sei erschienen.[115]

> ["O you dearest, most beloved father, you want to accept your son?—See, see! He makes so with his arms to me!" Marggraf exclaimed and extended his own (arms). Because he saw in the mirror his own figure, which he considered because of the similarity for the father's. . . . "I don't see anything from the old Master,"—answered Stoß, who thought the old Henoch appeared.]

Much as in the case of the diamond production, the finding of the father is here presented as a claim which originates in a productive self-delusion: Nikolaus recognizes himself as his own originator. By juxtaposing the servant's observation with his master's, the narrator presents this father-finding and self-creation as illusion. What Nikolaus sees in the mirror is not the father, but himself. The similarity between father and son, which the narrator mentions ironically ("der Ähnlichkeit wegen" [because of the similarity]), is the similarity to himself. It is a similarity that misses its referent: Nikolaus Marggraf becomes the father that he has been looking for.[116]

Nikolaus's biography is narrated as an attempt to find the original to the scars and halo that he apparently inherited. His autopoietic process is completed once he becomes an original himself. This process has started with the attempt to produce the diamond and climaxes at the moment when Nikolaus becomes an "original." Having come to money, Nikolaus establishes his own court and he can now also afford to pay artists to paint his portrait. In the final section of the novel, we find Nikolaus in Lukas-Stadt, a city known of its arts. Unfortunately, the local aristocracy has run out of money to support its artists. In this situation, Nikolaus Marggraf appears and invites the painters of the city to make a portrait of him. These painters are known for being copyists. They are divided into two big schools, the Dutch and the Italian school, and they name themselves after famous artists of the past whose style and work they copy. Nikolaus invites all of them, first the sixteen painters of the Dutch and then the sixteen painters of the Italian school, to portrait him:

> Hasencoppen [Nikolaus Marggraf] war leicht in die Mitte des Saals gesetzt, großen Spiegeln gegenüber—um ihn herum stellten sich die Tischchen der verschiedenen Meister, aber nur einige konnten ihn im Vollgesicht ergreifen—andere bloß im Dreiviertelprofil—mehrere im Halbgesicht—ein paar im Viertelgesicht, und die vielen hinter seinem Rücken hatten gar nichts von vorne zu sehen;—diesen aber waren jedoch Spiegel gegenübergehängt, so daß aus letzten wieder Vollgesichter und Dreiviertel- und Halbgesichter äußerst bequem herauszumalen waren.[117]

> [Hasencoppen (Nikolaus Marggraf) was easily put in the middle of the hall, in front of big mirrors—the little tables of the different masters were placed around him, but only a few could grasp his full face—others grasped him only in three quarters profile—several in half face—some in quarter face, and the many (placed) behind his back could not see anything from front;—but mirrors were hung across from them so that again from these (mirrors) they could paint full, three quarters, and half faces.]

Nikolaus is now surrounded by artists who are themselves copyists of the old painters, copyists of his face and copyists of his face's reflection. Hence somehow, in the midst of this copying of copies, Nikolaus Marggraf turns out to be an original.[118]

When the pictures of both schools are completed, they are shown in an exhibition. At the exhibition Nikolaus plans to make the first contact with "his" aristocratic family. Nikolaus hopes that the duke will recognize the similarity to his own face and will then accept Nikolaus as his son and heir. The narrator describes him entering the room that members of the court are visiting:

> Hier trat Graf Hacencoppen, der Fürst Nikolaus, mit seinem Gefolge ein. . . . Der Hof . . . wollte, wie es schien, den Eintritt des Inkognito-Fürsten nicht zu bemerken scheinen, welchem, gleichsam als 32 blasende Postillone, 32 stumme stille Ahnen vorausgegangen waren, wie ich seine 32 aufgehangene Gesichter nennen kann, da er mit seinem eignen Gesicht ihr Ahnherr ist, ob er gleich darunter (wie jeder Stammvater) auch manche ihm unähnliche Ahnen und nur 16 ganz veredelte aus der welschen Schule zählt.[119]
>
> [Lord Hasencoppen, duke Nikolaus, entered here with his entourage. . . . The court . . . did not want, as it seemed, to appear as if it had noted the incognito duke, who was—as thirty-two playing stage coach drivers—preceded by thirty-two silent ancestors, as I can call his thirty-two faces on the wall, because he is with his own face their progenitor, even though he counts among them (like any progenitor) several ancestors which did not resemble him and only sixteen entirely noble from the Latin (Welsch) school.]

Although the court does not take notice of Nikolaus Marggraf, in this room he becomes the aristocrat that his mother had claimed him to be. The images that are painted after his face become his ancestors ("Ahnen"). He is at the same time their progenitor ("Stammvater") as he is their heir. He is their precondition and at the same time they legitimize his identity. Finally, he has created his own past and thereby himself.

AUTOPOIESIS UNDER ATTACK

So far, it seems that the *Komet* celebrates self-creation as an aesthetic process independent of its natural grounding and modeled after the speculative system of paper money. Nevertheless, Jean Paul Richter had insisted in the *Entstehen* essay that creative experiments such as those of Koelreuter have only temporary effect, and that the novelty of his cross-bred hybrids had the status of an absolute exception. Thus far, I have highlighted how Marggraf's identity becomes possible precisely through an ignorance of natural conditions: the pockmarks become places for self-creation. However, as in the *Entstehen* essay and as with Dr. Katzenberger's insistence on the infertility of the monster, *Komet* shows how problematic it would be for Nikolaus Marggraf's self-creation to be without limits.

When Nikolaus enters Lukasstadt, in other words in the climax of his self-creation, he is confronted with the eerie and sinister figure of the

Leatherman, a man clothed in leather without flesh and with hair shaped like horns.[120] The leather covering his skeleton is an ape-skin, the Leatherman is thus a mixture between ape, death, and devil. Calling himself the "ruler [literally: duke] of the world" [*Fürst der Welt*], thereby alluding explicitly to the devil but also to Nikolaus's aristocratic ambitions, the Leatherman attacks Nikolaus's autopoietic project. He can see the problems of Nikolaus's autopoietic creation so clearly because his own project has many similarities with Marggraf's. The Leatherman's biggest problem is that he sees himself as "ruler of the world," but that he is constantly confronted with the fact that other people also declare themselves rulers: his aim for total autonomy is constantly confronted with similar claims. Pabst, Nikolaus Marggraf's host in Lukasstadt, explains the Leatherman's occasional paroxysm:

> In dergleichen Paroxysmen gerate er aber jedesmal, vorzüglich wenn große Herren in der Stadt eintreffen, die er sämtlich nicht ausstehen will, weil er allein der regierende Fürst der Welt in seiner ganz erbärmlichen Narrheit zu sein denkt.[121]
>
> [In such paroxysm he would get each time, in particular when great lords arrive in the city, who he all despises, because he—in his entire pitiful madness—thinks that he alone is the ruler of the world.]

The Leatherman mirrors Nikolaus Marggraf's agenda to create himself as a duke and, therefore, must consider him as competition.[122]

By casting the Leatherman as a devil figure, Richter envisions the Leatherman's attack on Nikolaus as being motivated by a similar project as Nikolaus's: in Judeo-Christian mythology, the devil figure is the angel who wanted to become autonomous from God and, therefore, like God. Drawing on this mythology, Richter considers the religious dimension of Nikolaus's self-creation. By presenting the Leatherman at the same time as an ape figure, Richter also suggests a kinship between the Leatherman's (and Nikolaus's) striving for autonomy and the evolutionary project. In the *Entstehen* essay Richter had criticized evolutionary models for the loss of a spiritual force in nature, and had done so with reference to the ape: "und zumal jetzo wären ein Paar Hochmenschen, gegen welche wir nur Untermenschen und Affen wären, eine Erlösung durch ein messianisches Paar"[123] [a couple of supermen, compared to which we were only subhumans and apes, would be—particularly now—a salvation by a messianic couple]. The Leatherman even places Nikolaus's project in an evolutionary context.

Considering the apes as the children of his empire, the Leatherman attacks the humans who claim to stand beyond the apes:

> Ihr wollt meine Reichskinder, die Affen, nachäffen, ihr Unteraffen; aber ihr versteht es schlecht—. . . Mein Affen sind klüger und lassen sich nicht, wie ihr, von euch regieren, nicht einmal von ihresgleichen. . . . Schälet einmal euere [sic] Haut ab und seht euch aufgedeckt und aufgemacht an: so hängen statt euerer [sic] Reize und Menschenmienen Gehirnkugeln und Herzklumpen und Magensäcke und Därme vor euch da und würmeln.[124]
>
> [You want to imitate the children of my realm, the apes, you underapes (Unteraffen); but you understand it badly—. . . . My apes are smarter and they don't accept to be governed by you, not even by individuals of their own. . . . Peel your skin once, and look at you uncovered and opened: then, instead of prettiness and human physiognomy, brain bowls and heart clumps and stomach sacks and intestines will hang in front of you.]

While in the *Enstehen* essay Richter had referred to the human as "Untermenschen" (subhumans) who are like apes when compared to the "Hochmenschen" (supermen), he lets the Leatherman in the *Komet* speak of the human as "Unteraffen" (underapes). While in the *Entstehen* essay Richter had described, and critiqued, a development from human to possible even higher-developed beings, the Leatherman focuses on the reversed development from human to ape, and possibly further to the Leatherman. Instead of a development from ape to a "messianic" couple, the Leatherman establishes a reversed hierarchy in which the ape represents a higher level than man.

The Leatherman accuses humans of trying to ridicule [*nachäffen*] the apes, while in fact they do not even understand the apes' nature. The ape-characteristic which the humans seem not to grasp is that the apes have an aversion against being governed. Neither do they want to be governed by the humans nor do they want to be governed by their fellow apes. Rather—the passage seems to imply—each of them is an autonomous ruler of itself. The Leatherman connects the reference to the apes' wish for autonomy to a comment on the material basis of human existence, thereby minimizing the human-ape gap. He does so by pointing to features which traditionally have been called upon to establish the human-ape distinction:

mimic expression and brain [*Menschenmienen und Gehirnkugeln*]. For the Leatherman, even the brain, since the eighteenth century the place where human superiority was located,[125] is nothing else than an accumulation of matter.[126] In the *Leatherman*, Richter created a figure which highlights the extent to which he considered evolutionary thought an attack on a divine dimension of human self-understanding. The Leatherman thus places Nikolaus's autopoietic production in close proximity to the evolutionary project and suggests that much as evolving matter must lead to "self-governing apes," Nikolaus's self-creation might lose the reference to the divine creator. In drawing attention to the material basis of the human being, Richter stresses that any autopoietic project ultimately risks the loss to a spiritual force and a force outside the autonomous state.

There are two ways in which the autopoietic totality which Richter sees as problematic could be kept in check. Either, there is a multiplicity of autonomous spheres and the very fact that there is a multiplicity guarantees the balance; or a hierarchy is established in which Nikolaus's autonomy is partial and embedded in a larger system. Both options appear in the *Komet*, but the second is ultimately given preference. Much as in the *Entstehen* essay, Richter locates the necessity for such hierarchy outside of rationally graspable parameters. After the Leatherman has attacked Nikolaus's autonomous project by juxtaposing it with his own and by stressing the material basis of existence, he surprisingly turns around to admit the necessity for a divine father. However, he does so only in what one could call a subconscious state, a kind of hypnotic sleep. Put into this sleep, this devil-like figure reveals the source of his problems by addressing himself to God:

> Ich habe in meinem Herzen dich, unendlicher Gott der Liebe, wieder, der in alle tausend tiefen Wunden der Menschen wärmend niedersieht und endlich die Wunde nimmt oder den Verwundeten. O Gott der Liebe, lasse dich fortlieben von mir, wenn ich erwache. Die schreckliche Stunde steht schon nahe, trägt mir meine Furienmaske entgegen und deckt sie auf mein Gesicht!—Vater der Menschen, ich bin ja auch dein Sohn und will dir ewig gehorchen; Vater, verlaß mich nicht, wenn das Glöckchen läutet.[127]

> [In my heart, I have you, eternal God of Love, again, who looks warmly into all thousand wounds of the human being, and who eventually takes the wound away or the wounded. O God of Love, please let me continue to love you when I wake up. The terrifying hour is close

> . . .!—Father of men, I am also your son and I will obey; Father, let me not alone, when the bell rings.]

The Leatherman describes what happens if the human being creates himself and loses his reference point, so that only in this dream-like, subconscious, situation can he refer again to the father God and appeal to him not to leave him alone when he wakes up and when he dies. The absolute isolation that Jean Paul Richter describes here is the isolation of the devil that has put himself above God. But only on this subconscious level can the Leatherman expresses his insight that autopoietic creation leads to total isolation from which there is no exit. Much as in the *Entstehen* essay, the spiritual origin of the world is not fully accessible to reason, but justified by the mere longing for it—which is expressed even by the very adversary of the divine, the devil figure.

Richter articulates here a problem which had occupied him since the 1790s. In his novel *Siebenkäs* (1796), published only two years after the publication of Erasmus Darwin's *Zoonomia*, Jean Paul Richter had already expressed the fear of being isolated in a fatherless world. In "Rede des todten Christus vom Weltgebäude herab, daß kein Gott sey" [Speech of the dead Christ from the top of the world], the following scene appears:

> Jetzo sank eine hohe edle Gestalt mit einem unvergänglichen Schmerz aus der Höhe auf den Altar hernieder, und alle Todten riefen: "Christus! ist kein Gott?" Er antwortete: "es ist keiner. . . . Ich stieg herab, so weit das Sein seine Schatten wirft, und schauete in den Abgrund und rief: Vater, wo bist du? aber ich hörte nur den ewigen Sturm, den niemand regiert. . . ." . . . Da kamen . . . gestorbenen Kinder . . . in den Tempel und warfen sich vor die hohe Gestalt am Altare und sagten: "Jesus! Haben wir keinen Vater?"—Und er antwortete mit strömenden Thränen: "wir sind alle Waisen, ich und ihr, wir sind ohne Vater."
>
> . . .
>
> so hob er [Christus] groß wie der höchste Endliche die Augen empor gegen das Nichts . . . und sagte: "starres, stummes Nichts! Kalte, ewige Nothwendigkeit! Wahnsinniger Zufall! . . . Wie ist jeder so allein in der weiten Leichengruft des All! Ich bin nur neben mir—O Vater! o Vater! wo ist deine unendliche Brust, daß ich an ihr ruhe?—Ach wenn jedes Ich sein eigner Vater und Schöpfer ist, warum kann es nicht auch sein eigner Würgeengel sein? . . ."[128]

> [And at this point a lofty, noble form, bearing the impress of eternal sorrow, came sinking down towards our group, and rested on the altar; whereupon all the dead cried out, "Christ! Is there no God?" He answered, "There is none. . . . I descended to where the very shadow cast by Being dies out and ends, and I gazed out into the gulf beyond, and cried, 'Father, where art Thou?' But answer came there none, save the eternal storm which rages on, controlled by none. . . ." . . . then the dead children . . . came into the temple, and fell down before the noble form . . . and cried, "Jesus, have we no Father?" He made answer, with streaming tears, "We are orphans all, both I and ye. We have no Father."
>
> . . . .
>
> He, sublime, loftiest of finite beings, raised his eyes towards the nothingness and boundless void, saying: "Oh dead, dumb, nothingness! Necessity endless and chill! . . . How every soul in this great corpse-trench of a universe is utterly alone? I am alone—none by me—O Father, Father! where is that boundless breast of thine, that I may rest upon it? Alas! if every soul be its own father and creator, why shall it not be its own destroying angel too? . . ."][129]

The optimism which has characterized Nikolaus Marggraf's endeavor to become his own father has as its obverse absolute desperation.[130] As in his comment on Darwin's idea of the "Fleischfädchen" in the essay on *Entstehen*, Jean Paul Richter motivates the necessity of the reference to the father with a psychological source: knowledge and fear of the absolute isolation in the moment of self-creation turns Jean Paul Richter against an evolutionary understanding of nature, in which matter is understood to develop without (spiritual) reference point. The idea of a human "superman," a "messianic couple," created via breeding is impossible for Jean Paul Richter, because it neglects a psychological necessity.

We can read the *Komet* as that satire of the notion of a superman that Jean Paul Richter had wanted to write according to his notes. As much as Jean Paul Richter believes in the power of nurture, he is also convinced that it cannot be transferred to the next generation, because it would eventually lead to the abandonment of the creator. Nurture can replace nature only as far as the individual's life goes, but it cannot change the finite structure of the world and the act of self-creation needs to be taken back in the moment of death. In Jean Paul Richter's world view, nature has not emerged in an evolutionary process. Rather, we have to assume that it is a stable entity, which serves as a reference point for a world that we can only access through nurture. He acknowledges that this reference point is psychologically motivated and leaves it open whether an empirical proof of this entity is possible.

Reading the *Komet* in light of socio-historical events has meant to see Nikolaus Marggraf's inability to obtain political power as a failure of bourgeois ambitions. Reading the novel in light of Richter's *Entstehen* essay and in light of his consideration of hybridity research suggests, however, that Nikolaus's inability to ascend the throne is a precondition for his fictitious existence. In light of evolutionary materialism, human autonomy is only possible as fiction and fiction is only possible as artificiality. Dennis Mahoney concludes that the double Nikolaus/Leatherman radically abandons "die Hoffnungen der Goethezeit auf die befreiende Macht der Kunst"[131] [the hope of the Goethe period for the liberating power of art]. Reading the *Komet* in light of evolutionary research suggests that art as artificial is indeed the space, and the only space, for freedom, precisely because it does not question God-created nature.

A TRAGELAPH OF THE FINEST SPECIES

In a letter to Schiller from June 10, 1795, Goethe described Richter's novel *Hesperus* as a "Tragelaph von der ersten Sorte" [a tragelaph of the first kind] and he repeated the reference in another letter to Schiller written only a week later.[132] By referring to the tragelaph, Goethe evoked a hybrid, deer-like, creature from Greek mythology. The word signaled that Richter had written a monstrous text which lacked the organic unity for which Goethe and Schiller were aiming at that moment. It is unlikely that Richter knew Goethe's letter to Schiller, but he certainly knew of Goethe's and Schiller's rejection of his literary projects and ambitions. The *Komet* can be understood as Richter's answer to Goethe's accusation of monstrous writing. If in the 1790s Goethe rejected Richter's *tragelaphic* literature, Richter articulated in the *Komet* a poetological program which endorses hybridity and stresses artificiality. Richter does not aim for realistic or even probable plots and characters, but stresses artificiality and surprising combinations of disparate elements.

The abandonment of reality which has been observed in Richter's characters has often been connected to his deep uneasiness vis-à-vis the bourgeois market economy, in which the maturation of the individual means his assimilation to economic structures. Richter's late work, according to Wulf Koepke, addresses "the problem of the artist and artistic creativity versus bourgeois capitalism." Koepke's stress on idealism as the "*conditio sine qua non* for creating a poetic world"[133] in Richter's work is well taken, but this idealism is thought of not in contradiction to capitalist society, but as modelled according to it. Much as the capital market is built on the model of speculation, so is Nikolaus Marggraf's life a speculation on a system for which there might be no realistic, natural, equivalent. Richter's work confirms a central thesis advanced by Richard T. Gray in *Money Matters* (2008), namely that "economic thought has a seldom recognized influence on the cultural regime" in eighteenth-century Germany. Refuting a long-standing dogma in German studies, Gray demonstrates that literature around 1800 does not develop in opposition to the economic paradigm, but is closely tied to it.[134] Referring to the economic theory of Goethe's brother-in-law Johann Georg Schlosser (1739–1799), Gray describes how in the last third of the eighteenth century a "'new' monetary economy" emerged in Germany "that knows no bounds and no limits, that is 'inexhaustible' . . . and grounded neither in nature nor in manual labor, but in the human power of imagination."[135] Richter was sensitive to the possibilities which this new monetary economy offers him for thinking through character formation in a context in which the model of organic autonomy becomes problematic in light of evolutionary thought. Nikolaus's creation of the diamond is not a transubstantiation of coal, but a speculation on such transubstantiation.

Once Richter recognized that the biological development does not provide a space any longer for thinking selfhood and aesthetics as autonomous spheres, he abandoned the organic model. If he referred to the biology of his time, he referred to those forms, which were in the understanding of the time precisely defined by their artificiality. Drawing on Koelreuter's research, Richter developed what one could call a *monstrous poetology*. Monstrous forms cannot (and even should not) reproduce in nature, as Dr. Katzenberger attests, but they are the forms which best describe art. They are closely linked to that idealism which Koepke has called in the quotation cited above the condition of Richter's work. It is, therefore, not astonishing that Dr. Katzenberger himself connects his monstrous specimen closely to idealism:

> Aber ich entsinne mich noch recht gut meiner Jugend, wo ich mehr *idealisierte* und weniger auf Erden als im Himmel wandelte, da weidete ich mich an geträumten, noch höhern Mißgeburten . . . ; da war es mir ein Leichtes, ganze ineinader hineingewachsenen Sessionen geboren und zu Kauf zu denken, die ich dann nach dem Ableben leicht in einem

> Spiritus-Glase bewahrte und bewegte nach Lust—oder einen Knaben mit einem angebornen vollständigen fleischernen Krönungshabit—oder einen tafelfähigen Edelmann mit zweiunddreißig Steißen besetzt—und doch sind das nicht ganz arkadische Träume.[136]
>
> [But I remember still quite well my youth, when I *idealized* more and when I walked less on earth than in the heavens, then I gazed on dreamt even higher monsters . . .; then, it was easy for me to imagine completely grown in (bodies) available for purchase, which I then, after death, kept easily in a glass with alcohol and moved as I pleased—or a boy with congenital complete coronation habit from flesh—or a table ready noble man equipped with thirty-two coccyx—and yet, these are not completely Arcadian dreams.]

In this passage, Dr. Katzenberger traces his fascination to his youth and admits to having dreamed of even more deformed monstrosities than the ones he possesses at the moment. The activity with which his imaginative production of monsters is connected is "idealisierte" [*idealizing*]. Drawing on Koelreuter's work and contemporary distinctions between nature and its artificial modification, it is now clear that Richter's idealism is understood precisely in opposition to any organic understanding which marked Goethe's endeavor in the 1790s when he called Richter's work a tragelaph.

We might also read the *Komet* as a text in which Richter negotiated questions of genre. In the *Vorschule zur Ästhetik* [preschool for aesthetics] (1812), Richter had distinguished an *Italian* and a *Dutch* type of the novel. While the Italian genre is defined by the classical and idealizing style and depicts its protagonist as hero, the Dutch genre is defined by the realistic depiction of everyday life and common people. A third genre, which stands between the Dutch and the Italian is defined as "German" and presents characters of the middle class. When the Dutch and the Italian school paint Nikolaus's portrait in the *Komet*, the Dutch artists focus on the realism of Nikolaus's pockmarks, while the Italian School depicts his idealist halo. The tripartite structure can also be mapped onto the Nikolaus's family: his mother, the beautiful Italian singer Mara, represents the Italian genre, Henoch Elias, the practical pharmacist, represents the Dutch School, while Nikolaus himself, the middle character with aristocratic ambitions, represents the German genre.

Establishing the *Komet* as genre of the middle ground illuminates Richter's *Genius* concept. In the epigraph to this chapter, Dr. Katzenberger offers an implicit definition of the genius in his analogy to the monstrous flowers produced by botanists, their exceptionality appears in their infertility. Three years after the publication of *Dr. Katzenberger*, 1812, Richter referred again to the hybridization experiments when defining the genius in the *Vorschule der Ästhetik*. He wrote:

> Im Genius stehen *alle* Kräfte auf einmal in Blüte; und die Phantasie ist darin nicht die Blume, sondern die Blumengöttin, welche die zusammenstäubenden Blumenkelche für neue Mischungen ordnet, gleichsam die Kraft voll Kräfte.[137]
>
> [In the genius all forces are blooming at the same time; and fantasy is in it not the flower, but the flower goddess, which orders the pollinating flower cups for new mixtures, so to say the force full of forces.]

The Genius is defined here as a figure which opens up a space so that new forms can be created. Fantasy arranges the genius' flowers in such a way that they can form new mixtures. Against the backdrop of Koelreuter's breeding experiments, we can now understand that the genius' flowers are precisely not of a totalizing unified origin, as an early *Geniekult* might have it, but of various kinds. The new forms that are created are thought of as hybrids, not as eternally recreating forms. We can read the Genius as a figure which opens up a space for cross-breeding. Nikolaus—like his nose, which became a place not only for tobacco cultivation but for entire tobacco planters—is then not only the product of a combination between *Poesie* and *Empirie*, but the place where new mixtures can be created. If we combine the reading of Nikolaus as the *German* genre with the reading of Nikolaus as Genius, we will realize once again that Richter comes to a definition of standard categories in the literature of the time that questions existing definitions of these categories. In Richter's reading, the Genius is not the autonomous entity that develops out of its natural internal origin, but a cross-bred figure that belongs to the middle ground of the bourgeois who speculates on the originality that still needs to be produced and that is only an effect of this speculation.

EPIGRAPH SOURCE

Jean Paul Richter, *Dr. Katzenbergers Badereise*, in *Jean Paul Richter Werke*, ed. Norbert Miller (München: Hanser, 1967), 4:128.

NOTES

1. James D. Watson and Francis H. C. Crick, "Molecular Structure of Nucleic Acids: A Structure of Deoxyribose Nucleic Acid," *Nature* 171 (April 25, 1953): 737–38.

2. David Dugan and Jeff Goldblum, *DNA: The Secret of Life. Pandora's Box*, produced and directed by David Glover (Princeton, NJ: Educational Broadcasting Corporation and Windfall Films: Films for the Humanities & Sciences, 2003), 4:45.

3. Ibid., 5:30. I am thanking Christian Weber for pointing me to this episode.

4. As Maximilian Bergengruen observes in his recent study *Schöne Seelen, groteske Körper: Jean Pauls ästhetische Dynamisierung der Anthropologie* (Hamburg: Meiner, 2003), for a long time it was the understanding that Jean Paul Richter's work can and should be read without reference to discourses outside the literary text (1). This view, of course, has changed dramatically in the last twenty-five years. Arguably the most innovative work in current Jean Paul Richter scholarship is devoted to the question of how to account for the many discourses which inform in particular Jean Paul Richter's excerpt booklets and how (or whether) to untangle the complex ways in which these discourses have found their way into his (literary) work. While Hans Esselborn's monograph *Das Universum der Bilder: Die Naturwissenschaft in den Schriften Jean Pauls* (Tübingen: M. Niemeyer, 1989), focused on the relationship between the "two cultures" literature and science, the latest studies problematize the very distinction between these two knowledge producing systems and shift the focus from content to structure within a theory of media (Annina Klappert, *Die Perspektiven von Link und Lücke: Sichtweisen auf Jean Pauls Texte und Hypertexte* [Bielefeld: Aisthesis, 2006]). Acknowledging that the discourses that inform Richter's literary texts do not enter these texts unmediated, I suggest in what follows that a detailed knowledge of these discourses can nevertheless contribute in crucial ways to a reading of Richter's essayistic and literary work.

5. The terms "bastard" and "hybrid" are used interchangeable around 1800 (cf. James L. Larson, *Interpreting Nature: The Science of Living Forms from Linnaeus to Kant* [Baltimore: Johns Hopkins University Press, 1994], 63).

6. Cf. Gottfried Reinhold Treviranus, *Biologie, oder Philosophie der lebenden Natur für Naturforscher und Aerzte* (Göttingen: Johann Friedrich Röwer, 1805), 3:477. In many ways, hybridization was understood as the opposite of incestuous reproduction. While incest brings together two too closely related individuals, hybridization unites two individuals who have too little in common. Both modes of reproduction were considered as producing infertile offspring. In *Biology*, Treviranus writes: "Einen andern, noch wichtigern Grund für die Existenz ursprünglicher Deformitäten geben die Bastarde der Thiere und Pflanzen. Diese sind den Mißgeburthen sehr nahe verwandt. Sie entstehen aus der Ungleichartigkeit des männlichen und weiblichen Saamens" [Another, even more important reason for the existence of initial deformations are the bastards of animals and plants. These are very closely related to monstrosities. They develop because of the dissimilarity of male and female seed] (Treviranus, *Biologie*, 3:447).

7. Staffan Müller-Wille and Vítezslav Orel, "From Linnaean Species to Mendelian Factors: Elements of Hybridism, 1751–1870," *Annals of Science* 64.2 (April 2007): 177.

8. Carl Linnaeus, quoted in ibid., 179.

9. Müller-Wille and Orel, "Linnaean Species," 175. The terms nature and nurture became popular in the early nineteenth century in particular in the context of the discussion around "Wolfskinder" and the question of the reproduction of the Genie (cf. Hans-Jörg Rheinberger and Staffan Müller-Wille, *Vererbung. Geschichte und Kultur eines biologischen*

Konzepts [Frankfurt am Main: Fischer- Taschenbuch-Verl., 2009], 99; and Nicolas Pethes, *Zöglinge der Natur: Der literarische Menschenversuch des 18. Jahrhunderts* [Göttingen: Wallstein, 2007], 62–99).

10. Linnaeus, quoted in Müller-Wille and Orel, "Linnaean Species," 180.

11. James L. Larson, *Interpreting Nature: The Science of Living Forms from Linnaeus to Kant* (Baltimore: Johns Hopkins University Press, 1994), 61.

12. Gregor Mendel, quoted in Müller-Wille and Orel, "Linnaean Species," 173.

13. Müller-Wille and Orel have argued that the novely of Mendel's approach was not so much that he stated the impossibility of drawing a "sharp line," but that he declared his work not focused on answering this question and instead focused on the transmission of individual traits. "Mendel's methodological decision to experiment with pea varities differing with respect to single character pairs only was not as dramatic as it seems in retrospect, but rather contiguous with the hybridists' tradition. Nevertheless, it decidedly shifted the level of theoretical analysis from the evolution of generic forms to the inheritance of individual characters, or rather, the inheritance of cellular dispositions for the formation of individual characters" (Müller-Wille and Orel, "Linnaean Species," 173–77).

14. Hans Kugler, "Kölreuter, Joseph Gottlieb," *Neue Deutsche Biographie* 12 (1979): 325–26.

15. So much so that in 1758/1759, the St. Petersburg Academy of Sciences held a prize competition on the question of whether or not there is plant sexuality. Even sixty years later, in his essay "Verstäubung, Verdunstung, Vertropfung" (Pollinating, Evaporating, Dropping) (1820), Goethe, while by and large accepting plant sexuality, still admitted (in erotically charged language) "gegen das Geschlechtssystem von Zeit zu Zeit erregten Zweifel" (from time to time raising [aroused/arousing] doubts regarding the sexual system of plants) (Johann Wolfgang Goethe, *Sämtliche Werke*, ed. Dieter Borchmeyer [Frankfurt a.M.: Deutscher Klassiker, 1987], series 1, 24:509).

16. Hermann Ziegenspeck, "Camerarius, Rudolph Jakob," *Neue Deutsche Biographie* 3 (1957): 107–8.

17. Müller-Wille and Orel, "Linnaean Species," 183.

18. Cf. Immanuel Kant, "Bestimmung des Begriffs einer Menschenrasse. Berliner Monatsschrift, Nov. 1785," in *Pädagogik*, ed. Wilhelm von Weischedel (Frankfurt am Main: Suhrkamp, 1993), 11:66.

19. Joseph Gottlieb Koelreuter, *D.J.G.K.'s Vorläufige Nachricht von einigen das Geschlecht der Pflanzen betreffenden Versuch und Beobachtungen* (Leipzig: Gleditsch, 1763), 58.

20. Dianne Morgan, *Kant Trouble: the Obscurities of the Enlightened* (London: Routledge, 2000), 137n22.

21. The degree to which Koelreuter understood his hybridization experiments (nevertheless) as creative acts is expressed in the following statement. "Weil ich schon lange von dem Geschlechte der Pflanzen überzeugt war, und an der Möglichkeit einer solchen widernatürlichen Erzeugung niemals gezweifelt hatte, so ließ ich mich auch durch nichts abhalten, Versuche hierüber anzustellen, *in der guten Hoffnung*, daß ich vielleicht einmal so glücklich seyn könnte, eine Bastartpflanze zuwegezubringen" [Since I was already for a long time convinced of the sexuality of plants, and I had never doubted the possibility of such unnatural reproduction, I let nothing stop me to conduct experiments on it, *expecting* that one day I would perhaps be so lucky to produce a bastard plant] (Koelreuter, *D.J.G.K.'s Vorläufige Nachricht* [1763], 38–39, italics mine). By using terminology used in

the context of child bearing ("in der guten Hoffnung"), Koelreuter appropriates female creativity for his activity.

22. Koelreuter, *D.J.G.K.'s Vorläufige Nachricht* (1761), 36–37.

23. Friedrich Schiller to Goethe (2 July 1796), quoted in Harmut Nonnenmacher, *Natur und Fatum: Inzest als Motiv und Thema in der französischen und deutschen Literatur des 18. Jahrhunderts* (Frankfurt am Main and New York: Peter Lang, 2002), 202–3. Liselotte Dieckmann translates "misconceptions of reason" (Johann Wolfgang Goethe and Friedrich Schiller, *Correspodence between Goethe and Schiller 1794–1805*, trans. Liselotte Dieckmann [New York: Peter Lang, 1994], 122).

24. Koelreuter, *D.J.G.K.'s Vorläufige Nachricht* (1761), 38.

25. Reinhart Koselleck, *Vergangene Zukunft. Zur Semantik geschichtlicher Zeit* (Frankfurt am Main: Suhrkamp, 1979).

26. Koelreuter, *D.J.G.K.'s Vorläufige Nachricht* (1761), 21–22, italics mine.

27. Ibid., 36.

28. Under the heading "Naturae Instinctus," the reader is referred to "Natur-Triebe," see Johann Heinrich Zedler, *Großes vollständiges Universal Lexicon aller Wissenschafften und Künste* (Leipzig and Halle: Zedler, 1749), 23:1225–26 and 1228.

29. It is noteworthy that the Zedler also offers a second notion of instinct by dinstinguishing between instincts before and after the fall. The entry states:

> Es müssen aber dies natürliche, von GOtt gegebene Lust-Triebe nicht mit andern, zwar auch natürlichen, aber nicht von GOTT gegebenen Trieben . . . verwirrte werden: man erkennet sie daran, daß jene einen GOtt gefälligen, auf alle Menschen zu allen Zeiten sich erstreckenden Zweck haben, diese aber entweder keinen GOttgefälligen . . .; oder nicht allen Menschen, zukommenden . . .; oder endlich nicht zu allen Zeiten sich ereignenden Zweck. . . . Darum nenneten die alten Welt-Weisen, gar nachdrücklich, die erste Art dieser Triebe, *prima naturalia*, die ersten natürlichen Triebe, um sie von den andern natürlichen, die sich erst nach Verderbniß und Fall des menschlichen Geschlechts eingefunden, zu entscheiden. (ibid., 1225)

> [However, these natural God given pleasure-drives should not be confused . . . with other drives which, though also natural, are not given by God: one can distinguish the two because the former have a purpose that pleases God and is true for all human beings in all times, the latter, however, either have no purpose that pleases God . . . or not all human beings common (purpose) . . . or they do not have a purpose that is present in all times. . . . That is why the ancient wise men, emphatically, called the first kind of these drives *prima naturalia*, the first natural drives, in order to distinguish them form the other natural (drives), which occured only after corruption and fall of the human race.]

Hence, we can observe in Zedler the attempt to distinguish between two kinds of nature, one original and one after the fall. Zedler's distinction demonstrates the theoretical difficulties which eighteenth-century thinkers encountered when confronting natural law philosophy with older, Christian notions of nature, including such concepts as instinct. Zedler solves the problem by suggesting two notions of nature, one good (conforming with natural law philosophy) and one distorted (conforming with Christian notions of sin). It is, therefore, not astonishing that the entry closes with the comment: "Sonsten ist noch eine Frage mitzunehmen: ob die Natur-Triebe den Grund des natürlichen Rechts abgeben können? davon aber schon gehandelt worden im Artickel: Natur-Rechts (Grund-Satz des)" (ibid., 1226) [Otherwise one more question is to be considered: whether the natural drives can be the basis of Natural Law? However, we have treated (this subject) already in the article on Natural Law (principle of)].

30. Lorraine Daston and Peter Gallison, *Objectivity* (New York: Zone Books, 2007), 30.

31. Koelreuter, *D.J.G.K.'s Vorläufige Nachricht* (1764), 3–4.

32. Erasmus. Darwin, *Zoonomia; or, The Laws of Organic Life* (London: Printed for J. Johnson, in St. Paul's Church-Yard, 1794–1796), 2:491.

33. Koelreuter, *D.J.G.K.'s Vorläufige Nachricht* (1764), 4–6. Koelreuter's reliance on alchemical vocabulary might seem metaphoric and anachronistic at first. However, as William R. Newman has demonstrated in *Promethean Ambitions: Alchemy and the Quest to Perfect Nature* (Chicago: University of Chicago Press, 2004), alchemical concepts played an important role in the conceptualization of breeding. While alchemic discussions around species transformation originated in the question of whether or not one species of metal can be transmuted into another, the Aristotelian species definition that it employed was not limited to species of metal and had as yet to be discovered impact on the conceptualization of species transformation in the context of the formulation of evolutionary theories. Within the context of alchemical thought, the scholastic philosopher Roger Bacon (ca. 1214–1292) had asked already whether or not grafted plants belong to a new species or not and eventually rejected the claim. In terms of animals, twelth-century Arabic philosopher Averroes (1126–1198) had debated whether a mouse produced by artificial means would be the same as a natural, sexually, generated mouse. Applying the answer to both the alchemical production of gold and to the artificially produced life form, Averroes came to the conclusion that "[n]o matter how closely the artificial product matched the properties of its natural exemplar, the two would be separated by an unbridgeable gulf" (Newman, *Promethean Ambitions*, 5–6). Averroes's discussion of the problem can be found here: Averroes. *Aristotelis de generatione animalium. Aristoteles opera cum Averrois comentariis.* (Venice: Junctae, 1562–1574; repr. Frankfurt: Minerva, 1962), 6:44v (in Newman, *Promethean Ambitions*, 6n15).

34. Müller-Wille and Orel, "Linnaean Species," 183.

35. Newman, *Promethean Ambitions*, 36–37. See also Manfred Ullmann, *Die Natur- und Geheimwissenschaften im Islam* (Leiden: Brill, 1972), in Newman, 37.

36. Newmanm, *Promethean Ambitions*, 38.

37. Avicenna, in Newmanm, *Promethean Ambitions*, 37.

38. Newmanm, *Promethean Ambitions*, 292.

39. Ibid.; Charles Darwin. *On the Origin of Species*, ed. Ernst Mayr (Cambridge, MA: Harvard University Press, 1964), 98, in ibid., 292.

40. Newmanm, *Promethean Ambitions*, 8.

41. Ibid., 6.

42. Cf. Götz Müller, *Jean Pauls Exzerpte* (Würzburg: Königshausen & Neumann, 1988).

43. Helmut Pfotenhauer, "Jean Pauls literarische Biologie. Zur Verschriftlichung von Zeugung und Tod, mit besonderer Berücksichtigung des Siebenkäs," in *Prägnanter Moment. Studien zur deutschen Literatur der Aufklärung und Klassik: Festschrift für Hans-Jürgen Schings*, ed. Peter-André Alt, Alexander Kosenina, Hartmut Reinhardt, and Wolfgang Riedel (Würzburg: Königshausen & Neumann, 2002), 461.

44. G. Müller, *Jean Pauls Exzerpte.*

45. Esselborn, *Das Universum der Bilder*, 3.

46. Götz Müller, *Jean Pauls Ästhetik und Naturphilosophie* (Tübingen: Niemeyer, 1983).

47. Gottfried Reinhold Treviranus, *Biologie, oder Philosophie der lebenden Natur für Naturforscher und Aerzte* (Göttingen: Johann Friedrich Röwer, 1802), 1:v.

48. Ibid.

49. Ibid., 1:vii.

50. Jean Paul Richter, *Frage über das Entstehen der ersten Pflanzen, Thiere und Menschen*, in *Jean Paul Richtler Sämtliche Werke. Historisch-kritische Ausgabe. Erste Abtheilung. Sechzehnter Band*, ed. Preußische Akademie der Wissenschaften (Weimar: Böhlau Nachfolger), series 1, 16:51.

51. Ibid., 69–70, italics mine.

52. Of course, Richter overlooks the fact that Erasmus Darwin also assumes a spiritual power in the beginning. For Darwin, the "living filament" is created by a father-like God who empowers it with all the potential that is later unfolded (Erasmus Darwin, *Zoonomia*, 2:513).

53. Richter, *Entstehen*, 53.

54. Cf. Timothy F. DeJager, *G. R. Treviranus (1776–1837) and the Biology of a World in Transition* (PhD diss., University of Toronto, 1991). Treviranus had expressed a similar view as Darwin:

> [E]rwägt man, daß die ganze lebende Natur ebenfalls bey ihrer Bildung stufenweise von Einfachern zum Zusammengesetztern fortgeschritten ist, so ist es klar, daß alles Leben nur von den niedern Stufen der Organisation zu den höhern gelangen kann. Diese müssen also durch jene bedingt seyn. Aber wie können sie dies anders seyn, als dadurch, daß der einfachere Organismus sich von Generation zu Generation immer mehr ausbildet? Wir glauben daher, daß die . . . Zoophyten der Vorwelt die Urformen sind aus welchen alle Organismen der höhern Classen durch allmählige Entwicklung entstanden sind (Treviranus, *Biologie* [1805], 3:225).
>
> [(I)f one considers that the entire animated nature progressed in its formation also gradually from simpler to more complex forms, then it is clear that all life can develop only from lower levels of organization to higher ones. Hence, the latter need to be conditioned by the former. But how can they be like that if not in such a way that the simpler organism develops more and more from generation to generation? We thus believe that the . . . zoophytes of prehistoric time are the basic forms from which all organisms of the higher classes have emerged by means of gradual development.]

55. Richter, *Entstehen*, 63. Treviranus had written "Selbst der Mensch wird vielleicht einst vergehen und verwandelt werden. Aber regelmäßig war von jeher der Gang der Natur bey allen Veränderungen; regelmäßig wird er bleiben bis ans Ende der Zeiten, und nicht ohne Grund läßt sich vermuthen, daß die Natur noch nicht die höchste Stufe der Organisation in dem Menschen erreicht hat, sondern in ihrer Ausbildung noch weiter fortschreiten und noch erhabenere Wesen, noch edlere Gestalten einst hervorbringen wird" (Treviranus 1805, 3:226).

56. Richter, *Entstehen*, 63.

57. Berend, "Anmerkungen," in *Jean Pauls Sämtliche Werke* (Weimar: Böhlaus Nachfolger, 1938), series 1, 16:480n63.

58. Richter, *Entstehen*, 69, italics in original.

59. Johann Friedrich Blumenbach, Über den Bildungstrieb und das Zeugungsgeschäfte (Göttingen: Dieterich 1781; repr. 1791) 74–75.

60. Johann Friedrich Blumenbach, *An Essay on Generation* (London: T. Caddell, Strand, 1792) 56.

61. Richter, *Entstehen*, 64.

62. Ibid., 64.

63. Jean Paul Richter, *Dr. Katzenbergers Badereise*. In *Jean Paul Richtler Werke*, edited Norbert Miller (München: Hanser, 1967), 4:128–29.

64. Richter, *Entstehen*, 66–68, italics mine.

65. Maximilian Rankl explains Richter's stating of a concepts such an eternal God and soul as a reaction to the temporalization and decentralization after Newton (*Jean Paul und die Naturwissenschaft* [Frankfurt am Main: Peter Lang, 1987], 154).

66. Richter, *Entstehen*, 59.

67. Cf. Lily C. Kay, *Who Wrote the Book of Life? A History of the Genetic Code* (Stanford, CA: Stanford University Press, 2000).

68. Linnaeus, Bonnet, and Buffon are, according to Esselborn, the biologists which are most frequently quoted in Jean Paul Richter's work. Jean Paul Richter, however, is not necessarily uncritical of Bonnet. See for example the parodistic comment by Leibgeber: "ich wandle hier mit einem Säetuch umhangen, worin die Sämerei aller Völker liegt, auf und ab und trage das Repetorium . . . des ganzen Menschengeschlechts. . . . Denn *Bonnet*, der im Magen mit steckt, wird, wenn er herausgehoben wird, sich niedersetzen und es auf seinem Schreibpult dartun, daß alles ineinanderstecke, eine Parenthese und Schachtel in der andern" [I am strolling here up and down with a seed cloth hanging around me, which contains the seeds of all peoples, and I am carrying the repertory . . . of the entire human race. . . . Because Bonnet, who is also hidden in the stomach, will, once he is taken out of it, sit down and present it at his desk that everything is encapsulated, one parenthesis and encapsulation in the other] (in Esselborn, *Das Universum der Bilder*, 103). Jean Paul Richter's critique of Bonnet, however, concentrates not on preformationism, but on Bonnet's conceptualization of the soul. Cf. Bergengruen, *Schöne Seelen*, 35. See also Lorin Anderson, *Charles Bonnet and the Order of the Known* (Boston: D. Reidel, 1982).

69. For a comprehensive account of the chain of beings see Arthur O. Lovejoy, *The Great Chain of Being: A Study of the History of an Idea* (Cambridge, MA: Harvard University Press, 1961).

70. Petra Feuerstein-Herz, *Die große Kette der Wesen: Ordnungen in der Naturgeschichte der Frühen Neuzeit* (Wiesbaden: Harrassowitz, 2007), 24–25.

71. Richter, *Entstehen*, 72.

72. Richter, quoted in Ralf Berhorst, *Anamorphosen der Zeit: Jean Pauls Romanästhetik und Geschichtsphilosophie* (Tübingen: Niemeyer, 2002), 32.

73. In the *Sudelbuch*, Lichtenberg makes a similar point about the anthropomorphism of knowledge when referring to creation: "Wenn einmal ein höheres Wesen sagte wie die Welt entstanden sei, so möchte ich wohl wissen ob wir im Stande wären es zu verstehen. Ich glaube nicht. Von Entstehung würde schwerlich etwas vorkommen, denn das ist bloßer

Anthropomorphismus." (Lichtenberg, quoted in Esselborn, *Das Universum der Bilder*, 311n92. [If a higher being were to tell us how the world came into existence I would like to know whether we would be in a position to understand him. I believe not. There would be hardly any mention of the act of coming into existence, for that is mere anthropomorphism (Lichtenberg, *The Waste Books*, trans. and intro. R. J. Hollingdale [New York: New York Review Books, 1990], 184.)]

74. In the end it is only God who can know: "Geht nicht durch die ganze Schöpfung eine unermessliche Kette—wo die ersten und hintersten Glieder einander zum Erstau[n] en unähnlich sehen, deren Ähnlichkeit und Verbindung aber nur der Vater aller Geister entdekken kan?" [is there not an infinite chain through the entire creation—where the first and the last members look astonishingly dissimilar, of which, however, only the father of all spirits can discover their similiarity and connection?] (quoted in Esselborn, *Das Universum der Bilder*, 273).

75. In a letter to Wernlein, Richter says "es giebt für iedes Subjekt keine andre Wahrheit als die gefühlte" [for each subject there is no other truth than the felt one] (quoted in Bergengruen, *Schöne Seelen*, 9).

76. Götz Müller, "Die Einbildungskraft im Wechsel der Diskurse. Annotationen zu Adam Bernd, Karl Philipp Moritz und Jean Paul," in *Jean Paul im Kontext*, ed. Wolfgang Riedel (Würzburg: Königshausen & Neumann, 1996), 162.

77. Berhorst also refers to Herders "Abhandlung über den Ursprung der Sprache" (1772) [On the Origin of Language], and a Jacobi as source for Richter's instinct notion (Berhorst, *Anamorphosen der Zeit*, 46).

78. Richter, quoted in Berhorst, *Anamorphosen der Zeit*, 46. Similar, Richter says "Unser Irthum daß wir uns unser ganz bewust zu sein glauben, da doch das, wodurch das Bewustsein entsteht, nicht in diesem sein kan" [our misapprehension to believe to be fully conscious of ourselves, while that which conditions consciousness cannot be in it] (quoted in Götz Müller, "Die Einbildungskraft," 161).

79. Dorothee Sölle, *Realisation: Studien zum Verhältnis von Theologie und Dichtung nach der Aufklärung* (Darmstadt: Luchterhand, 1973), 208.

80. Immanuel Kant, *Kritik der Urteilskraft*, ed. Wilhelm Weischedel (Frankfurt: Suhrkamp, 1974), 10:347, italics mine.

81. Georg Christoph Lichtenberg, *Schriften und Briefe*, ed. Wolfgang Promies (München: C. Hanser, 1968), 1:710.

82. Lichtenberg, *Waste Books*, 140–41. Hollingdale seems to omit "Ähnlichkeiten bis auf den Reim hinaus" (similarities down to the rhyme).

83. Richter, *Entstehen*, 68.

84. Koelreuter, *D.J.G.K.'s Vorläufige Nachricht* (1763), 58–59.

85. Daniel Purdy, quoted in Elliott Schreiber, *The Topography of Modernity: Karl Philipp Moritz and the Space of Autonomy* (Ithaca, NY: Cornell University Press, 2012), 17.

86. Martha Woodmansee, *The Author, Art, and the Market: Rereading the History of Aesthetics* (New York: Columbia University Press, 1994), 27; Schreiber, *Topography of Modernity*, 17.

87. Schreiber, *Topography of Modernity*, 156.

88. Karl Philipp Moritz, *Schriften zur Ästhetik und Poetik*, ed. Hans Joachim Schrimpf (Tübingen: Niemeyer, 1962), 122.

89. Karl Philipp Moritz, "On the Artistic Imitation of the Beautiful," In *Classic and Romantic German Aesthetics*, ed. J. M. Bernstein (Cambridge: Cambridge University Press), 142.

90. Manfred Engel, *Der Roman der Goethezeit. Anfänge in Klassik und Frühromantik: Transzendentale Geschichten* (Stuttgart: J. B. Metzler, 1993), 1:313.

91. Helmut Müller-Sievers, *Self-Generation: Biology, Philosophy and Literature around 1800* (Stanford: Stanford University Press, 1997), 4.

92. As Eduard Berend has noted, the writing of the *Komet* overlapped with the writing on the *Entstehen* essay (cf. Berend, "Einleitung," in *Jean Pauls Sämtliche Werke*, Der Komet [Weimar: Böhlaus Nachfolger, 1937], 1, 15:xviii).

93. Cf. Robert J. Richards, *The Romantic Conception of Life: Science and Philosophy in the Age of Goethe* (Chicago: University of Chicago Press, 2002), 515–18.

94. Karl Christian Gottlieb Sturm, *Ueber Raçen, Kreuzungen und Veredlung der landwirthschaftlichen Hausthiere* (Eberfeld: Büschler, 1825), 30.

95. Uwe Schweikert, "Jean Paul und die Restauration," *Jahrbuch der Jean Paul Gesellschaft* 10 (1975): 141–49.

96. Dennis F. Mahoney, *Der Roman der Goethezeit, 1774–1829* (Stuttgart: Metzler, 1988), 150.

97. Berend, "Einleitung," 1, 15:xi.

98. Hans Heinrich Borcherdt, *Der Roman der Goethezeit* (Urach: Port, 1949), 470.

99. Jean Paul Richter, *Komet*, in *Jean Paul Richtler Werke*, ed. Norbert Miller (München: Hanser, 1967) 4:576.

100. Ibid., 580.

101. In fact, the pocks were so hotly debated in the eighteenth century precisely because they lost their status as a determining fate. New vaccination methods made it possible to prevent the full outbreak of the disease. This was so much so the case that Justus Möser (1720–1794) in his *Patriotische Phantasien* chose the voice of a "young matron" through which we learn of the demographic challenge posed by pock vaccination which the matron expresses with Malthusian logic: "[D]ie jetzige [Zeit] . . . arbeitet beständig an Stamm und Namen, und doch soll jeder Stamm von unendlichen Sprößlingen erschöpft werden. . . . Sie klagt, daß ihr die Kinder täglich mehr kosten, und tadelt gleichwohl ihre Vorfahren, welche in glücklichen Zeiten die Hälfte davon an den Blattern sterben ließ" [(O)ur current (time) . . . works continually on stem and name, and nevertheless one says that every stem is exhausted by infinite offspring. . . . It (our time) complaints that the children cost every day more and nevertheless it rebukes its ancestors because (even) in happy times they let half of them die from the pocks] (Möser, *Sämtliche Werke*, Historisch-kritische Ausgabe in 14 Bänden, ed. Ludwig Schirmeyer and Werner Kohlschmidt. [Hamburg: Gerhard Stalling, 1943–1954] 7:59–61). If we follow Möser's matron, pockmarks and hereditary traits have to be seen as pointing in opposite directions. While the pocks tend to destroy genealogy by killing its carrier, traits are a sign that transmission along the genealoical line takes place. See Schreiber/Mathys. For the cultural significance of vaccination see also Johannes Türk, *Die Immunität der Literatur* (Frankfurt am Main: M. S. Fischer, 2011).

102. Richter, *Komet*, 4:577–78.

103. Eduar Berend connects the protagonist's search for the father among aristocrats to an anecdote which Goethe related in *Dichtung and Wahrheit*. Goethe reported that as a child he found a certain joy about the doubt regarding the identity of his biological grandfather and that he searched among contemporary princes for signs of facial similarities to establish whether one of them could be considered his kin. The Swedish writer Atterbom reports that Richter had referred to this anecdote in a conversation with him (Berend, "Einleitung" 1, 15:xxii).

104. Richter, *Komet*, 4:655.

105. Ibid., 4:583.

106. Koelreuter, *D.J.G.K.'s Vorläufige Nachricht* (1761), 38; and 36–37.

107. Richter, *Komet*, 4:658.

108. For a discussion of money as metaphor for the circulation of signs in Jean Paul Richter's work see Caroline Pross, *Falschnamenmünzer: Zur Figuration von Autorschaft und Textualität im Bildfeld der Ökonomie bei Jean Paul* (Frankfurt am Main: Peter Lang, 1997).

109. Richter, *Komet*, 4:810.

110. Ibid., 4:779.

111. Ibid., 4:743

112. Ibid., 4:720.

113. Similarly, Marggraf calls the money that he produces in one of his many alchemic attempts "*Prinz*metall." The narrator comments in a footnote: "Prinzmetall wird bekanntlich aus Kupfer und Zink (nebst wenigem Zinn) verfertigt. Dabei überseh' ich indes nicht, daß er [Marggraf] versteckt genug die gewonnenen Diamanten für Metall zu seinem Prinzwerden erklärt" [as is well-known princmetal is produced by combining copper and zinc (together with a little tin). I am not overlooking that he (Marggraf) declares, hidden enough, the obtained diamonds for metal for his inauguration as prince] (ibid., 4:793).

114. Further confirmation of the close connection between diamond production and French Revolution can be found in the following quote. Marggraf says to Stößer: "Jetzo sind vollends die Zeiten, wo mancher Thron, weil alles unten um ihn herum rebelliert, für Geld zu haben ist, und ich kann Königen, die ihren abstehen, dafür vielleicht etwas bieten, wenn es dort im faulen Heinze [Ofen der Diamantproduktion] zu etwas kommt" [Now we have the times when a few thrones—because everything is in rebellion below them—can be bought and I can perhaps offer kings, who want to relinquish theirs, something for it, if something is produced in the oven] (ibid., 4:748). The following quote alludes to a similar situation. Marggraf says: "Wie würd' es mit allen gekrönten Häuptern beiderlei Geschlechts aussehen, wenn ich ihre Kronjuwelen so gemein machte wie Haarnadeln?" [How would it look like for all crowned heads of both sexes, if I made their jewels as common as hairpins?] (ibid., 794) And a last quote to confirm our claim about the close connection between diamond production and political discourse: "Kurz er [Marggraf] wollte wirklich einen *Regenten* machen. Nun ist freilich ein Regent leichter in einem Fürstenhause als in einem faulen Heinze [Ofen] oder in einer voltaischen Säule zu machen; aber es ist ja hier offenbar die Rede bloß von jenem großen Diamant, 'Regent' oder auch 'Pitt' genannt, welchen zuletzt Pitts Feind und Opfer, Bonaparte, besessen haben soll" [In short, he (Marggraf) really wanted to produce a regent. Of course a regent is easier to produce in a royal house than in an oven or in a voltaic column; however, we seem to speak here only of that big diamond, called "Regent" or also "Pitt," of which one says that it was owned in the end by Pitt's enemy and victim, Bonaparte] (ibid., 4:811). Bergengruen alludes to the fact that for Johann Georg Büsch, the most famous economic thinker of Jean Paul Richter's times, circulation of money is "Triebfeder" behind the wealth of the bourgeois society. Johann Georg Büsch. *Abhandlung von dem Geldumlauf in anhaltender Rücksicht auf die Staatswirtschaft und Handlung*. 2 Vols. Hamburg, Kiel 1780. 61 (in Bergengruen, *Schöne Seelen*, 62).

115. Richter, *Komet*, 4:812.

116. The close relationship between chemical and biological production is further stressed in the following comment by the narrator: "Fleiß und Mühe . . . hab' ich daher

mehr, als ich sagen will, darauf verwandt, um Marggrafs kleinstes Wörtchen aus seiner Tisch-Rede über präexistierende Keime, Samentierchen, Muttertrompeten, Geburtstühle und Geburtzangen des glänzendsten Sohns des Jahrhunderts—so nennt' ich mit Recht den Diamant . . . redlich darzureichen" [I have thus put more effort into it than I want to say in order to present Marggraf's smallest word from his table speech about preexisting germs, amoeba, fallopian tubes, birthing chairs and forceps of the century's most brilliant son—that is how I call the diamond] (ibid., 4:790–91). At the same time that Nikolaus creates himself as his father, he also changes his understanding of kinship with this father. Nikolaus addresses his followers with the following words: "Liebe Getreue! Wir tun euch hiemit zu wissen, daß Wir die fürstliche Würde, die Uns Gott längst durch Unsere Geburt verliehen, fortan öffentlich durch zweckdienliche Mittel behaupten können und wollen, wohlwissend, daß es der Vorsehung . . . gefallen, Uns im dürftigen . . . Stand . . . erziehen zu lassen, um Uns . . . von aller Überhebung einer höhern Geburt—über welche der Fürst so gar leicht die Verwandtschaft mit andern, gleichfalls wie er von demselben Adam abstammenden Menschen vergißt—" [Dear followers! We let you know herewith that from now on, with help of convenient means, we are able to and want to publicly claim the aristocratic dignity which God has given to us a long time ago by birth, in full knowledge that destiny . . . liked to educate us among the lower . . . ranks . . . so that we don't (become arrogant) in light of higher birth—because of which the prince so easily forgets the kinshp with others who also originate from Adam—] (ibid.,4:815). The quote suggests that Nikolaus reinterprets his relationship to the claimed father as a relationship that connects him also with all other human beings.

117. Ibid., 4:938–40.

118. Although the Italian and the Dutch School have an opposing approach to their object—the first idealizing, the second realistic representation—Nikolaus Marggraf manages to praise both of them for their relationship to the original. First, he addresses the Dutch School: "Er [Nikolaus] brachte vor, wie sehr gerade ihre Schule [die Niederländische] den Kenner befriedige, der sich oder jemand anders malen lasse, weil er von ihnen doch eigentlich kein Scheinbild seiner selbst erhalte, sondern ein wahres, nichts Hineingepinseltes, nichts Vertuschtes, sondern gerade nur das, was er selber sei.—Und eben dieses Selbst sei es ja, was der Liebende im fremden Bildnis allein aufsuche.—Niemand werde sich einen schönern Vater wählen, als sein wirklicher sei, und ebenso geh' es mit dessen Bildnis" [He (Nikolaus) expressed how much precisely their School (the Dutch) satisfied the expert who let himself or somebody else paint because from them he would not receive an illusionary image of himself, but a true one, not fantasy, nothing hidden, but only precisely that what he himself is.—And precisely that self (he said) would be the only thing that the loving person would search in someone else's image.—Nobody would choose a more beautiful father than the real one is, and the same is true for his image] (ibid., 4:939). Then, he addresses the Italian School in such a way that his praise resembles that of the Dutch School: "Mit Vergnügen konnte man das fürstliche Zartgefühl bemerken, daß Nikolaus die welsche Schule ganz mit demselben leutseligen Anteil wie die Schule des vorigen Tages behandelte. . . . So schickte er auch, ehe er und alle sich setzten, wie tags vorher einige kurze Anreden voraus und tat dar: Kunst als solche veredle stets; sie sei kein bloßes Silhouettenbrett des Gesichts oder eine englische Kopiermaschine der Gestalten, sondern eine selber gebärende Madonna" [With pleasure one could note the noble gentleness of the fact that Nikolaus treated the Italian School entirely with the same affable sympathy as the School of the previous day. . . .

Hence, like on the previous day, he gave a few short speeches before he and everybody else sat down and said: art as such ennobles always; she is not mere silhouette of the face or an English copy machine of characters, but a madonna who gives birth herself] (ibid., 4:953).

119. Ibid., 4:985.

120. Ibid., 4:911.

121. Ibid., 4:965.

122. Ibid., 4:974.

123. Richter, *Entstehen*, 63.

124. Richter, *Komet*, 4:999–1000.

125. Albrecht Koschorke describes how, during the eighteenth century, the supposed location of the human soul moved upwards to the head (Koschorke, *Körperströme und Schriftverkehr: Mediologie des 18. Jahrhunderts* [München: Fink, 1999], 130–34).

126. The question of whether humans and apes were to be considered as related was greatly debated in the eighteenth century. Londa Schiebinger comments: "Anthropocentricism continued to dominate natural history long after geocentrism had been displaced in astronomy. Copernicus dethroned humans from the center of the universe, banishing them to a small planet circling the sun. Humankind nevertheless remained the chief of creation . . . Linnaeus captured these presumptions in the term *Primates*—literally, 'of the first rank'—a term coined in 1758. Though human might be primates, they were no longer uniquely situated just below angels: apes, monkeys, lemurs, and bats were also considered primates." "Comparative anatomists were struck by similarities in form and figure between apes and humans. Edward Tyson, in his pioneering *Orang-Outang, sive Homo Sylvestris; or, The Anatomy of a Pygmie Compared with that of a Monkey, an Ape, and a Man* (1699), documented forty-eight ways in which his 'pygmie' resembled a human, alogn with thirty-four ways in which it differed. The French naturalist Buffon characteristically remarked in the eighteenth century that 'if figure alone is regarded, we might consider this animal [generic orang] as the first among apes, or the last among men.' Montaigne, by contrast, rejected the human-animal dualism and suggested that there was more difference between any two given men than between any given man and a beast. Nonetheless the great apes held a very special place in the European imagination. They seemed to confirm the notion of hierarchy and continuity in nature. Humans—part brut, part angel—were thought to link the mortal world to the divine." "Carl Linnaeus jolted the scholarly world when, in 1735, he placed humans in the same order with monkeys and sloths. . . . Concerning his radical reordering of nature, Linnaeus remarked: 'I know full well what great difference exists between man and beast when viewed from a moral point of view: man is the only creature with a rational and immortal soul . . . If viewed, however, from the point of view of natural history and considering only the body, I can discover scarcely any mark by which man can be distinguished from the apes . . . Neither in the face nor in the feet, nor in the upright gait, nor in any other aspect of his external structure does man differ from the apes'" (Schiebinger, *Nature's Body: Gender in the Making of Modern Science* [Boston: Beacon Press, 1993; repr. New Brunswick, NJ: Rutgers University Press, 2004], 79–80). Linnaeus's remark suggests that the invention of the nature-nurture split has to be seen in connection with the discussion on the man-ape divide. If the body, the natural equipment, of man was comparable to the ape, then man's moral superiority could not be based on his bodily equipment, rather it had to be placed in an entity independent of nature: nurture.

127. Richter, *Komet*, 4:1004.

128. Jean Paul Richter, "Rede des todten Christus vom Weltgebäude herab, daß kein Gott sey" (*Siebenkäs*), in *Jean Paul Richtler Sämtliche Werke. Historisch-kritische Ausgabe. Erste Abtheilung. Sechster Band*, ed. Kurt Schreinert (Weimar: Böhlau Nachfolger, 1928; repr. 1977), 250–51.

129. Jean Paul Richter, *Flower, Fruit, and Thron Pieces; Or, the Wedded Life, Death, and Marriage of Firmian Stanislaus Siebenkaes*, trans. Alexander Ewing (London: George Bell and Sons, 1897), 262–64.

130. Helmut Pfotenhauer has alluded to the fact that *Rede des toten Christus* has a predecessor in *Des todten Shakespear's [sic] Klage unter todten[sic] Zuhörern in der Kirche, daß kein Gott sei* (1790). He points out that it is significant that it is the author, the creator of imaginative worlds, who articulates the experience of isolation (Pfotenhauer, "Jean Pauls literarische Biologie. Zur Verschriftlichung von Zeugung und Tod, mit besonderer Berücksichtigung des Siebenkäs," in *Prägnanter Moment. Studien zur deutschen Literatur der Aufklärung und Klassik: Festschrift für Hans-Jürgen Schings*, ed. Peter-André Alt, Alexander Kosenina, Hartmut Reinhardt, and Wolfgang Riedel [Würzburg: Königshausen & Neumann, 2002], 474).

131. Mahoney, *Der Roman der Goethezeit*, 153.

132. Friedrich Schiller and Johann Wolfgang von Goethe, *Der Briefwechsel Schiller and Goethe*, ed. Emil Staiger (Frankfurt am Main: Insel, 1977), 110.

133. Wulf Koepke, "Bildung and the Transformation of Society: Jean Paul's *Titan* and *Flegeljahre*," in *Reflection and Action: Essays on the Bildungsroman*, ed. James N. Hardin (Columbia: University of South Carolina Press, 1991), 230; and 248.

134. Richard T. Gray, *Money Matters: Economics and the German Cultural Imagination, 1770–1850* (Seattle: University of Washington Press, 2008), 4. On the close relationship between money and selfhood see Fritz Breithaupt, *Der Ich-Effekt des Geldes* (Frankfurt a. M.: Fischer, 2008).

135. Gray, *Money Matters*, 149.

136. Richter, *Dr. Katzenberger*, 199, italics mine.

137. Jean Paul Richter, *Vorschule der Ästhetik*. In *Jean Paul Richtler Werke.*, ed. Norbert Miller (München: Hanser, 1963), 5:56, italics in original.

CHAPTER FOUR

Habe das Blatt zwanzig mal mit süßen liebevollen Blicken der Vaterfreude angekuckt. —*E. T. A. Hoffmann, October 26, 1803*

[*I have looked twenty times at the page with sweet loving glances of fatherjoys.*]

Oneiros der Traumgott hat mir einen Roman inspiriert, der in lichten Farben hervorbricht. . . . Das Büchlein heißt: Die Elixiere des Teufels. —*E. T. A. Hoffmann, March 24, 1814*

[*Oneiros, the god of dreams, inspired me with a novel that burst forth in luminous colors. . . . The little book is called: . . . The Elixirs of the Devil.*]

From Blood to Trauma

PATER SEMPER INCERTUS EST?

WHEN TAKING the Berlin subway in the beginning of this century, the rider could enjoy not only a variety of colorful advertisements promoting the use of condoms, but also a number of more subdued posters promising certainty regarding fatherhood by means of genetic testing. The availability of the so called "Vaterschaftstest" (paternity test), made possible by a simple probe from the child's and father's skin cells, promised to put an end to the age-old *pater semper incertus est.* Not an insignificant deed, considering that the *Ärztezeitung* (Journal for Physicians) has estimated that 5–10 percent of infants in Germany (an equivalent of thirty-five thousand to seventy thousand births) are born in relationships where the male partner is not the child's biological father.[1] But the new tests also threatened the naturalness with which in previous centuries fathers had accepted children born in marriage as their own and with which children accepted the mother's husband as father.

In the years after paternity tests became available and affordable for the average (potential) father, they were hotly debated in the public arena and in legal circles. One of the most pressing legal questions was whether such tests, when conducted in secret, could violate the child's personal rights. In 2005, the Federal Supreme Court (Bundesgerichtshof) decided in the affirmative and since February 2010 the tests are prohibited in Germany unless they are done with the knowledge and agreement of both parents and the child (respectively the child's legal guardian). In the decade leading up to these decisions, the effects of the tests were discussed in the press in touching personal stories.

In April 2002, the German weekly *Die Zeit* ran a story on a woman in her thirties, Stefanie Schmied, who had doubted since childhood whether her mother's (former) husband was the biological father.[2] The paternity test, which Schmied commissioned in agreement with her estranged parents, brought certainty that this was not the case. The new birth certificate that Schmied received now lists the father as "unknown." Schmied has finally obtained the truth for which she had hoped so long. However, as the article's author Lotta Wieden observes, even with this truth in hand, and after fifteen years of psychiatric treatment, Stefanie Schmied seems to have difficulties in launching a new beginning. Years shaped by anorectic behavior and self-mutilation, including a suicide attempt lie behind her. Wieden states laconically "Warum Stefanie schmied zu einem psychiatrischen Fall wurde, einem mit der Diagnose schwere Persönlichkeitsstörung, kann heute niemand genau erklären" [The reasons for which Stefanie Schmied became a case for psychiatry—with the diagnosis of severe personality disorder—nobody can really explain today]. Referring to work done by Wolfgang Krauss, a specialist in identity formation, Wieden can only witness that the wish for certainty about the biological origin has increased to the same degree to which other concepts of identity formation, such as profession and social relations, have lost their significance.

The debate in Germany on the paternity test has highlighted a number of questions: How does biological certainty regarding the origin matter? How does the knowledge about biological kinship shape the parent-child relationship? And what does it contribute to identity formation? While the urgency with which biology imposes itself might be new, this chapter explores how questions surrounding biology and selfhood have been anticipated by writers around 1800 who were confronted with a new significance of biological paternity in the context of heredity studies.

My interest in this chapter is with E. T. A. Hoffmann's conceptualization of the father origin in his novel *Elixiere des Teufels* (Devil's Elixirs) (1815–1816) not as a biological certainty but as a heredity and identity that we imagine. Taking as a starting point the protagonist's mental problems, the first section of this chapter looks into accounts of insanity around 1800. The question pursued here is how contemporary studies on heredity shaped the perception of insanity in Hoffmann's literary reflection. More precisely, my question is how literature engaged with contemporary convictions of a biological basis of insanity? Turning, in a second step, to a close reading of the *Elixiere* the chapter pays particular attention to the textual accounts of the father in the protagonist's narrative. Here, the chapter zeroes in on the lack of information about the father with particular interest for the epistemological implications of this absence. In a third section, the chapter asks for the consequences of Hoffmann's account of the origin for conceptualizations of gender. The chapter concludes with reflections on the formation of selfhood and the function of the novel once the publications of Charles Darwin's *Origin of Species* (1859) and the *Descent of Man* (1871) had made it virtually impossible to think the human outside natural history.

Encouraged by entries in Hoffmann's diaries in which the fear of *Ich-Spaltung* is mentioned frequently,[3] scholarly works on the *Elixiere* so far have focused almost exclusively on the topic of the double. The protagonist's *Doppelgänger*—his half brother—has been read as projection and personification of the protagonist's guilt-driven and suppressed unconscious. The focus on the synchronic dimension of kinship (i.e., on the relationship between the half-brothers) and its metaphorical reading has enabled scholars to reveal Hoffmann's critique of the unified and self-governing identity of subjectivity and has often provoked—justifiably so—statements that render the author a predecessor of psychoanalytic thought.[4] In a recent reconsideration of the protagonist's pathology, Susan E. Gustafson has traced the problems to the female figures in his life. In her discussion of the *Elixiere*, she points out that "pathology throughout Hoffmann's tales is principally traced from mother to child."[5] While confirming Gustafson's comprehensive account of Hoffmann's indebtedness to a tradition that associates female imagination and melancholy with monstrous births, this chapter also complicates her claim by examining the significance of contemporary theories of heredity for the novel and thereby demonstrates the prevalence of the *patrilineal* line and *male* imagination. By doing so, this chapter shows that the focus on the mother and the double has enabled us to see Hoffmann's problematization of desire and selfhood, but it has also obscured Hoffmann's even more radical critique of perception, knowledge production, and the underlying gender politics. By shifting the attention from the synchronic to the diachronic dimension of kinship, from the horizontal axis of siblings to the vertical axis of father and son, the epistemological dimension of the *Elixiere* can be revealed.[6]

A brief plot summary will be helpful here. The *Elixiere* are the confessions of the monk Medardus. His father died at the moment the monk was born, and Medardus's biography develops as the search for his father's past. It turns out that his father and all of his ancestors are burdened by a hereditary drive toward incest, which often leads to murder. In each family member, the strength of this drive correlates with the number of blood connections to the original progenitor, so that it is particularly strong in those descendants who are the offspring of one or several incestuous relationships. At the origin of the family tree, Medardus discovers the patriarch Francesko whose intercourse with a devilish woman, it seems, provides the primal scene for all his offspring. The discovery of the progenitor's original sin coincides in the novel with the protagonist's firm conviction that he has found his identity, and provides him with an explanation for his own problems: insanity, alcoholism, rape, and murder. Medardus spends the rest of his life in a monastery where part of his penance involves writing his confessions, the novel itself. Because neither Medardus nor his siblings have children, the family now dies out.[7]

Much has been said about Hoffmann's preoccupation with the question of insanity and its treatment in contemporary medicine.[8] Wulf Segebrecht (1978) in particular has stressed the significance of Bamberg, where Hoffmann lived while conceptualizing the *Elixiere* (1808–1813), in shaping medical discourses, and has given us a detailed account of Hoffmann's personal encounters with progressive Bambergian physicians like Adalbert Friedrich Marcus. Among the scientific literature that had the greatest influence on Hoffmann's formulation of insanity, Johann Christian Reil's *Rhapsodien über die Anwendung der psychischen Curmethode auf Geisteszerrüttungen* [Rhapsodies on the Application of Psychological Methods of Cure to the Mentally Disturbed] (1803), Pinel's *Traité medico-philosophique sur l'aliénation mentale, ou la manie* [Medico-philosophical Treatise on Mental Alienation] (1801), and Cox's *Practical Observation on Insanity* (1804) are often quoted. The references to these sources have helped us better to understand and contextualize the cases of mental aberrations that Hoffmann depicts in his works.

At the same time, none of these texts gives an in-depth account and theory of hereditary transmission of insanity, which could have served as a model for Medardus's medical history and biography. Thus, it is astonishing that in Hoffmann scholarship little attention has been paid to that physician who was internationally recognized as *the* authority on questions of heredity: Erasmus Darwin. Although Darwin is sometimes mentioned as a source that Hoffmann might have known indirectly, a direct connection has never been established. Not much use has been made of a short remark by Peter Vodosek (1973) in an article about the lending library of Hoffmann's friend, wine dealer, and publisher Carl Friedrich Kunz. Kunz, whom Hoffmann met in 1809, supplied Hoffmann with alcohol, but also with reading material from his library and he was among the first who learned of Hoffmann's plan to write the *Elixiere*.[9] Vodosek has demonstrated that Kunz possessed not only an extensive selection of national and international fiction, but also a significant collection of scientific treatises. More importantly, Vodosek reports that among the books listed in Kunz's catalogue is Erasmus Darwin's *Zoonomia*.[10] Hoffmann was intimately familiar with Kunz's book holdings and was involved in preparing the catalogue for the latter's lending library, for which Hoffmann also wrote the preface which was published as a supplement in the *Fränkische Merkur*.[11] Given Hoffmann's strong interest in scientific literature in general and questions of reproduction and genealogy in particular—see the Spallanzani figure in the *Sandmann*[12] and the question of ancestry in *Das Majorat*[13]—it seems likely that Hoffmann studied Kunz's copy of Darwin's *Zoonomia*.

Translated into German only two years after its original publication, Darwin's work was widely quoted as an authority on the question of evolution and hereditary transmission of insanity in scientific treatises by authors like Reil and the phrenologist Johann Gaspar Spurzheim.[14] Hoffmann refers to Spurzheim's work in his report of the Schmolling murder case for which he had to write a report in his function as councilor at the Supreme Court in Berlin.[15] Embracing the idea of a hereditary cause of insanity, Spurzheim declares:

> Many physicians speak of diseases of the mind; others admit both mental and corporeal causes of insanity; a few acknowledge only corporeal diseases, and with the latter I decidedly agree. . . . Considerable diversity of opinion has prevailed, whether insanity be hereditary or not; and much is said on both sides. With Darwin . . ., I consider insanity as hereditary.[16]

In this assessment, the recognition of hereditary transmission of insanity goes hand in hand with the recognition of the material basis of mental capacities—and the question of whether the animal-human distinction so forcefully conceptualized by Descartes is still justified (see figures 4.1,

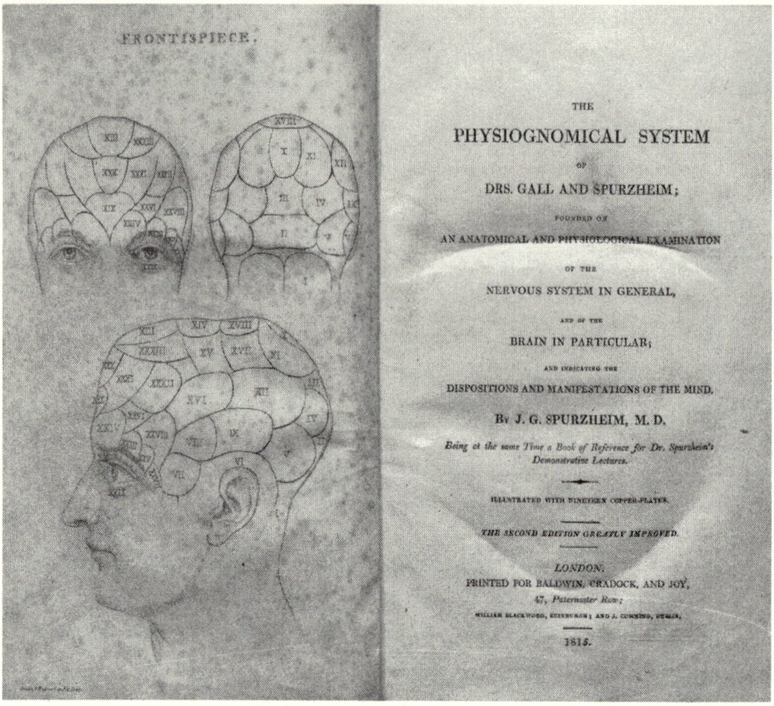

fig. 4.1. Organs of the human brain. Courtesy of Thomas Fisher Rare Book Library, University of Toronto. Johann Gaspar Spurzheim. The physiognomical system of Drs. Gall and Spurzheim. *London: Baldwin, Cradock, and Joy, 1815.*

4.2, and 4.3). Among early nineteenth-century researchers on insanity, Erasmus Darwin was conceived of as the man who provided the theory for an understanding of insanity as a corporeal and hereditary disease.

PRODUCTIVE PERCEPTION

On the surface the plot-line of *Elixiere des Teufels* overlaps with several of Darwin's statements and seems to confirm his theory of insanity as a heritable disease via close connection between parent and child so that the child is determined by the progenitor's habit, hereditary transmission via the patrilineal line,[17] negative traits stimulated by environmental influences like wrong nutrition (e.g., alcoholism or inappropriate nutrition supplied by the mother during pregnancy),[18] and appearance and disappearance of these negative traits within four generations.[19] The novel's most direct reflection of such determinism appears in the Pope's materialistic interpretation of the doctrine of original sin when Medardus relates his tragic life story to him:

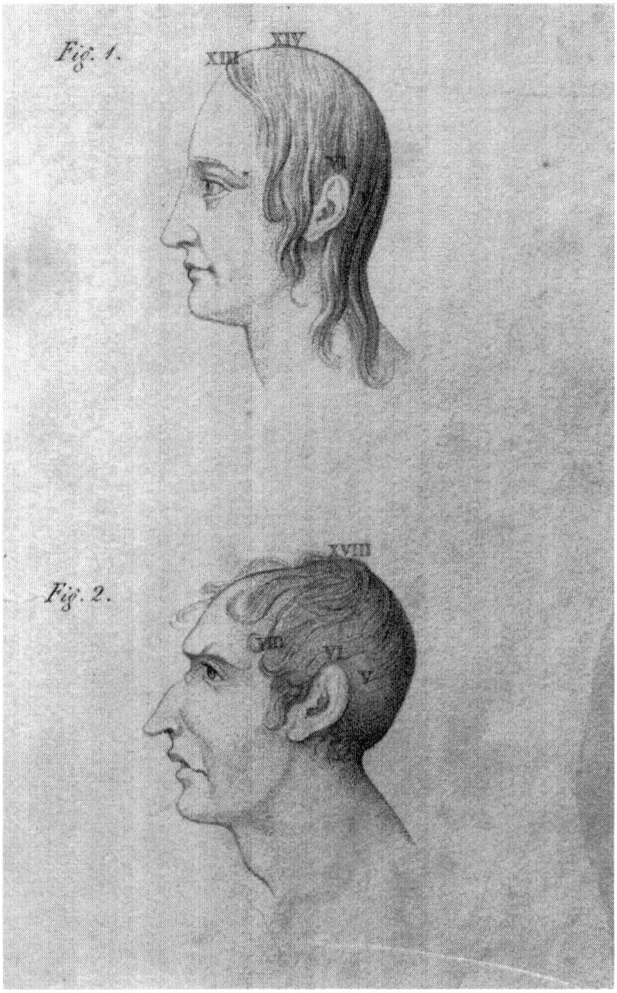

fig. 4.2. Organ of combativeness (V) and destructiveness (VI) in two individuals. Courtesy of Thomas Fisher Rare Book Library, University of Toronto. Johann Gaspar Spurzheim. The physiognomical system of Drs. Gall and Spurzheim. *London: Baldwin, Cradock, and Joy, 1815.*

fig. 4.3. Organ of philoprogenitiveness (love of offspring) (II), combativeness (V), and benevolence (XIII) in horse and dog. Courtesy of Thomas Fisher Rare Book Library, University of Toronto. Johann Gaspar Spurzheim. The physiognomical system of Drs. Gall and Spurzheim. *London: Baldwin, Cradock, and Joy, 1815.*

> [Pabst:] "Glaubt Ihr, daß der Wein . . . Euch zu den Freveln trieb, die Ihr beginget?"—[Medardus:] "Wie ein von giftigen Dünsten geschwängertes Wasser gab er Kraft dem bösen Keim, der in mir ruhete, daß er fortzuwuchern vermochte!"—Als ich dies erwidert, schwieg der Papst einige Augenblicke, dann fuhr er mit ernstem in sich gekehrtem Blick fort: "Wie, wenn die Natur die Regel des körperlichen Organism auch im geistigen befolgte, daß gleicher Keim nur gleiches zu gebären vermag? . . . Wenn Neigung und Wollen,—wie die Kraft, die im Kern verschlossen, des hervorschießenden Baumes Blätter wieder grün färbt—sich fortpflanzte von Vätern zu Vätern, alle Willkür aufhebend? . . . Es gibt Familien von Mördern, von Räubern! . . . Das wäre die Erbsünde . . .!"[20]

> [(Pope:) "Do you believe that it was the wine . . . which drove you to these crimes?" (Medardus:) "Like water teeming with foul odours it gave strength to the seed of evil latent with me, so that it multiplied beyond all knowledge." The Pope was silent for a few moments. Then he continued with a serious expression: "What if Nature were to follow in the realm of the spirit the physical law by which an organism can only reproduce its own kind; if propensity and desire, like the in-dwelling power which makes the leaves green, were to be handed down from father to son, obliterating free will? There are whole families of robbers and murderers. This would then be the original sin, an eternal, ineradicable curse on a guilty house, for which no sacrifice could atone."][21]

Yet despite the novel's preoccupation with the father figure and his deterministic influence on the family, it starts with an absence of the father and a lack of information about him. Medardus opens his confessions:

> *Nie* hat mir meine Mutter gesagt, in welchen Verhältnissen mein Vater in der Welt lebte; rufe ich mir aber alles das in's Gedächtnis zurück, was sie mir schon in meiner frühesten Jugend von ihm erzählte, so muß ich wohl glauben, daß es ein mit tiefen Kenntnissen begabter und lebenskluger Mann war.[22]

> [My mother *never* told me of the circumstances in which my father lived, but when I call to mind the stories she told about him in my childhood years, I cannot help thinking that he must have been a man of great experience and profound knowledge.][23]

This first sentence contains two contradictory statements. Although Medardus says first that his mother never told him about his father, he then states that she told him about his father early in his life. The gap between these contradictory clauses can only be bridged by an imaginative process, which is presented as an act of recollection ("rufe ich mir aber alles das in's Gedächtnis zurück"). Only in this process of imaginative recollection can the father emerge. Thus, the perception of the father figure turns out to be a productive process, for which the mother functions as a catalyst that provides the medium, the narratives.[24]

Three decisive elements characterize this first sentence: First, the necessity for the son to refer to the father to explain his identity; second, the father's absence, which raises the suspicion that the son is the actual producer of the father's image; and third, the mother figure that serves as a "medium" (she provides the narrations) to enable the production of the father's image. The product of the son's imaginative recollection—a process that I want to call *productive perception*—is a positive father figure that stands as "ein mit tiefen Kenntnissen begabter und lebenskluger Mann" at the end of the first sentence. And yet, the second sentence alters this father image, as Medardus continues:

> Eben aus diesen Erzählungen und einzelnen Äußerungen meiner Mutter, über ihr früheres Leben, die mir erst später verständlich worden, weiß ich, daß meine Eltern von einem bequemen Leben, welches sie im Besitz vieles Reichtums führten, herab sanken in die drückendste bitterste Armut, und daß mein Vater, einst durch den Satan verlockt zum verruchten Frevel, eine Todsünde beging, die er, als ihn in späten Jahren die Gnade Gottes erleuchtete, abbüßen wollte, auf einer Pilgerreise nach der heiligen Linde im weit entfernten kalten Preußen.[25]

> [From these stories, and from various remarks which my mother made about her earlier life and which I only later came to understand, I know that my parents sank from a life of plenty to a state of the most abject poverty. I learned too that my father had been led by Satan to commit a heinous crime, a deadly sin. When, in his old age, the grace of God shone upon him, it became his desire to do penance by making a pilgrimage to the shrine of the Holy Linden in cold, far distant Prussia.][26]

At the end of the autobiography's first sentence, the father was a "talented" and "experienced" man, a positive father-figure and potential role model. In the second sentence, however, this father turns into a character that is haunted by his own sin. Hence, the protagonist's investment in the past, searching for the absent father, produces an imaginative process, which develops from a positive outlook to a negative situation. In the third sentence, this culminates in the father's death which coincides with the protagonist's birth:

> In der heiligen Linde erkrankte mein Vater, und je weniger er die vorgeschriebenen beschwerlichen Andachtsübungen seiner Schwäche unerachtet aussetzen wollte, desto mehr nahm das Übel überhand; er starb entsündigt und getröstet in demselben Augenblick, als ich [Medardus] geboren wurde.[27]

> [In the monastery of the Holy Linden he fell sick, but in spite of his weakness he would not relax the rigorous devotional discipline imposed upon him, and his condition became more and more serious. He died, at peace in the knowledge of his salvation, at the moment I was born.][28]

The temporal simultaneity of the father's death and the son's birth seems to suggest a causal relationship. The father's death appears to be a precondition of the son's existence. Hence, as in the first sentence of the confession, there is a curious connection between the fact that the formation of the son's identity apparently requires the reference to the father *and* that this reference is only possible as a reference to the dead, and therefore absent, father. Furthermore, although Medardus informs the reader that the father's sins are redeemed at the moment of the son's birth, the reader soon learns that the son's biography unfolds like a repetition of the father's sinful life. If we consider that the father is a product of the son's *productive perception*, we have to suspect that these sins have less to do with the father's biological heritage than with the son's imagination of them. By exposing the process of biological inheritance as a process of imaginative recollection, Hoffmann questions the assumed determinism and, simultaneously acknowledges the power of imagination, triggered by the mother's narratives.

The elements described in the first three sentences of the confessions—an absent father–origin only graspable in an act of imaginative recollection via an apparently distorting mother–medium (leading to death and repetition of the cycle)—form a structural unit from which the novel evolves and to which it returns again and again. Preceding Medardus's confessions, the novel has a preface. This preface has rarely been discussed in detail, yet it provides in its structure a key to the understanding of the entire novel.[29] In this preface, the fictitious editor describes how he discovered Medardus's autobiography in the monastery where the protagonist lived during the last days of his life. For the prior of this monastery, Medardus's confessions have no basis in reality and are "das regellose Spiel der erhitzten Einbildungskraft"[30] [the caprice of an inflamed imagination].[31] However, although the Church fears the text's haunting and seducing influence, the editor hopes that the reader will appreciate their publication.[32] If the reader is open for an engagement with the text, the editor promises that a whole new world will open up for him.[33] He addresses the reader directly:

> Entschließest du dich aber aber, mit dem Medardus, als seist du sein treuer Gefährte, durch finstre Kreuzgänge und Zellen—durch die bunte—bunteste Welt zu ziehen, und mit ihm das Schauerliche, Entsetzliche, Tolle, Possenhafte seines Lebens zu ertragen, so wirst du dich vielleicht an den mannigfachen Bildern der Camera obscura, die sich dir aufgetan ergötzen.—Es kann auch kommen, daß das gestaltlosscheinende [sic], so wie du es schärfer ins Auge fassest, sich dir bald deutlich und rund darstellt. Du erkennst den verborgenen Keim, den ein dunkles Verhängnis gebar, und der, zur üppigen Pflanze emporgeschossen, fort und fort wuchert in tausend Ranken, bis eine Blüte, zur Frucht reifend, allen Lebenssaft an sich zieht, und den Keim selbst tötet.[34]

> [But if you decide to accompany Medardus through gloomy cloisters and cells, through the lurid episodes of his passage through the world, and to bear the horror, the fear, the madness, the ludicrous perversity of his life as if you were his faithful companion—then, maybe, you will derive some pleasure from those glimpses of a camera obscura which have been vouchsafed to you. It may even be that, as you look more closely, what seemed formless will become clear and precise; you will come to recognise (sic) the hidden seed which, born of a secret union, grows into a luxuriant plant and spreads forth in a thousand tendrils, until a single blossom, swelling to maturity, absorbs all the life-sap and kills the seed itself.][35]

The editor states that the identification of the reader with the protagonist is the precondition for the emergence of images depicting an absent reality. The editor also mentions that these images are the source of entertainment for the reader ("ergötzen"), thereby highlighting an atmosphere of superficiality. However, it turns out that in the moment when the reader engages with these images of the camera obscura, they obtain a quality that goes beyond mere representation.[36] In this moment the images of the camera obscura—these flat media of representation—obtain a new reality, a three dimensional quality. They become "clear and round" ("deutlich und rund"). While the editor first indicates that this new quality might be an effect of the way the images are presented or imagined rather than of their genuine qualities ("sich dir . . . darstellt"), he speaks in the next sentence of a "germ" ("Keim"), thereby switching from a language of representation and superficiality to that of nature and reality. A vocabulary of cognition ("erkennst") and of origin and meaning ("Keim, den ein dunkles Verhängnis gebar") further supports the tendency away from the supposedly superficial world of the image towards a dimension of depth and origin.

Yet it remains unclear of what the "germ" is in the origin. At first sight, it might seem as if the reader discovers in the representation its origin, hence the reality to which the images of the camera obscura refer. If this were the case, the second part of the sentence should describe the germ's growth in past tense, following the development from germ (original) to its representation (images of the camera obscura). Instead, however, in mid sentence, the editor switches to present tense (in German also used for future), suggesting that the plant's growth is still and always going on ("fort und fort wuchert") and thereby surpasses the representation from which its germ sprang. We have to conclude that in the act of perception, the reader has not only discovered the germ belonging to the representation, but has created it, thereby unleashing a threatening dynamics. Hence, it turns out that the uncovering of the hidden germ is at the same time the act of production of this germ and its proliferating growth. The investment in the medium of representation seems to have created a *second nature* where the first one was missing. In other words, the editor has reversed the relationship between origin and representation: Not the origin has created an imitation but the imitation has triggered a perception process that posits an origin.

The germ's growth is described by Hoffmann with words like "voluptuous" ("üppig"), "proliferating" ("wuchert"), and "to kill" ("tötet"). This vocabulary creates an atmosphere of impending danger. Indeed, the growth of the germ develops a self-destructive dynamic. It kills the germ from which it grew by replacing it with a new germ. The threatening growth that only finds an end by its continuation seems to confirm the Church's warning about the deteriorating and uncontrollable products of a medium that has no basis in reality. However, a close look at the text shows that Hoffmann insists that the problem does not lie in the fact that the representation has no basis in reality or distorts the reality to which it refers, but in the fact that the reader assumes and posits such an origin. If the germ is "born by a somber fate," then, this fate is not some unknown force, instead: The germ is born in the moment in which it is recognized by the reader and precisely this recognizing—the process that I have called *productive perception—is* the "somber fate." *Because* we posit the germ, its growth deteriorates.

A GERM THAT PROLIFERATES

In the two passages already discussed (preface and opening of the confession), the importance of a *productive perception*, an imaginative process, emerges as a dominant theme. When we juxtapose these two passages, an astonishing structure of shared elements emerges, indeed a structure of five steps: First, we note a moment of absence. The Church claims that there is no reality corresponding to the confessions and the opening of the autobiography is marked by the absence of the father.[37] The second step provides an engagement with this absence, an active perception process triggered by this absence. In the preface, the reader is promised entrance to a new reality if he identifies with the protagonist's life. The opening paragraph of the autobiography shows the protagonist's active engagement with the missing knowledge about the father as the precondition for the emergence of a father image.[38] The active perception process produces, in the third step, a representation of that which was absent in step one. In other words, absence is replaced by representation in such a way that the representation and the creation of its object coincide. This representation comes into being with help of a "catalyst." In the preface, this catalyst function is provided by the camera obscura, creating images. In the confessions, the mother serves as a catalyst of the narrations.[39] Via the comparison with the mother, the camera obscura and the reality that is depicted by it obtain a gendered quality. If the mother's narrations refer to a father figure, then, the projections of the camera obscura refer to a "father" reality. The connection between the two media or "production machines" is highlighted by their metonymic connection: the dark chamber. If the mother's uterus— in German of the time literally called "Mutter"[40]—projects the father in the son, the "camera obscura" projects reality in the image. The fourth step then is defined by a new engagement of the observer (protagonist or reader) with the medium of representation. This engagement reveals, or rather produces, a reality that is independent of the reality to which the representation referred in the first place. I have called this new reality a second nature. In the preface, the reader discovers a germ. In the opening paragraph, the protagonist announces that his mother is pregnant with him.[41] In step five, this second nature develops a threatening dynamics. The father, who turns out to be burdened by his owns sins, dies in the moment of the son's birth and the germ develops an uncontainable growth that leads eventually to its self-destruction. In both cases, the lines of developments end by being replaced with a new being.[42]

Hoffmann further questions the biological determinism propagated in the plot-line but undermined already in the first three sentences by demonstrating its structural relatedness to religious belief. When Medardus, as a young monk, is given the task of taking care of the monastery's reliquary, he expresses his doubts regarding the authenticity of the relics. In the long conversation which follows, an older monk, Cyrillus, voices the conviction that it is belief, not the authenticity of the material which brings about the miracle attributed to the reliquary:

> Was ist jenes Stückchen Holz— . . . —man sagt aus dem Kreuz Christi sei es gehauen . . .; aber den Gläubigen, der ohne zu grübeln, sein ganzes Gemüt darauf richtet, erfüllt bald jene überirdische Begeisterung, die ihm das Reich der Seligkeit erschließt, das er hienieden nur geahnet . . . Ja, diese in ihm erweckte höhere geistige Kraft wird selbst Leiden des Körpers zu überwinden vermögen, und daher kommt es, daß diese Reliquien jene Mirakel bewirken, die, da sie so oft vor den Augen des versammelten Volks geschehen, wohl nicht geleugnet werden können.[43]

> [What is that bit of wood. . . . People say that it was hewn from the cross of Christ. . . . But the believer does not brood over those things: he is absorbed with the thought of that supernatural power which will open to him a realm of bliss which he only dreamed of here below . . . this spiritual strength will of itself be able to overcome the sufferings of the body, and that is why the relics work those miracle which happen so often before the eyes of the assembled multitude that they can scarcely be denied.][44]

Hence, Cyrillus admits that the claimed metonymic connection between the divine and the relic is doubtful, but he states that the important process derives from the believer's engagement with the medium that fills the absence. Furthermore, in a striking parallel to the structure of the first paragraph, Cyrillus's own engagement with the medium, the reliquary develops a threatening dynamic. It turns out that the most precious of the monastery's relics, the "Elixiere," has destructive powers. Showing Medardus the box in which these elixirs are kept, Cyrillus explains:

> [H]ierinnen, lieber Bruder Medardus! ist die geheimnisvollste wunderbarste Reliquie enthalten, die unser Kloster besitzt. . . . Ich kann die Kiste nicht ohne inneren Schauer anrühren, es ist als sei darin ein böser Zauber verschlossen, der, gelänge es ihm, den Bann der ihn umschließt und wirkungslos macht, zu zersprengen, Verderben und heillosen Untergang jedem bereiten könnte, den er ereilt.—Das was darinnen enthalten, stammt unmittelbar von dem Widersacher her. . . . Ich will mich lieber Bruder Medardus gänzlich enthalten, in dieser höchst mystischen Sache nur irgend eine Meinung zu äußern . . . , sondern lieber getreulich dir das erzählen, was die, über jene Reliquie vorhandenen Dokumente davon sagen.[45]

> [This, dear Brother Medardus, is the most marvellous (sic) most mysterious relic that our monastery possesses. . . . I cannot touch the chest without shuddering, for it is as if it contained a demon which, if it succeeded in bursting the bonds which now render it powerless, would bring irrevocable disaster upon everyone it overtook. The contents of the chest were prepared by Satan himself. . . . I will refrain, dear Brother, from expressing any personal opinion on this most mysterious affair . . . rather I will tell you faithfully what the relevant documents say about the relic][46]

Cyrillus, who just admitted the reliquaries' lack of authenticity, states now that he has immediate reactions to the described reliquary. While before, he mentioned the believer's engagement, he makes now an effort to distance himself from any subjective involvement. Without hesitations or doubts, he refers to documentary evidence that confirms the authenticity of the elixirs—which turn out to be the elixirs which were offered by the devil to St. Antonius in the desert. Thus, he himself becomes an example of the mechanisms that he has described. His engagement with the absence has created a second nature, the authentic reliquary. As in the structure described before, this second nature develops a threatening dynamics: The elixirs derive, as Cyrillus declares, directly from the devil ("stammt unmittelbar von dem Widersacher her") and threaten the mental stability of everybody who comes in contact with them. Just as the imagination of the father figure has lead first to a role model and then to sin and death, the imaginative process connected to the relics leads first to miracles and eventually to the devil himself. In both examples, the imaginative process seems to get distorted by a medium, which is defined by a problematic materiality leading to evil.

The pattern discussed here is repeated a fourth time, this time on the aesthetic level. After the opening paragraph about the father figure that we have already discussed, Medardus describes a pleasant childhood in the convent, where his father died. When Franz decides to take the oath (and to become Medardus), his mother is enthusiastic because she is convinced that only then the father's sins are truly overcome. However, already early on, Medardus is confronted with sexual desires and thoughts of hubris that threaten his faithfulness to the monastic oath. These transgressions come to a climax after Medardus is given the responsibility over the reliquary. As mentioned, he is warned that one of the reliquaries, the "elixirs of Saint Anthony," has devilish powers if opened. However, the young monk cannot resist the temptation. Repeatedly, he drinks from the elixirs, a deed that only increases his hubris and sexual desire. The prior of the monastery believes he can cure Medardus's problems by sending him to fulfill tasks in Rome. However, on his way to Rome, Medardus drifts from his way, rapes, kills, is put in prison, and spends time in a lunatic asylum. Only after many detours does Medardus return to his original path. Shortly before arriving in Rome, he stays overnight at a monastery where he receives a book that, he is promised, will solve the riddle of his life. Medardus opens the book and discovers a number of pictures that resemble images of his childhood and although the book is written in an almost illegible hand, Medardus recognizes it as a document connected to his own life. He writes:

> Das, was der Maler auf den letzten Seiten des Buchs in kleiner, kaum lesbarer bunt gefärbter Schrift zusammen getragen hatte, waren meine Träume, meine Ahnungen, nur deutlich, bestimmt in scharfen Zügen dargestellt, wie ich es niemals zu tun vermochte.[47]

> [The barely legible writing (of the painter) in brightly-coloured ink on the final pages described all my dreams and forebodings, but with such clear, sharply-defined outlines as I could never have achieved.][48]

Similar to the media before (narrations and images), this book serves as a screen for Medardus's projections. Medardus's comment on the book repeats the editor's description of the novel: The amorphous and two-dimensional page creates a life of its own and like the beginning of Medardus's confessions started with an apodictic "Nie," so do the pages of this book start with three hyphens, an empty space. The book tells the story of a certain Francesko, who turns out to be the progenitor of the family to which Medardus belongs (Medardus is mentioned on the last page of the book).

Here, towards the end of his long pilgrimage to Rome, Medardus reads the biography of the family's progenitor whose evil deeds, it seems, have led to the family's decline. The figure Medardus discovers at the origin of his family tree was an artist. This artist once tried to paint the image of Saint Rosalia. However, he was only able to complete the image after getting intoxicated with the "Syrakuser Wein"[49] [Syracusan wine] offered by his pagan friends—a wine which the friends praise as "Wein aus dem Keller des heiligen Antonius"[50] [wine from the cellars of St. Antonius] (the reference connects this episode to Medardus drinking the elixirs). In a moment between sleeping and being awake, Francesko drinks the wine that his friends offered him, and the female figure in his painting becomes alive:

> So wie der Berg Vesuv in wildem Brausen verzehrende Flammen aussprüht, so tobte es jetzt in Feuerströmen heraus aus Francesko's Innern. Alle heidnischen Geschichten, die er jemals gemalt, sah er vor Augen, als ob sie lebendig worden, und er rief mit gewaltiger Stimme: "Ach du mußt kommen, meine geliebte Göttin, du mußt leben und mein sein, oder ich weihe mich den unterirdischen Göttern!" Da erblickte er Frau Venus dicht vor dem Bilde stehend, und ihm freundlich zuwinkend.[51]
>
> [As Mount Vesuvius roars and sends forth its destructive flames, so streams of fire shot through Francesco's soul. All the pagan subjects he had ever painted appeared as living forms before his eyes, and he cried out fiercly: "You too must come, my beloved goddess. You too must live and be mine, or I shall deliver myself up to the gods of the underworld." Then he saw Venus standing close to the picture, beckoning gently to him.][52]

Again and again, the Venus figure draws back, but, finally, on the third day, she stays. While Medardus turns into an image, the image becomes alive:

> . . . am dritten Tag, als er [Francesko], *wie eine erstarrte Bildsäule*, vor dem Bild stand, ging die Tür seines Gemachs auf, und es rauschte hinter ihm wie mit weiblichen Gewändern. Er drehte sich um und erblickte ein Weib, das er *für das Original seines Bildes* erkannte.[53]

> [On the third day, while he was standing before the picture motionless as a statue, the door of his room opened and he heard a rustling sound behind him as of a woman's dress. Turning around, he saw alive before him, in all her lovelines, the woman of his picture.][54]

Full of admiration, the artist falls to his knees:

> Das fremde Weib richtete ihn aber lächelnd auf und sagte ihm, daß sie ihn schon damals, als er in der Malerschule des alten Leonardo da Vinci gewesen, als ein kleines Mädchen oftmals gesehen und eine unsägliche Liebe zu ihm gefaßt habe.[55]

> [The woman lifted him up, a gentle smile on her lips, and told him that, when she had been a little girl and he a student at Leonardo's academy, she had often see him and cherished even then an all-consuming love for him.][56]

Like Medardus's imagination of the father figure, the artist's imagination is presented as an act of recollection: the beautiful woman is a recollection of Francesko's childhood.

The artist—like Pygmalion—falls passionately in love with this woman, who shortly thereafter becomes pregnant. However, while giving birth, her beautiful figure turns into a distorted and abject body mass and she dies in childbirth. The neighbors interpret her transfiguration as a sign of her diabolic origin.[57] In a vision, the artist hears a heavenly voice condemning him and all his descendants never to find peace as long as they reproduce: "Nicht Gnade, nicht Ruhe im Grabe soll er [Francesko] finden, so lange der Stamm, den sein Verbrechen erzeugte, fortwuchert, in frevelicher Sünde!" [He shall find no grace, no rest in the grave so long as his stem continues to multiply in sin and wickedness!].[58]

All the elements observed in the previous two examples are represented here once again: absence, engagement with this absence, representation of this absence with help of a medium, investment in the representation and materialization into a (second) reality, threatening development and death (coinciding with the birth of a new being, which will repeat the cycle). However, we witness one crucial change compared to the life cycle that Medardus narrated: while in the opening paragraph the narrator posited an origin that was defined as male and thought to come into being by a female, polluting medium, in this last scene the absent reality is feminine

(Saint Rosalia). As a consequence, the implicit gender politics change: the medium and the origin to which it refers cannot be differentiated any longer in terms of purity and pollution. Moreover, it is the medium that delivers the original. When the woman on the canvas comes alive, the narrator comments that the artist "erblickte ein Weib, das er für das Original seines Bildes erkannte"[59] [he saw a woman whom he recognized as the original of his picture]. Hence, where the difference between original and medium is dismissed or even reversed, the idea of a medium that pollutes the origin no longer makes sense. The text suggests that Hoffmann rejects not only the idea of a determining past, but also the gender politics linked to this idea, namely the assumption that the woman is to be blamed for the pollution of the pure father origin. Instead, Hoffmann stresses that all that is accessible to us is the medium and that this medium is the product of a progenitor, who in turn, is a product of the male descendant's imagination.[60]

Thus, if from Aristotle to Erasmus Darwin men blamed women for distorting the father origin, Hoffmann exposes this argument as an attempt at avoiding the source of the problem: *male* imagination.[61] The text suggests that it is not due to some polluting powers and inherited traits that the progenitor and all his descendants become rapists and murderers. Rather, their mental problems spring from their attempt to imagine the absent origin which comes to determine them. The primal scene to which Medardus refers to explain his life reveals the nature of his progenitor's and family's incestuous drive in its most extreme form: the progenitor has intercourse with the product of his own imagination. He and his descendants are caught in a vicious cycle of an imagined world which they take for the real. For Hoffmann, it is not problematic that these imaginative projections occur (their pervasive nature actually hints to an anthropological constant), but that they are given ontological status. Positing the origin as real is the moment and reason for its degeneration.[62]

Hoffmann further highlights the mechanisms of imagination by referring to the biblical prohibition against the production of carven images. The passage of the Ten Commandments in the Deuteronomy has the following formulation:

> You shall have no other gods before Me. You shall not make for yourself an idol, or any likeness of what is in heaven above or on the earth beneath or in the water under the earth. You shall not worship them or serve them; for I, the Lord your God, am a jealous God, visiting the iniquity of the fathers on the children, and on the third and the fourth generations of

> those who hate Me, but showing loving kindness to thousands, to those who love Me and keep My commandments.[63]

While the Pope tried to explain original sin by referring to its corporeal basis, the fact that in this novel an artist fails, when working on a work commissioned by the Church, suggests that the Church's own tradition of depicting the divine is at the source of the problem. However, a close look at the biblical passage shows that it overlaps in important features with the biological model. Both models warn of pollution by sensual engagement, and, as in Darwin's treatise, in the Ten Commandments, the violation of the prohibition against images is directly connected to a curse of several generations. By inviting the reader to draw the connections between original sin (understood as violation against the prohibition of graven images) and the biological concept of inheritance, Hoffmann questions the empiricist basis of the scientific claim, on which scholars such as Darwin insists. Hoffmann shows that Darwin's claims rely heavily on cultural concepts and that Darwin's assertion of the material basis of his claim must falter because of its reliance on these concepts. Hence, Hoffmann returns the scientific model to its imaginative source.

Hoffmann's conflation of biological, religious, and aesthetic discourses has far-reaching consequences not only for the understanding of the formation of the self as an autopoietic process, but also for Hoffmann's assessment of the newly emerging scientific disciplines and the basis of their truth claims. Erasmus Darwin's *Zoonomia* opens up a space for Hoffmann to rethink reproduction as creation, but he pushes further still by showing the potential of the theory to undermine the material foundation on which its truth claim rests.[64] While scientists like Erasmus Darwin were beginning to pride themselves in being able to "inlist [sic] Imagination under the banner of Science"[65] by working empirically, Hoffmann questions the confidence in the material reality, which he exposes as a fiction filling an absence. To be clear, Hoffmann's criticism of a truth claim, which is made with reference to empirical data does not imply a rejection of scientific explanation in general. His intensive preoccupation with medical literature in preparation for the Schmolling case shows that he valued medical science and tried hard to get insight from its latest developments.[66] Rather, Hoffmann urges us to keep in mind that our accounts of reality can only have the character of a model.

Against the backdrop of this assessment of reality, it now becomes clear why the "living filament" of Darwin's treatise acquires new meaning in Hoffmann's *Elixiere*. In the last sentence of the programmatic preface, the fictitious editor contemplates:

> Nachdem ich die Papiere des Capuziners Medardus recht emsig durchgelesen . . ., war es mir auch, als könne das, was wir insgemein Traum und Einbildung nennen, wohl die symbolische Erkenntnis des geheimen *Fadens* sein, der sich durch unser Leben zieht, es festknüpfend in allen seinen Bedingungen, als sei der aber für verloren zu achten, der mit jener Erkenntnis die Kraft gewonnen glaubt, jenen Faden gewaltsam zu zerreißen, und es aufzunehmen, mit der dunklen Macht, die über uns gebietet.[67]
>
> [After I had with great diligence read through the papers of Medardus the Capuchin—which was extremely difficult because of his minute and barely legible monastic handwriting I came to feel that what we call simply dream and imagination might represent the secret thread that runs through our lives and links its varied facets; and that the man who thinks that, because he has perceived this, he has acquired the power to break the thread and challenge that mysterious force which rules us, is to be given up as lost.][68]

Darwin's "living filament," which Jean Paul had rendered as "Fleischfädchen" (thread of flesh) (see Chapter 3), reoccurs here as *Lebens-Faden* (thread of life). Hoffmann makes clear that this "Faden" might be the product of "Einbildung" (imagination), but it is the only one that can give unity and meaning to our life and establish our identity. The danger lies only in forgetting that the meaning that we give to our lives is one of many possible explanations, and in assuming that we have found access and control of the life-governing power.[69]

For Hoffmann, the close connection between living filament and imagination raises two fundamental and connected questions: can there be a biological approach to explain the self and its moral behavior? And what is the methodological foundation of scientific claims about the self? A close reading of the *Elixiere* suggests that Hoffmann argues against a biological determination of human psychological and moral behavior. He is suspicious about scientific findings which claim ontological status with reference to an empirical methodology, ignorant of the imaginative conditions predating the material observation and covering the epistemological vacuum. Consequently, Hoffmann rejects attempts to apply biological models to the self.

LITERATURE AS SCIENCE OF THE SELF

Under these conditions, literature receives two new functions. First, since an independent psychological discipline had not yet been established, the novel, rather than the life sciences becomes, for Hoffmann, the forum for discussing psychological questions—and it is not surprising that the emergence of psychology as an independent discipline is so closely connected to the reception of Hoffmann's oeuvre.[70] The novel obtains a second new function as a corrective to, and exploration of, truth claims that are increasingly being made with reference to their empirical basis, a practice that soon becomes standard in nineteenth-century sciences. Hence, Hoffmann does not suggest that literature is omnipotent and has access to truth; however, he sees in the novel a sphere that, by providing alternative views, can offer a critique of science. Moreover, the novel can provide an exploratory space for scientific theories, which allows for further speculation on their implications and potential.

Writing in a climate of increasing rejection of the *Elixiere*, in the middle of the nineteenth century, Friedrich Hebbel acknowledges Hoffmann's innovative approach to the novel. In his diary, he comments:

> [S]eine Elixiere des Teufels sind und bleiben ein höchst bedeutendes Buch, . . . so wunderbar angelegt und mit solcher Konsequenz durchgeführt, daß, wenn es noch keine Gattung gibt, der Darstellungen dieser Art angehören, das Buch eine eigne Gattung bilden wird.[71]

> [(H)is Devil's Elixirs are and remain the most important book, . . . so beautifully outlined, and executed with such consequence that—if there is no genre yet to which works of this kind belong—the book will form its own genre.]

By establishing the novel as a genre which investigates the fundamental premises of increasingly divergent models of human understanding, Hoffmann's *Elixiere* marks a crucial departure from previous understandings of the relationship between science and literature. If, as Richards has argued convincingly, the "fundamental organizing principle in the philosophy of the Romantics . . . was the aesthetic-epistemic tenet of the complementarity of the poetic and scientific conceptions of nature" and both "aesthetic intuition and scientific comprehension drove down to the type, to the underlying force that gave form to the surface of things,"[72] then Hoffmann questions the certainty that underlying forces can be found and contingency can be overcome. Hence, in the *Elixiere* Hoffmann introduces a new possibility of conceptualizing the relationship between science and art. As we have seen, literature becomes an independent discipline in which questions about the sciences can be asked.

This new understanding of the relationship between the disciplines becomes perhaps most obvious where Hoffmann's vocabulary seems closest to Goethe's. In an often quoted letter to Herder from May 17, 1787, Goethe, writing from Italy, announces his conception of the *Urpflanze*:

> Ferner muß ich Dir vertrauen daß ich dem Geheimnis der Pflanzenzeugung und Organisation ganz nahe bin und daß es das einfachste ist was nur gedacht werden kann. . . . Den Hauptpunkt, wo der Keim steckt, habe ich ganz klar und zweifellos gefunden, alles Übrige seh' ich auch schon im Ganzen und nur noch einige Punkte müssen bestimmt werden. . . . Dasselbe Gesetz wird sich auf alles übrige Lebendige anwenden lassen.[73]

> [Further, I have to tell you that I am very close to (the discovery of) the secret of plant reproduction and organization and that it is the simplest thing that can be conceived. . . . I have found the central point, where the germ is hidden, in total clarity and without doubt; I see everything else also already in its entirety and I have to determine only a few more points. . . . The same law will be applicable to all other living nature.]

Goethe elaborates further, stressing that he does not describe "malerische oder dichterische Schatten und Scheine" [painterly or poetic shadows and appearances], but "eine innerliche Wahrheit" [an inner truth].[74] The lexical spectrum ranging from plant reproduction and germ to painterly appearance and the recognition of inner truths resonates in the *Elixiere*. In the programmatic preface already quoted, the fictitious editor promises the reader new insights if he is willing to identify with the protagonist.[75] However, while Goethe describes the recognition of the "Keim" [germ] of a law-guided development belonging to a nature that is understood or even defined by an economic balance that never degenerates,[76] Hoffmann describes a germ which develops a life-threatening quality. While Goethe suggests that the observer can eventually penetrate to the essential structures of life, Hoffmann creates an air of superficiality and turns the attention to the process of perception itself. He does so in such a way that the germ and its threatening growth turn out to be intrinsically linked to that process of perception that Goethe famously referred to as an immediate seeing.[77]

However, Goethe too had already engaged with the problem of perception precisely at the place and moment of his botanical discovery. It is noteworthy that Palermo, the place of Goethe's discovery of the *Urpflanze* (see diary entry from April 17, 1787),[78] is also the place where Goethe locates his own *Rosalia* experience. In October 1788, Goethe published the account of this experience in Wieland's *Deutsche Merkur* in an essay entitled "Rosaliens Heiligtum" (Rosalia's Shrine):

> Ich setzte mich auf eine Bank, gegen dem Altar über, und hörte ihnen [den Geistlichen] eine Weile zu; alsdenn begab ich mich wieder zum Altare, kniete nieder, und suchte das schöne Bild der Heiligen noch deutlicher gewahr zu werden, und überließ mich der reizenden Illusion der Gestalt und des Orts. . . . Ich habe nachher manchmal mit mir selbst darüber gescherzt, und das Vergnügen, das ich dort empfunden, mehr einer glücklichen Stimmung und einigen Gläsern guten Sicilianischen Weins, als den Gegenständen selbst zuschreiben wollen. . . .[79]
>
> [I sat down on a bench, across from the altar, and listened a while to [the clergy], then I went again to the altar, knelt down, and tried to see the Saint's beautiful image in greater detail, and abandoned myself to the delightful illusion of the figure and the place. . . . Later, I have sometimes made fun of myself and have attributed the pleasures which I have enjoyed there more to a happy atmosphere and a few glasses of good Sicilian wine than to the objects themselves. . . .]

The constellation of discovery of the *Urpflanze*, Rosalia, Sicilian wine, and the reflection on the perception process seem to resurface in the *Elixiere* and suggest that the novel can be read as a comment on Goethe's Palermo experience and its theorization.[80]

Towards the end of the eighteenth century, at the very moment when scientific models emerge, literature assumes more and more a correlative function, distinguishing itself from the scientific enterprise and demonstrating the potential limits of scientific truth claims. By establishing the novel as critical space vis-à-vis science, Hoffmann takes part in the formulation of disciplines and their functions. I suggest that the *Elixiere des Teufels* should be considered one of the first inquiries into the relationship between the humanities and the sciences, in which each of them is considered an independent discipline with its unique tasks and methods. Hoffman anticipates a differentiation of value spheres which will be explored further only almost a century later by thinkers like Wilhelm Dilthey (1988) and later C. P. Snow (1959).[81]

Aside from a short spark of interest in the first decades of the twentieth century, few scholars in the twentieth century have placed the development of the self as it appears in the novel in relation to biological notions of heredity and evolution.[82] However, since the mid-1990s, we can observe a growing number of publications that have begun to address the relationship of individual and society as depicted in literature from a Darwinian perspective (i.e., Charles Darwin). Proponents of this idea argue for a paradigm shift away from "irrational" postmodernism towards a "rational" and empirically proven theory which is based on concepts developed in the natural sciences and applies them to literature.[83] If today literary historians like Joseph Carroll aim for a return to a methodology before the differentiation of disciplines by asking for a Darwinian approach to literature and culture, it might be timely to consider Hoffmann's epistemological legacy in order to better understand the implications and challenges of such an enterprise. The *Engführung* of scientific and religious discourses in the *Elixiere* suggests that Hoffmann contemplates the danger that in modernity science replaces religion as that vantage point from where truth claims are made.

CODA

When the first scientific works that described corporeal and mental qualities as the result of hereditary transmission appeared around 1800, they challenged the claim of the self to independence and singularity. While the Enlightenment had posited a potentially autonomous individual within a hierarchically structured nature, new findings in medicine and biology indicated determinism on the level of the individual and the elimination of inter-species boundaries on the level of the group. The claim of an autonomous self had been developed in connection with new religious concepts supported by Pietism and emerging ideas of childhood,[84] but it had also received decisive impulses from scientific explorations into embryology, which suggested the embryo's self-regulated development from an amorphous substance by means of a *Bildungstrieb* (formative drive). The biological term of this formative drive was lively debated in philosophical and literary texts and accompanied the emergence of a new genre, the *Bildungsroman*, which promised to provide an ideal program any individual could follow to develop a free and independent self.[85] The reference to the biological concept of self-regulation provided a model for totality and telos that the fragmented social world could not offer any longer.

As the concept of the *Bildungstrieb* had triggered new investigations into autonomy, temporality, and epistemology, new findings regarding heredity and evolution forced a reconsideration of selfhood and *Bildung* in literary texts. If the model of the formative drive had seemed to justify the promotion of a self-regulation, the discovery and exploration of hereditary patterns and evolutionary potentials challenged the promise of this autonomy on three levels. First, it pointed to the biological determinism that conditions selfhood. Second, it undermined the boundary between subject and other by pointing to kinship relation. And, third, it suggested the loss of the teleological framework that had formed the condition of possibility of selfhood as striving towards self-contained autonomy. Confronted with these challenges, three literary responses were possible in principle. First, the novel could react by giving in to the powerful biological model and by developing notions of identity that incorporated the idea of heredity. Second, it could challenge the very concept of biological determination and its significance for questions of selfhood. Third, the novel could choose to understand itself as a space in which knowledge about the science could be discussed. Considering the novels discussed in this book, we can see that Hoffmann's *Elixiere* responds to all three of these options. Hoffmann takes seriously the proposition of a hereditary transmission of insanity—but only to eventually dismiss it. By doing so, he elevates the novel to a meta-sphere in which questions about scientific propositions can be raised. Discussing the Mignon episode in Goethe's *Wilhelm Meister*, we found greater willingness to admit determining forces which cannot be overcome by acts of will. However, this determinism was ultimately considered of only very limited significance for notions of selfhood. Here too, the novel functions as a space where findings in the sciences are taken up quickly—or even anticipated—and then tested, as it were, for their relevance in the literary realm. Finally, Jean Paul Richter is most fascinated—and repulsed—by the materialism of evolutionary models of nature. Selfhood as celebration of autopoeisis becomes possible precisely because a clear border between nature and nurture is established. While Hoffmann in the *Elixiere* highlights the imaginary, and thereby cultural, conditions of what we understand when we speak of nature, for Richter the imaginary projection of selfhood becomes possible precisely as a counter model to a nature that is (projected) outside culture.

In the course of the nineteenth century the novel of formation explores the range of the three principal options vis-à-vis biology outlined above. While biological considerations never determine characters and novels entirely, considerations on determinism occupy an important place in many of the novels. Examining determinism is important because it can help us to understand the changes that happen in the conceptualization of selfhood under the pressure of evolutionary thought. At the origin of the *Bildungsroman* we find a novel that is already deeply invested in questions of determinism. In Karl Philipp Moritz's *Anton Reiser*, we encounter an individual whose existence and painful biography is explained to a great extent as a result of determining forces: "Unter diesen Umständen wurde Anton geboren, und von ihm kann mit Wahrheit sagen, daß er von der Wiege an unterdrückt war." [Anton was born in these circumstances and one can say in fact that he was suppressed from the cradle.][86] The moment of repression *is* the moment of the self.[87] However, the promise of autonomy is possible precisely because these forces are considered historical and, therefore, can be overcome—even though only by turning moments of repression via external forces into moments of repressions by forces governed by the self. Once determinism is understood as biological force such *Selbstermächtigung* (self-authorization) by means of internalization becomes much more difficult. Émile Zola's *Les Rougon-Macquart* with its programmatic subtitle *Histoire Naturelle et Sociale d'une Famille* (1871–1893) presents protagonists that are determined by family and heredity. But such strictly deterministic interpretations occupy only a short moment in the history and reception of evolutionary thought.[88] Further investigations are necessary to describe better the relationship between the novel and biological understandings of heredity as it evolved in the nineteenth century and beyond.

Of course, not all novels which are interested in formations of the self can or should be connected to what happens in the life sciences. However, the impact of life sciences and evolutionary thought on significant segments of our literary landscape remains powerful as we approach our times (Franz Kafka, Gottfried Benn, Ernst Jünger, W. G. Sebald, Durs Grünbein, etc.). A look at Thomas Mann's novels reveals how intimately questions of selfhood are bound to—or are discussed in—biological contexts after Charles Darwin's *Origin of Species* (1859) and the *Descent of Man* (1871) had made it necessary to reconsider the status of the human as animal. Mann's writing covers an immense area of topics related to heredity and evolution, to name just a few: Hanno's decline in the *Buddenbrooks*, which neither cod-liver oil nor summer vacation can cure; Hans Castorp's vision in the *Zauberberg* (Magic Mountain) of a body that is not individual any longer but an organism constituted out of a myriad of cells; and in *Krull* the visit to the natural history museum which confronts the protagonist with his prehistoric past.

In all three instances, selfhood is in crisis. It succumbs to family genealogy, is undermined by evolutionary developments, or disintegrates in cells and particles. Faced with these options, selfhood is only possible as claim: While Hanno dies and Castorp falls ill, Krull's self survives as trickster.

It is not surprising then that Krull's character has so much in common with Jean Paul Richter's picaresque Nikolaus Marggraf in the *Komet*. Like Marggraf who searches for signs of his aristocratic origin among all aristocrats travelling through town, Krull searches for signs of noble birth:

> Erinnert der geneigte Leser sich wohl, wie ich in eigener Frühzeit, aus Neugier nach den Ursprüngen meiner auffallenden Wohlschaffenheit, unter allerlei Vorfahrenbildnissen nach ersten Hinweisen auf mein Selbst mich forschend umsah?[89]

> [Does the kind reader perhaps recall how in my earlier days I had been at pains to search among the pictures of my forebears for some hint or indication of the source of my own striking physical perfection?][90]

Krull makes these deliberations in the context of his visit to the natural history museum. And it is here that we need to note decisive differences between Nikolaus and Felix. For Nikolaus the claim of selfhood was possible as projection: he recognizes himself as father (and nobility) in the mirror. Krull, however, discovers an origin that is strangely self and other:

> Ganz fühlte ich mich zurückversetzt in jene Beschäftigung [der Suche nach Ahnen], als ich nun dringlichen Auges und klopfenden Herzens das aus grauester Ferne auf mich Abzielende besichtigte. Du mein Gott, was hockte da klein und beflaumt in scheuer Gruppe beisammen, als beriete man sich in schnalzender, gurrender Vor-Sprache, wie auf dieser Erde, die man beherrscht von weit günstiger ausgestatteten, stärker bewaffneten Wesen vorgefunden, ein Durchkommen, ein Auskommen sei?[91]

> [I now felt completely re-immersed in that activity as, with probing and beating heart, I saw what had been striving toward me from the grey reaches of antiquity. Good God, what were those small, shaggy creatures squatting together in timid groups as though conferring in some cooing and hissing pre-language about the means of surviving and prospering on an earth already possessed by much better-equipped and more strongly armed creatures?][92]

Searching for noble lineage, Krull discovers himself as savage. In language reminicsent of Arnold Gehlen's *Der Mensch. Seine Natur und seine Stellung in der Welt* (Man. His Nature and Place in the World) (1940), the human is depicted here as lacking natural survival tools. The human being is threatened by the powerful natural world that surrounds him. His formation is not search for nobility but a necessity in order to survive. Like for Rotpeter in Kafka's *Bericht für eine Akademie* (A Report to an Academy) (1917) becoming human is a survival technique. Humanity is both an overcoming of that feeble nature and an expression of it. This survival becomes possible because of a pretense of strength by means of the tool, the artifice, where nature is lacking. When Krull reads natural history as developing towards him as the crown of evolution ("auf mich Abzielende" [striving toward me]), this belief is at the same time projection *and* an expression of a natural condition.[93]

EPIGRAPH SOURCE

Entry in Hoffmann's journal from October 26, 1803 (Hoffmann 1:336); and Letter to Carl Friedrich Kunz from March 24, 1814 (Hoffmann 6:24). Kunz was the publisher, bookseller, and wine dealer that Hoffmann met in Bamberg, where he was music director from 1808–1813. Hoffmann profited enormously from Kunz's collection of scientific literature. See Walther (1994) and Segebrecht (1978). For the translation of the letter to Kunz see E. T. A. Hoffmann, *Selected Letters*, ed. and tran. Johanna C. Sahlin (Chicago: University of Chicago Press, 1977), 225.

NOTES

1. Sandra-Valeska Bruhns, "Ein paar Härchen sorgen für Gewissheit," *Die Zeit*, February 13, 2007.

2. Lotta Wieden, "Vater Unbekannt," *Die Zeit*, April 18, 2002.

3. See entries January 6, 1804, and November 6, 1809 (E. T. A. Hoffmann, *Die Elixiere des Teufels. Werke 1814–1816*). In *Sämtliche Werke in sechs Bänden*, ed. Hartmut Steinecke and Wulf Segebrecht (Frankfurt am Main: Deutscher Klassiker Verlag, 2003) 2.2:564.

4. See Natalie Reber, *Studien zum Motiv des Doppelgängers bei Dostojevskij und E. T. A. Hoffmann* (Gießen: Wilhelm Schmitz, 1964); Karin Cramer, "Bewußtseinspaltung in E. T. A. Hoffmanns Roman Die Elixiere des Teufels," *Mitteilungen der E. T. A. Hoffmann-Gesellschaft* 16 (1970): 8–18; Horst S. Daemmrich, *The Shattered Self: E. T. A. Hoffmann's Tragic Vision* (Detroit: Wayne State University Press, 1973); Susanne Olson, "Das Wunderbare und seine psychologische Funktion in E. T. A. Hoffmanns Die Elixiere des Teufels," *Mitteilungen der E. T. A. Hoffmann-Gesellschaft* 24 (1978): 34–50; Sarah Kofman, "Vautour Rouge: Le double dans Les Élixirs du diable d'Hoffmann," in *Mimesis*, ed. Sylviane Agacinsky et al. (Paris: Flammarion, 1975), 95–163; and Detlef Kremer, "Flüchtige Familien-Phantome: 'Die Elixiere des Teufels,'" in *Romantische Metamorphosen: E. T. A. Hoffmanns Erzählungen* (Stuttgart: Metzler, 1993), 245–59.

5. Susan E. Gustafson, "The Cadaverous Bodies of Vampiric Mothers and the Genealogy of Pathology in E. T. A. Hoffmann's Tales," *German Life and Letters* 52, no. 2 (1999): 247.

6. Commentators have often described the shift from a feudal society in which a patriarch reigns over a household to a bourgeois society in which the mother dominates the nuclear family as a weakening of masculinity and a devaluation of lineage in the wake of the French Revolution. See Friedrich Kittler, "Der Dichter, die Mutter, das Kind: Zur romantischen Erfindung der Sexualität," in *Romantik in Deutschland: ein interdisziplinäres Symposion*, ed. Richard Brinkmann (Stuttgart: Metzler, 1978), 102–14; and Peter Uwe Hohendahl, "Die Krise der Männlichkeit im späten 18. Jahrhundert. Eine Problemskizze," *Zeitschrift für Germanistik* 12, no. 2 (2002): 275–86. The claim of a crisis of masculinity has been contested only rarely; see Martha Helfer, "The Male Muse of Romanticism: The Poetics of Gender in Novalis, E. T. A. Hoffmann, and Eichendorff," *German Quarterly* 78, no. 3, (July 2005): 299–319, on the dominance of the male muse in Romantic poetic production and for the significance of the father in the legal discourse, see Sara P. Eigen, "A Mother's Love, a Father's Line: Law, Medicine, and the 18th-Century Fictions of Patrilineal Genealogy," in *Genealogie als Denkform in Mittelalter und Früher Neuzeit*, ed. Kilian Heck and Bernhard Jahn (Tübingen: Niemeyer, 2000), 87–107. Dorothea von Mücke has argued that the literary genres of the eighteenth century are conceived primarily as instruments for disciplining the male bourgeois subject. However, by declaring that "with Rousseau's novel [*Julie*] the focus of the family romance shifts from an 'ideal' father/daughter paradigm to one of a tender attachment to a mother/lover figure" and that "the paternal authority is replaced by a maternal agency in the service of a new social order, the modern state," (Mücke, *Virtue and the Veil of Illusion: Generic Innovation and the Pedagogical Project in Eighteenth-Century Literature* [Stanford: Stanford University Press, 1991], 16), she overlooks that this shift is accompanied by a new significance of the progenitor who overshadows the mother/son relationship. For a discussion of eighteenth-century gender roles in the generational process and the new family concept see Edward Shorter, *The Making of the Modern Family* (New York: Basic Books, 1975); Reinhard Sieder, *Sozialgeschichte der Familie* (Frankfurt am Main: Suhrkamp, 1987); Marc Shell, *Children of the Earth: Literature, Politics, and Nationhood* (Oxford and New York: Oxford University Press, 1993); and Suzanne Desan, *The Family on Trial in Revolutionary France* (Berkeley: University of California Press, 2004).

7. We can observe vivid interest for questions of lineage in texts by Romantic authors. Jochen Strobel discusses the fact that Medardus belongs to an aristocratic family ("Die Ahnenprobe des Medardus," *E. T. A. Hoffmann Jahrbuch* 13 [2005]: 29–46). Referring to Hoffmann's *Das Majorrat*, Achim von Arnims *Die Majorratsherren*; and Adam Müller's *Elemente der Staatskunst* (among others), Strobel argues against the claim that German Romanticism was bourgeois and speaks of a new significance of aristocracy in the post-Revolution decades. See also Peter Philipp Riedl, "Die Zeichen der Krise. Erbe und Eigentum in Achim von Arnims 'Die Majoratsherren' und E. T. A. Hoffmanns 'Das Majorrat,'" *Aurora* 52 (1992): 17–50.

8. For a comprehensive overview of Hoffmann's sources see James M. McGlathery, *Mysticism and Sexuality: E. T. A. Hoffmann* (Las Vegas: Peter Lang, 1981).

9. Hartmut Steinecke, "Die Elixiere des Teufels [Kommentar], E. T. A. Hoffmann," in *Sämtliche Werke in sechs Bänden*, ed. Hartmut Steinecke and Wulf Segebrecht (Frankfurt am Main: Deutscher Klassiker Verlag, 1988), 2:546; and Wulf Segebrecht, "Krankheit und

Gesellschaft," in *Romantik in Deutschland: Ein interdisziplinäres Symposion*, ed. Rolf Dieter [Richard] Brinkmann (Stuttgart: Metzler, 1978), 280–81.

10. Peter Vodosek, "Eine Leihbibliothek der Goethe-Zeit. Das 'Königl Privilegierte neue LeseInstitute' des Carl Friedrich Kunz zu Bamberg," *Jahrbuch des Wiener Goethe-Vereins* 77 (1973): 130.

11. In his memoirs, Carl Friedrich Kunz writes: "Ehe wir (d. i. Kunz und Hoffmann. Der Verf.) zu dem Entschluß gelangten, irgend ein Werk der neuern Literatur zum Lesegegenstand zu wählen, dauerte es lange. Stundelang wählten wir oft in meiner Bibliothek unter mehrern tausend Bänden, ohne etwas zu finden, wie die Hoffmann'sche Stimmung es wünschte . . ." [It often took a long time until we (Kunz and Hoffmann) came to the decision to select any of the newer literary pieces as reading material. Often we searched many hours in my library among several thousand volumes, without finding anything that would correspond to Hoffmann's mood] (Vodosek, "Eine Leihbibliothek," 110). Hoffmann worked on parts of the catalogue in November 1812 and on December 9, he wrote the preface for the catalogue; see (Hoffmann, *Späte Prosa. Briefe. Tagebücher und Aufzeichnungen. Juristische Schriften. Werke 1814–1822*, in *Sämtliche Werke in sechs Bänden*, ed. Hartmut Steinecke and Wulf Segebrecht [Frankfurt am Main: Deutscher Klassiker Verlag, 2004], 6:1111).

12. See also Stefani Engelstein, "Reproductive Machines in E. T. A. Hoffmann," in *Body Dialects in the Age of Goethe*, ed. Marianne Henn and Holger A. Pausch (Amsterdam: Rodopi, 2003), 169–93.

13. Hoffmann's work is preoccupied with genealogy and the question of an inherited evil. In *Das Majorat* (1817) the characters blame the ancestors for all their misery in life (Hoffmann 3:284). The transmission of insanity from parent to child shapes not only E. T. A. Hoffmann's *Elixiere*, but it also surfaces in Hoffmann's *Der goldene Topf* (1813–1814). In view of his daughter's wedding, Konrektor Paulmann mentions to his future son-in-law: "Ich will indesssen glauben, daß es die Liebe ist die Euch in dem Gehirn spukt, das gibt sich aber bald in der Ehe, sonst wäre mir bange, daß auch Sie in einigen Wahnsinn verfallen verehrungswürdiger Hofrat, und würde dann Sorge tragen wegen der Deszendenz die das Malum der Eltern vererben könnte" (Hoffmann, *Der Goldene Topf*, in *Sämtliche Werke in sechs Bänden*, ed. Hartmut Steinecke and Wulf Segebrecht [Frankfurt am Main: Deutscher Klassiker Verlag, 2003] 2.1:315). [I will suppose, however, that it is love which haunts your brain. Marriage will soon set all this to rights. Otherwise I should be apprehensive lest you too were falling into some degree of madness, most honored Privy Councillor. Then what would become of the future branches of the family, inheriting the *malum* of their parents? (Hoffmann, *The Golden Pot*, in *Tales of Hoffmann*, trans. Christopher Lazare [New York: Grover Press, 1946], 212).]

14. Johann Christian Reil, *Rhapsodien über die Anwendung der psychischen Curmethode auf Geisteszerrüttungen* (Halle: Curt, 1803; repr. Amsterdam: Bonset, 1968), 17; and Johann Gaspar Spurzheim, *Observations on the Deranged Manifestations of the Mind; or, Insanity* (Boston: March, Capen & Lyon, 1833; repr. Gainesville, FL: Scholars Facsimiles and Reprints, 1970).

15. Cf. Hoffmann, *Späte Prosa*, in *Sämtliche Werke*, ed. Steinecke and Segebrecht, 6:336, 6:702. Daniel Schmolling, an apprentice in the tobacco industry, had killed his girlfriend Henriette Lehne, although, shortly before the murder, he had been very affectionate to her. His case spurred a lively debate about whether or not Schmolling should be held

responsible for his deed or not, considering a possible psychological deformity. Hoffmann read a wide range of medical literature as well as Immanuel Kant's *Anthropologie* and Moritz's *Magazin der Erfahrungsseelenkunde* in preparation for the case. In his report, he argued that medicine can only make statements on the physical aspect of the human being (Hartmut Mangold, *Gerechtigkeit durch Poesie: Rechtliche Konfliktsituationen und ihre literarische Gestaltung bei E. T. A. Hoffmann* [Wiesbaden: Deutscher Universitäts Verlag, 1988], 80). In his view, only a very restricted number of mental problems belonged in this area. This confirms the thesis and shows that the *Elixiere's* argument is parallel to Hoffmann's professional activities and does *not* fulfill a compensating function for the author. Eventually, Hoffmann dismissed appeals to consider Schmolling not responsible for his deeds. The Schmolling case remained important for German discussions on mental soundness and guilt and was taken up again by Georg Büchner when preparing for *Woyzeck;* see Hoffmann's *Juristische Arbeiten*; Mangold, *Gerechtigkeit*, 71–93; and also Friedhelm Auhuber, *In einem fernen dunklen Spiegel. E. T. A. Hoffmanns Poetisierung der Medizin* (Opladen: Westdeutscher Verlag, 1986).

16. Spurzheim, *Observations*, 96.

17. "I conclude, that the embryon is produced solely by the male, and that the female supplies it with a proper nidus, with sustenance, and with oxygenation. . . ." (Erasmus Darwin, *Zoonomia*, 1:489).

"The living filament is a part of the father, and has therefore certain propensities, or appetencies, which belong to him; which may have been gradually acquired during a million of generations, even from the infancy of the habitable earth. . . ." (1:526).

18. "as a scrophulous race is frequently produced among the poor from the deficient stimulus of bad diet, or of hunger; and among the rich, by a deficient irritability from their having been long accustomed to too great stimulus, as of vinous spirit" (ibid., 1:502; 1:511).

"In some cases by the nutriment originally deposited by the mother the filament acquires parts not exactly similar to those of the father, as in the production of mules and mulattos. In other cases, the deficiency of this original nutriment causes deficiencies of the extreme parts of the fetus, which are last formed, as the fingers, toes, lips. In other cases, a duplicature of limbs are caused by the superabundance of this original nutritive fluid, as in the double yolks of eggs, and the chicken from them with four legs and four wings" (ibid., 1:526).

19. ". . . hereditary diseases, as scrophula, epilepsy, insanity, have their origin produced in one or perhaps two generations; as in the progeny of those who drink much vinous spirit; and those hereditary propensities cease again, as I have observed, if one or two sober generations succeed; otherwise the family becomes extinct." (ibid., 1:527)

20. Hoffmann, *Die Elixiere des Teufels*, 2.2:300.

21. Hoffmann, *The Devil's Elixirs*, trans. Ronald Taylor (New York: John Calder, 1963), 273.

22. Hoffmann, *Die Elixiere des Teufels*, 2.2:15, italics mine.

23. Hoffmann, *Devil's Elixirs*, 3, italics mine.

24. One could argue that there is no gap between the first and second part of the sentence because the protagonist differentiates between "Verhältnisse" [conditions] and qualities of the father. However, in the following third sentence (see below), the protagonist, referring to the mother's narrations, describes exactly these "Verhältnisse." Hence, the contradiction as outlined above is confirmed.

25. Hoffmann, *Die Elixiere des Teufels*, 2.2:15.

26. Hoffmann, *Devil's Elixirs*, 3.

27. Hoffmann, *Die Elixiere des Teufels*, 2.2:15.

28. Hoffmann, *Devil's Elixirs*, 3.

29. One of the exceptions is Friedrich A. Kittler, "Die Laterna magica der Literatur: Schillers und Hoffmanns Medienstrategien," *Athenäum. Jahrbuch für Romantik* 4 (1994): 219–37. See also Kittler, "Das Phantom unseres Ichs und die Literaturpsychologie: E. T. A. Hoffmann, Freud, Lacan," in *Romantikforschung seit 1945*, ed. Klaus Peter (Königstein/Ts.: Verlagsgruppe Athenäum, 1980), 335–56.

30. Hoffmann, *Die Elixiere des Teufels*, 2.2:11.

31. Hoffmann, *Devil's Elixirs*, 1.

32. Altrud Dumont claims that the discovery of the individual and its depiction coincides with the discovery of the abnormality of the human psyche. On the basis of this claim, Dumont sees a strong correspondence between Anton Reiser and Medardus. Cf. Dumont, "Die Einflüsse von Identitätsphilosophie und Erfahrungsseelenkunde auf E. T. A. Hoffmanns 'Elixiere des Teufels,'" *Zeitschrift für Germanistik*. 1 (1991): 37–54.

33. For the purpose of convenience and likelihood, I have addressed reader and editor here with the masculine pronoun. Of course, this does not mean that reader and editor might not be feminine.

34. Hoffmann, *Die Elixiere des Teufels*, 2.2:12.

35. Hoffmann, *Devil's Elixirs*, 2.

36. On the function of the camera obscura see Stiegler (1995) and Kaminski (2001).

37. "die sonderbaren Visionen des Mönchs . . . das regellose Spiel der erhitzten Einbildungskraft" (Hoffmann, *Die Elixiere des Teufels*, 2.2:11). "Nie hat mir meine Mutter gesagt, in welchen Verhältnissen mein Vater in der Welt lebte" (ibid., 2.2:15). "the monk's strange visions . . . the caprice of an inflamed imagination" (Hoffmann, *Devil's Elixirs*, 1). "My mother never told me of the circumstances in which my father lived" (ibid., 3).

38. "Entschließest du dich aber, mit dem Medardus, als seist du sein treuer Gefährte, durch . . . die bunte—bunteste Welt zu ziehen. . . ." (Hoffmann, *Die Elixiere des Teufels*, 2.2:12). "rufe ich mir aber alles, das in's Gedächtnis zurück" (ibid). "But if you decide to accompany Medardus . . . through the lurid episodes of his passage through the world . . ." (Hoffmann, *Devil's Elixirs*, 2). "but when I call to mind the stories" (ibid. 3).

39. "so wirst du dich vielleicht an den mannigfachen Bildern der camera obscura, die sich dir aufgetan, ergötzen" (Hoffmann, *Die Elixiere des Teufels*, 2.2:12). "was sie mir schon in meiner frühesten Jugend von ihm erzählte" (ibid., 2.2:15). "then, maybe, you will derive some pleasure from those glimpses of a camera obscura which have been vouchsafed to you" (Hoffmann, *Devil's Elixirs*, 2). "she told me about him in my childhood years" (ibid., 3).

40. Cf. Jacob Grimm and Wilhelm Grimm, *Deutsches Wörterbuch* (Leipzig: S. Hirzel, 1854–1913), vol. 6.

41. "Es kann auch kommen, daß das gestaltlosscheinende, so wie du es schärfer ins Auge fassest, sich dir bald deutlich und rund darstellt. Du erkennst den verborgenen Keim. . . ." (Hoffmann, *Die Elixiere des Teufels*, 2.2:12). "so muss ich wohl glauben, daß es ein mit tiefen Kenntnissen begabter lebenskluger Mann war. . . . Auf der beschwerlichen Wanderung dahin [zur heiligen Linde] fühlte meine Mutter nach mehreren Jahren der Ehe zum erstenmal, daß diese nicht unfruchtbar bleiben würde . . ." (ibid., 2.2:15). "It may even be that, as you look more closely, what seemed formless will become clear and precise; you will recognize the seed . . ." (Hoffmann, *Devil's Elixirs*, 2). "I cannot help thinking that he must

have been a man of great experience and profound knowledge. . . . During the wearisome journey my mother felt for the first time in the many years of her marriage that she would not remain childless . . ." (ibid. 3).

42. "Du erkennst den verborgenen Keim, den ein dunkles Verhängnis gebar, und der, zur üppigen Pflanze emporgeschossen, fort und fort wuchert in tausend Ranken bis eine Blüte zur Frucht reifend, allen Lebenssaft an sich zieht, und den Keim selbst tötet" (Hoffmann, *Die Elixiere des Teufels*, 2.2:12). "Eben aus diesen Erzählungen und einzelnen Äußerungen meiner Mutter, über ihr früheres Leben, die mir erst später veständlich worden, weiß ich, daß meine Eltern von einem bequemen Leben herabsanken in die drückendste bitterste Armut, und daß mein Vater, einst durch den Satan verlockt zum verruchten Frevel, eine Todsünde beging . . . er starb entsündigt und getröstet in demselben Augenblick, als ich geboren wurde" (ibid., 2.2:15). "you will come to recognise the hidden seed which, born of a secret union, grows into a luxuriant plant and spreads forth in a thousand tendrils, until a single blossom, swelling to maturity, abosrbs all the life-sap and kills the seed itself" (Hoffmann, *Devil's Elixirs*, 2). "From these stories, and from various remarks which my mother made about her earlier life and which I only later came to understand, I know that my parents sank from a life of plenty to a state of the most abject poverty. I learned too that my father had been led by Satan to commit a heinous crime, a deadly sin . . . He died, at peace in the knowledge of his salvation, at the moment I was born" (ibid., 3).

43. Hoffmann, *Die Elixiere des Teufels*, 2.2:33–34.

44. Hoffmann, *Devil's Elixirs*, 21.

45. Hoffmann, *Die Elixiere des Teufels*, 2.2:34.

46. Hoffmann, *Devil's Elixirs*, 21–22.

47. Hoffmann, *Die Elixiere des Teufels*, 2.2:275.

48. Hoffmann, *Devil's Elixirs*, 250.

49. Hoffmann, *Die Elixiere des Teufels*, 2.2:281.

50. Ibid., 2.2:282.

51. Ibid.

52. Hoffmann, *Devil's Elixirs*, 257.

53. Hoffmann, *Die Elixiere des Teufels*, 2.2:283, italics mine.

54. Hoffmann, *Devil's Elixirs*, 258.

55. Hoffmann, *Die Elixiere des Teufels*, 2.2:284.

56. Hoffmann, *Devil's Elixirs*, 258.

57. Death and distortion of the female body in the moment of desire or childbirth and her association with the devil is a recurrent topic in Hoffmann's oeuvre, see for example *Ignaz Denner* (Hoffmann, *Die Elixiere des Teufels*, 3:51), *Das öde Haus* (188), and *Vampyrismus* (1130). See also Gustafson (1999). For an excellent account of the close connection between sexual drive, imagination, and evil in Christian thought see Barbara Maria Stafford, *Body Criticism: Imaging the Unseen in Enlightenment Art and Medicine* (Cambridge, MA: MIT Press, 1991), 211–79.

58. Hoffmann, *Elixiere des Teufels*, 2.2:286. Hoffmann, *Devil's Elixirs*, 260.

59. Hoffmann, *Die Elixiere des Teufels*, 2.2:283.

60. By proposing this argument, I complicate Martha Helfer's paradigmatically new reading of Romantic texts as shaped by "male muses" and an exclusion of female voices. Discussing Hoffmann's *Goldene Topf*, Helfer states that the "recursive process of male *autopoiesis* is jeopardized by a malevolent female force, the Äpfelweib, who is explicitly linked

to the biblical Fall story from the start of the text. The Äpfelweib, as her name suggests, is not just any woman or the stock evil witch, but Woman herself, who must be killed off as a precondition for the male poet to write" (Helfer, "The Male Muse," 309). While my interpretation of the *Elixiere* confirms Helfer's claim of a male *autopoiesis* in Romantic poetology, I argue that Hoffmann's texts are shaped by an awareness of the underlying gender politics and takes a meta-critical stance towards it by showing that "Woman herself " is a product of "the self-positing male subject" (300).

61. For a detail discussion of the history of monstrous imagination Marie Hélène Huet, *Monstrous Imagination* (Cambridge, MA: Harvard University Press, 1993), in particular 13–123.

62. Erasmus Darwin's reinvention of the father's creative powers, vis-à-vis a merely procreative woman, proved to be extremely attractive for Romantic authors who were searching for ways to conceptualize their artistic endeavors as different from classical models. Furthermore, in a time in which more and more women took to writing, male authors like Wilhelm Humboldt, "Über den Geschlechtsunterschied (1794)," in *Schriften zur Anthropologie und Geschichte*, ed. Andreas Flitner, 4th ed. (Darmstadt: Wissenschaftliche Buchgesellschaft, 2002), 1:268–95; and Friedrich Schlegel (*Lucinde*) referred to such biological models to relegate women to childbirth. Schlegel's narrator for example describes himself as fathering the novel-son *Lucinde*: "was soll mein Geist seinem Sohne geben . . .?"(2) [what should my mind give to my son . . .?], and at the same time limits the pregnant Lucinde to her motherly duties: "Denke, daß eine neue Ordnung der Dinge für Dich beginnt. . . . Nun ist aber etwas da, worauf Du immer Rücksicht nehmen, worauf Du alles beziehen wirst" (223) [Remember that you're now beginning a new order of things. . . . But now something else exists which you'll always have to take into consideration, around which your whole world will turn (Friedrich Schlegel, *Lucinde and the Fragments*, trans. Peter Firchow [Minneapolis: University of Minnesota Press, 1971], 112).]. Hoffmann's critique of the self's biologization, therefore, also implies a critique of the exclusion of women from artistic production on the basis of (claimed) sexual differences and their contribution to reproduction.

63. *New American Standard Bible*, Deuteronomy 5, 7–10.

64. This reinterpretation of reproduction also explains why in Hoffmann's *Der goldene Topf* the becoming artist, Anselmus, is a copyist. Reproducing the text, Anselmus creates the original, namely Archivarius Lindhorst's genealogy (Hoffmann, *Elixiere des Teufels*, 2.1:229–83).

65. Erasmus Darwin, *The Botanic Garden: A Poem, in Two Parts Containing* "The Economy of Vegetation" *and* "The Loves of the Plants"*: With Philosophical Notes* (London: Jones & Company, 1824). Darwin's "The Loves of the Plants" (1789), here "advertisement," no page number given) forms the second part of Erasmus Darwin's *The Botanic Garden* (1789), an immensely popular work, both in Britain and in other countries. Goethe, for example, mentions it in a letter to Schiller from January 26–27, 1798 (Goethe, *Sämtliche Werke. Briefe, Tagebücher und Gespräche*, ed. Dieter Borchmeyer, series 2, 4:486–88).

66. The question of determinism was not only discussed in biology, but also highly debated in the legal sphere, where the answer had far-reaching consequences for criminal law: The culprit's fate was highly depending on whether freedom of will was assumed or not. While Samuel Pufendorf (*De jure naturae et gentium*, 1781); Karl Ludwig Grolman (*Begründung des Criminalrechts*, 1799); and Carl August Tittmann (*Handbuch der Strafrechtswissenschaft* [Halle: Bei Hemmerde und Schwetschke,1806]) assumed that the human being is

able to decide freely on his actions, Christian Thomasius first developed a deterministic legal theory (*Grundlehren des Natur- und Völckerrechts* [Hall im Magdeburgischen: Renger, 1709]). Johann Jakob Schmauß (*Neues Systema des Rechts der Natur* [Göttingen: Verlegts Abram Bandenhoecks seel. Witwe, 1754]); and Karl Ferdinand Hommel (*Über die Belohnung und Strafe nach türkischen Gesetzen*, 2nd ed. [Bayreuth, Leipzig, Lübeck 1772]) were to follow him in this argumentation. The controversy found perhaps its most prominent adversaries in Kant whose system was built on the assumption of a free will and Paul Anselm von Feuerbach (*Lehrbuch des gemeinen in Deutschland geltenden Peinlichen Rechts* [Gießen: Georg Friedrich Heyer, 1801]) who assumed the opposite. A summary of these debates can be found in Eberhard Schmidt, *Einführung in die Geschichte der deutschen Strafrechtspflege*, 3rd ed. (Göttingen: Vandenhoeck & Ruprecht, 1965); and Hermann Conrad, *Deutsche Rechtsgeschichte: Neuzeit bis 1806* (Karlsruhe: C. F. Müller, 1962), vol. 2.

67. Hoffmann, *Elixiere des Teufels*, 2.2:12, italics mine.

68. Hoffmann, *Devil's Elixirs*, 2.

69. For a discussion of the ambivalence of the "dunkle Macht" see Rainer Pabst, *Schicksal bei E. T. A. Hoffmann: zur Erscheinungsform, Funktion und Entwicklung eines Interpretationsmusters* (Köln: Böhlau, 1989). However, I disagree with Pabst's claim that Hoffmann's interest in the notion of unity and coherence is driven by the wish for deeper insight. Rather, it is Hoffmann's wish to show the construction of this unity.

70. It is no coincidence that Sigmund Freud—arguably the author who establishes the psychological discipline—relied in significant aspects of his work on Hoffmann. The connection between the *Sandmann* and the "Unheimliche" is explicit and has been discussed frequently, but equally striking is the connection between the father-son structure developed in *Totem und Tabu: Einige Übereinstimmungen im Seelenleben der Wilden und der Neurotiker* (Leipzig, H. Heller, 1913); and Hoffmann's oedipal tale in the *Elixiere* where the father's death is the condition of possibility of the son's existence and the reason for the father's subsequent idolization (albeit negative). However, while Freud tells the primal scene as a chronologically organized narrative, it is an important detail of the *Elixiere* that the father's death *coincides* with the son's birth—and is recounted by the mother. As a self-referential gesture this absence at the origin not only allows alternative (maternal) tales of the individual's development, but it also turns Hoffmann's story in a meta-critical device. See also the ample references to the *Elixiere* in psychological texts around 1900; Freud's *Das Unheimliche* (257); or Otto Rank's "Der Doppelgänger," in *Psychoanalytische Beiträge zur Mythenforschung: Gesammelte Studien aus den Jahren 1912 bis 1914* (Leipzig and Vienna: Internationaler Psychoanalytischer Verlag, 1919), 267–354; and Carl Gustav Jung's *Das Unbewußte im normalen und kranken Seelenleben* (1925) (Zurich: Rascher, 1926), as quoted in Steinecke, "Die Elixiere des Teufels," 2:574.

71. Hoffmann, *Die Elixiere des Teufels*, 2.2:571–72.

72. Robert J. Richards, *The Romantic Conception of Life: Science and Philosophy in the Age of Goethe* (Chicago: University of Chicago Press, 2002), 329. For a similar approach see Astrida Orle Tantillo, *The Will to Create: Goethe's Philosophy of Nature* (Pittsburgh, PA: University of Pittsburgh Press, 2002). One might also consider Clifford Geertz's notion of "Thick Description" a variation of this approach. On methodological questions in the context of writing cultural history see his "Thick Description: Toward an Interpretive Theory of Culture," in *The Interpretations of Cultures: Selected Essays* (New York: Basic Books, 1973), 3–30.

73. Johann Wolfgang Goethe, *Sämtliche Werke: Briefe, Tagebücher und Gespräche*, ed. Dieter Borchmeyer (Frankfurt am Main: Deutscher Klassiker Verlag, 1987), series 2, 15:393–94; punctuation as given.

74. Ibid., series 2, 15: 394.

75. Hoffmann, *Die Elixiere des Teufels*, 2.2:12.

76. Goethe, *Sämtliche Werke*, ed. Dieter Borchmeyer, series 1, 24:233–34.

See also Richards, *The Romantic Conception of Life*, 416.

77. See Goethe's account of the discussion of the problem with Schiller in "Glückliches Ereignis" (12: 88–89).

78. Diary entry from April 17, 1787, Goethe, *Sämtliche Werke: Briefe*, 15:455.

79. Goethe, *Sämtliche Werke: Briefe, Tagebücher und Gespräche*, ed. Dieter Borchmeyer, series 2, 3:163–64.

80. Medardus's Rosalia experience comes surprisingly close to Goethe's account: "Ein Altar in unserer Kirche war der heiligen Rosalia geweiht, und ihr herrliches Bild in dem Moment gemalt, als sie den Märtyrer Tod erleidet.—Es war meine Geliebte . . . Da lag ich [Medardus] stundenlang, wie von verderblichem Wahnsinn befangen, niedergeworfen auf den Stufen des Altars . . ." (Hoffmann, *Die Elixiere des Teufels*, 2.2:52). [There was an altar in our chapel dedicated to Saint Rosalia, above which hung a wonderful painting of her at the moment of martyrdom. It was my beloved . . . As though in the terrible grip of madness, I lay prostrate for hours on the steps of the altar . . . (Hoffmann, *Devil's Elixirs*, 39).]

81. Wihelm Dilthey, *Introduction to the Human Sciences: An Attempt to Lay a Foundation for the Study of Society and History*, trans. and intro. Ramon J. Betanzos (Detroit: Wayne State University Press, 1988); and C. P. Snow, *The Two Cultures and the Scientific Revolution. The Rede Lecture 1959* (New York and London: Cambridge University Press, 1961).

82. See Elisabeth Kutzer, *Zum Stammbaumroman in der neueren Literatur* (PhD diss., Universität Leipzig, Leipzig: Gerhardt, 1929); and A. E. Zucker, "The Genealogical Novel, a New Genre," *PMLA* 43, no. 2 (June 1928): 551–60.

83. Joseph Carroll, "Pluralism, Poststructuralism, and Evolutionary Theory," *Academic Questions* 9, no. 3 (Summer 1996): 43–57; and his "Evolution and Literary Theory" *Human Nature* 6, no. 2 (June 1995): 119–34.

84. See Charles Taylor, *Sources of the Self: The Making of the Modern Identity* (Cambridge, MA: Harvard University Press, 1989); Carolyn Steedman, *Strange Dislocations: Childhood and the Idea of Human Interiority, 1780–1930* (Cambridge, MA: Harvard University Press, 1995); Larry Wolff, "When I Imagine a Child: The Idea of Childhood and the Philosophy of Memory in the Enlightenment," *Eighteenth-Century Studies* 31, no. 4 (Summer 1998): 377–401; Corey Roberts, "German Pietism and the Genesis of Literary Aesthetics: The Discourse of 'Erfahrung' in the 1700," *Deutsche Vierteljahrsschrift* 78, no. 2 (June 2004): 200–28; and Wolfram Malte Fues, "Der Pietismus im Roman der deutschen Aufklärung," *Deutsche Vierteljahrsschrift* 78, no. 4 (December 2004): 535–49.

85. On the question of whether or not the actual novels lived up to this promise, see also chap. 2.

86. Karl Philipp Moritz, *Anton Reiser* (Stuttgart: Reclam, 1986), 12.

87. See also Schreiber, *Topography of Modernity*.

88. Elisabeth Kutzer has described such novels in the late 1920s under the rubric of the genre *Stammbaumroman* (genealogical novel). She defines the genre using the following criteria: Focus on different generations of one family, characters are defined by their family

and their function in this family, family is a political entity, decisive influence of progenitor on the fate and moral behavior of descendants (focus on vertical direction). See Kutzer, *Zum Stammbaumroman.*

89. Thomas Mann, *Bekenntnisse des Hochstaplers Krull* (Frankfurt a. M.: S. Fischer, 2012), 352.

90. Thomas Mann, *Confessions of Felix Krull: Confidence Man*, trans. Denver Lindley (New York: Knopf, 1955), 304.

91. Mann, *Bekenntnisse*, 352.

92. Mann, *Confessions*, 304–05.

93. It is fitting that in the first concrete mentioning of the novel in *Notizbuch* 7 (notebook 7), Mann remarks: "*Zum Hochstapler* Er markiert in gefährlichen Augenblicken Husten-Anfälle, aus dem Instinkt" [*On the Parvenue* In dangerous moments, he fakes a coughing fit, by instinct]. Mann quoted in Thomas Sprecher and Monica Bussmann, *Bekenntnisse des Hochstaplers Felix Krull. Memoiren erster Teil. Kommentar* (Frankfurt a. M.: S. Fischer, 2012), 17.

BIBLIOGRAPHY

Adelung, Johann Christoph. *Grammatisch-kritisches Wörterbuch der hochdeutschen Mundart.* vol. 1. Leipzig: Breitkopf und Compagnie, 1793.

Agamben, Giorgio. *Homo sacer: Die souveräne Macht und das nackte Leben.* Frankfurt am Main: Suhrkamp, 2002.

Allgemeine Encyclopädie der Wissenschaften und Künste. Edited by Johann Samuel Ersch and Johann Gustav Gruber. vol. 4. Leipzig: Johann Friedrich Gleditsch, 1819.

Allgemeines Landrecht für die Preußischen Staaten. 4 vols. Berlin: Pauli, 1796.

Allgemeines Landrecht für die Preußischen Staaten. February 5, 1794. http://www.smixx.de/ra/Links_F-R/PrALR/PrALR_II_1.pdf (accessed March 9, 2012).

Ammerlahn, Hellmut. "Wilhelm Meisters Mignon—ein offenbares Rätsel. Name, Gestalt, Symbol, Wesen und Werden." *DVJS* 42, no. 1 (1968): 89–116.

———. "Puppe-Tänzer-Dämon-Genius-Engel: Naturkind, Poesiekind und Kunstwerdung bei Goethe." *German Quarterly* 54, no. 1 (January 1981): 19–32.

Anderson, Lorin. *Charles Bonnet and the Order of the Known.* Boston: D. Reidel, 1982.

Apel, Karl-Otto. *Die Idee der Sprache in der Tradition des Humanismus von Dante bis Vico.* Bonn: H. Bouvier, 1963.

Ariès, Philippe. *Centuries of Childhood: A Social History of Family Life.* New York: Vintage Books, 1965.

Aristotle. *Generation of Animals.* Translated by A. L. Peck. Cambridge, MA: Heinemann and Harvard University Press, 2000.

Arnaud, Georg. *Anatomisch-Chirurgische Abhandlung über die Hermaphroditen, aus dem Französischen übersetzt, nebst Kupfertafeln.* Straßburg: Amand König, 1777.

Arnim, Achim von. "Die Majorrathsherren." In *Erzählungen.* Edited by Gisela Henckmann. 211–251. Stuttgart: Reclam, 1991.

Atsma, Aaron J. *Theoi Project.* 2000. http://www.theoi.com/Protogenos/Phanes.html (accessed March 29, 2012).

Augustine. *City of God.* Translated by Marcus Dods. Edinburgh: T & T Clark, 1913.

Auhuber, Friedhelm. *In einem fernen dunklen Spiegel. E. T. A. Hoffmanns Poetisierung der Medizin.* Opladen: Westdeutscher Verlag, 1986.

Baßler, Moritz. "Goethe und die Bodysnatcher." In *Von der Natur zur Kunst zurück*, edited by Moritz Baßler et al., 181–97. Tübingen: Niemeyer, 1997.

Baer, Karl Ernst von. *Über Entwicklungsgeschichte der Thiere. Beobachtung und Reflexion*. Edited and introduction by Olaf Breidbach. Königsberg: Bornträger, 1828–1837. Reprint Hildesheim, Zürich, and New York: Olms-Weidmann, 1999.

Baudelaire, Charles. *The Painter of Modern Life and other Essays*. Translated and edited by Jonathan Mayne. London: Phaidon Press, 1964.

Bauer, Volker. "Die gedruckte Ahnentafel als Ahnensformular: zur Interferenz von Herrschafts-, Wissens- und Medienordnung in der Universalgenealogie des 17. Jahrhunderts." In *Die Ahnenprobe in der Vormoderne*, edited by Elizabeth Harding and Michael Hecht, 125–56. Münster: Rhema, 2011.

Bäumer, Konstanze. "Wiederholte Spiegelungen-Goethes 'Mignon' und die 'Neue Melusine.'" In *Goethes Mignon und ihre Schwestern*, edited by Gerhart Hoffmeister, 113–33. New York: Peter Lang, 1993.

Bayertz, Kurt, Jürgen Kroll, and Peter Weingart. *Rasse, Blut und Gene: Geschichte der Eugenik und der Rassenhygiene in Deutschland*. Frankfurt am Main: Suhrkamp, 1988.

Behrens, Rudolf. *Umstrittene Theodizee, erzählte Kontingenz: Die Krise teleologischer Weltdeutung und der französische Roman (1670–1770)*. Tübingen: Niemeyer, 1994.

Berend, Eduard. "Anmerkungen." In *Jean Pauls Sämtliche Werke*, series 1, 16:477–508. Museum. Weimar: Böhlaus Nachfolger, 1938.

———. "Einleitung." In *Jean Pauls Sämtliche Werke*, series 1, 15:v-lxvi. Der Komet. Weimar: Böhlaus Nachfolger, 1937.

Bergen, Joachim Christian. *Anleitung für die Landwirthe zur Verbesserung der Viehzucht*. Berlin and Stralsund: G. A. Lange, 1781.

Bergengruen, Maximilian. *Schöne Seelen, groteske Körper: Jean Pauls ästhetische Dynamisierung der Anthropologie*. Hamburg: Meiner, 2003.

Berhorst, Ralf. *Anamorphosen der Zeit: Jean Pauls Romanästhetik und Geschichtsphilosophie*. Tübingen: Niemeyer, 2002.

Berman, Marshall. *All That is Solid Melts Into Air: The Experience of Modernity*. New York: Simon & Schuster, 1982.

Bischoff, Christian Heinrich Ernst. *Darstellung der Gallschen Gehirn- und Schädel-Lehre. Nebst Bemerkungen über diese Lehre von Dr. C. W. Hufeland*. Zweite, vermehrte und verbesserte Auflage. Berlin: In Commission bei L. W. Wittich, 1805.

Blessin, Stefan. *Goethes Romane: Aufbruch in die Moderne*. Paderborn: Schöningh, 1996.

Blumenbach, Johann Friedrich. *Beyträge zur Naturgeschichte Erster Theil*. Göttingen: Dieterich, 1790.

———. *An Essay on Generation* (Über den Bildungstrieb). London: T. Cadell, Strand, 1792.

———. *Über den Bildungstrieb und das Zeugungsgeschäfte*. Göttingen: Dieterich 1781. Reprint 1791.

Borcherdt, Hans Heinrich. *Der Roman der Goethezeit*. Urach: Port, 1949.

Börne, Ludwig. *Sämtliche Schriften*. Edited by Inge Rippmann and Peter Rippmann, vol. 1. Düsseldorf: Joseph Melzer, 1964.

Breidbach, Olaf. "Einleitung." In *Über Entwicklungsgeschichte der Thiere: Beobachtungen und Reflexion*, by Karl Ernst von Baer, v-xxiii. Hildesheim, Zürich and New York: Olms-Weidmann, 1999.

Breithaupt, Fritz. *Der Ich-Effekt des Geldes*. Frankfurt a.M.: Fischer, 2008.

Brown, Jane K. "The Theatrical Mission of the *Lehrjahre*." In *Reflection and Action: Essays on the Bildungsroman*, edited by James N. Hardin, 142–62. Columbia: University of South Carolina Press, 1991.

Bruhns, Sandra-Valeska. "Ein paar Härchen sorgen für Gewissheit." *Die Zeit*. February 13, 2007.

Brunner, Otto, ed. "Das 'ganze' Haus und die alteuropäische 'Ökonomik.'" In *Neue Wege der Sozialgeschichte*, 33–61. Göttingen: Vandenhoeck & Ruprecht, 1956.

Buffon, Georges Louis Leclerc. *Buffon's Natural history, containing a theory of the earth, a general history of man, of the brute creation, and of vegetables, minerals, &c. &c.* 10 vols. London: J. S. Barr, 1792.

Carey, Nessa. *The Epigenetics Revolution: How Modern Biology is Rewriting our Understanding of Genetics, Disease, and Inheritance*. London: Icon Books, 2011.

Carroll, Joseph. "Pluralism, Poststructuralism, and Evolutionary Theory." *Academic Questions* 9, no. 3 (Summer 1996): 43–57.

———. "Evolution and Literary Theory." *Human Nature* 6, no. 2 (June 1995): 119–34.

Carter, Harold B. *His Majesty's Spanish Flock: Sir Joseph Banks and the Merinos of George III of England*. London: Angus & Robertson, 1964.

Carus, Friedrich August. *Nachgelassene Werke. Sechster Theil. Ideen zur Geschichte der Menschheit*. Leipzig: Iohann Ambrosius Barth und Paul Gotthelf Kummer, 1809.

Chaouli, Michel. *The Laboratory of Poetry: Chemistry and Poetics in the Work of Friedrich Schlegel*. Baltimore: Johns Hopkins University Press, 2002.

Conrad, Hermann. *Deutsche Rechtsgeschichte: Neuzeit bis 1806*. vol. 2. Karlsruhe: C. F. Müller, 1962.

Cook, Harold J. "Time's Bodies: Crafting the Preparation and Preservation of Naturalia." In *Merchants and Marvels: Commerce, Science, and Art in Early Modern Europe*, edited by Pamela H. Smith and Paula Findlen, 223–47. London: Routledge, 2002.

Cox, Joseph Mason. *Practical Observations on Insanity. . . :* London: Baldwin, 1804. Reprint 2nd corrected and enlarged ed. Philadelphia: Fry Kammerer for Thomas Dobson, 1811.

Cramer, Karin. "Bewußtseinspaltung in E. T. A. Hoffmanns Roman *Die Elixiere des Teufels*." *Mitteilungen der E. T. A. Hoffmann-Gesellschaft* 16 (1970): 8–18.

Daemmrich, Horst S. *The Shattered Self: E. T. A. Hoffmann's Tragic Vision*. Detroit: Wayne State University Press, 1973.

Dann, Otto. "Verbrecher aus Infamie" [Kommentar]. In *Werke und Briefe*, by Friedrich Schiller, vol. 7, 991–97. Frankfurt am Main: Deutscher Klassiker Verlag, 2002.

Darwin, Charles. *The Descent of Man, and Selection in Relation to Sex* (1871). Reprint Princeton, NJ: Princeton University Press, 1981.

———. *The Origin of Species by Means of Natural Selection; or, The Preservation of Favoured Races in the Struggle for Life*. 6th ed., with additons and corrections. London: John Murray, 1891. Reprint edited by Ernst Mayr. Cambridge, MA: Harvard University Press, 1964.

Darwin, Erasmus. *The Botanic Garden: A Poem, in Two Parts Containing* "The Economy of Vegetation" *and* "The Loves of the Plants": *With Philosophical Notes*. London: Jones & Company, 1824.

———. *Zoonomia; or, The Laws of Organic Life*, 2 vols. London: Printed for J. Johnson, in St. Paul's Church-Yard, 1794–1796.

Daston, Lorraine, and Peter Galison. *Objectivity*. New York: Zone Books, 2007.

Davidson, Jenny. *Breeding: A Partial History of the Eighteenth Century*. New York: Columbia University Press, 2009.

DeJager, Timothy F. *G. R. Treviranus (1776–1837) and the Biology of a World in Transition*. PhD diss., University of Toronto, 1991.

Delcourt, Marie. *Hermaphrodite: Myths and Rites of the Bisexual Figure in Classical Antiquity*. Translated by Jennifer Nicholson. London: Studio Books, 1961.

Desan, Suzanne. *The Family on Trial in Revolutionary France*. Berkeley: University of California Press, 2004.

Deutsche Encyclopädie oder allgemeines Real-Wörterbuch aller Künste und Wissenschaften von einer Gesellschaft Gelehrten. Edited by Heinrich Martin Gottfried Köster and Johann Friedrich Roos. 23 vols. Frankfurt am Main: Varrentrapp Sohn und Werner, 1778–1807.

Die Bibel. Die Gute Nachricht des Alten und Neuen Testaments. Edited by Deutsche Bibelgesellschaft. Stuttgart: Vandenhoeck & Ruprecht GmbH, 1982.

Dieterichs, Joachim Friedrich Christian. *Von der Zucht der Hunde, ihren vorzüglichen Rassen und den wichtigsten Krankheiten derselben*. Leipzig: Baumgärtner, 1831.

Dewitz, Hans-Georg, and Gerhard Neumann. "Kommentar zu 'Der Mann von fünfzig Jahren.'" In *Johann Wolfgang Goethe. Sämtliche Werke*, edited by Hendrik Birus, Dieter Borchmeyer, and Hans-Georg Dewitz et al., vol. 10, 1104–13. Frankfurt am Main: Deutscher Klassiker Verlag, 1989.

Dilthey, Wilhelm. *Introduction to the Human Sciences: An Attempt to Lay a Foundation for the Study of Society and History*. Translation and introduction by Ramon J. Betanzos. Detroit: Wayne State University Press, 1988.

Dugan, David, and Jeff Goldblum. *DNA: The Secret of Life. Pandora's Box*. Produced and directed by David Glover. Princeton, NJ: Educational Broadcasting Corporation and Windfall Films: Films for the Humanities & Sciences, 2003.

Dumont, Altrud. "Die Einflüsse von Identitätsphilosophie und Erfahrungsseelenkunde auf E. T. A. Hoffmanns 'Elixiere des Teufels.'" *Zeitschrift für Germanistik*. 1 (1991): 37–54.

Egger, Irmgard. "'Verbinden mehr als Trennen': Goethe und die plastische Anatomie." *Germanisch-romanische Monatsschrift* 51, no. 1 (2001): 45–53.

Eigen, Sara P. "A Mother's Love, a Father's Line: Law, Medicine, and the 18th-Century Fictions of Patrilineal Genealogy." In *Genealogie als Denkform in Mittelalter und Früher Neuzeit*, edited by Kilian Heck and Bernhard Jahn, 87–107. Tübingen: Niemeyer, 2000.

Eissler, Kurt Robert. *Goethe: eine psychoanalytische Studie 1775–1786*. Translated by Peter Fischer and Rüdiger Scholz, edited by. Rüdiger Scholz with Wolfram Mauser and Johannes Cremerius. 2 vols. Basel and Frankfurt am Main: Stroemfeld/Roter Stern, 1983–1985.

Elsaghe, Yahya A. "'Anni demunt': Die drei Paraphrasen des Mannes." *Zeitschrift für Deutsche Philologie* 112, no. 4 (1993): 509–28.

Engel, Manfred. *Der Roman der Goethezeit. Anfänge in Klassik und Frühromantik: Transzendentale Geschichten*. vol. 1. Stuttgart: J. B. Metzler, 1993.

Engelstein, Stefani. *Anxious Anatomy: The Conception of the Human Form in Literary and Naturalist Discourse*. Albany: State University of New York Press, 2008.

———. "Reproductive Machines in E. T. A. Hoffmann." In *Body Dialects in the Age of Goethe*, edited by Marianne Henn and Holger A. Pausch, 169–93. Amsterdam: Rodopi, 2003.

Esselborn, Hans. *Das Universum der Bilder: Die Naturwissenschaft in den Schriften Jean Pauls.* Tübingen: M. Niemeyer, 1989.

Étienne, Servais. *Le genre romanesque en France depuis l'apparition de la "Nouvelle Heloïse" jusqu'aux approches de la révolution.* Brüssel: M. Lamertin, 1922.

Feiler, Johann. *Über angeborene menschliche Missbildungen im Allgemeinen und Hermaphroditen insbesondere. Ein Beitrag zur Physiologie, pathologische Anatomie, und gerichtlichen Arzneiwissenschaft.* Landshut: Philipp Krüll, Universitätsbuchhändler, 1820.

Feuerbach, Paul Anselm von. *Lehrbuch des gemeinen in Deutschland geltenden peinlichen Rechts.* Gießen: Georg Friedrich Heyer, 1801.

Feuerstein-Herz, Petra. *Die große Kette der Wesen: Ordnungen in der Naturgeschichte der Frühen Neuzeit.* Wiesbaden: Harrassowitz, 2007.

Fick, Monika. *Das Scheitern des Genius: Mignon und die Symbolik der Liebesgeschichten in "Wilhelm Meisters Lehrjahre."* Würzburg: Königshausen & Neumann, 1987.

Fincke, Johann Heinrich. *Schreiben über ungekünstelte und sichere Verfeinerung aller groben Wolle aus eigener Erfahrung und Thathandlungen aufgestellt.* Celle: Ernst August Richter, 1790.

Fink, Gonthier-Louis, Gerhart Baumann, and Johannes John, "Kommentar," in *Johann Wolfgang Goethe. Sämtliche Werke nach Epochen seines Schaffens.* Edited by Karl Richter et al. München: Carl Hanser, 1991. 17:1131.

Fischer, Johann Leonhard. *Anweisungen zur praktischen Zergliederungskunst. Nach Anleitung des Thomas Pole.* Leipzig: Weygandsche Buchhandlung, 1791.

Fleming, Paul. "The Crisis of Art: Max Kommerell and Jean Paul's Gestures." *Modern Languages Notes* 115, no. 3 (April 2000): 519–43.

Forster, E. M. *The Longest Journey.* Edinburgh: William Blackwood and Sons, 1907.

Forster, Georg. "Beschluss der im vorigen Monat angefangenen Abhandlung des Herrn G. R. Forsters über die Menschenrassen." *Teutscher Merkur* (November 1786): 150–66.

———. "Noch etwas über die Menschenrassen. An Herrn Dr. Biester." *Teutscher Merkur* (October 1786): 57–86.

Foucault, Michel. *Herculine Barbin: Being the Recently Discovered Memoirs of a Nineteenth-Century French Hermaphrodite.* Translated by Richard McDougall. New York: Pantheon Books, 1980.

———. *The History of Sexuality.* Translated by Robert Hurley. 3 vols. New York: Vintage, 1990.

———. *The Order of Things: An Archeology of the Human Sciences.* New York: Vintage Books, 1973.

Francis, Richard C. *Epigenetics: The Ultimate Mystery of Inheritance.* New York: W. W. Norton, 2011.

Frank, Johann Peter. *System einer vollständigen medicinischen Policey. Erster Band. Von Fortpflanzung der Menschen und Ehe-Anstalten, von Erhaltung und Pflege schwangerer Mütter, ihrer Leibesfrucht und der Kind-Betterinnen in jedem Gemeinwesen.* Mannheim: Schwan und Götz, 1784.

Freud, Sigmund. *Totem und Tabu: Einige Übereinstimmungen im Seelenleben der Wilden und der Neurotiker.* Leipzig, H. Heller, 1913.

———. "Das Unheimliche." In *Psychoanalytische Studien an Werken der Dichtung und Kunst.* by Sigmund Freud, 99–138. Leipzig: Internationaler Psychoanalytischer Verlag, 1924.

Friedenthal, Richard. *Goethe: His Life and His Times.* New York: Transaction Publishers, 2010.

Fues, Wolfram Malte. "Der Pietismus im Roman der deutschen Aufklärung." *Deutsche Vierteljahrsschrift* 78, no. 4 (December 2004): 535–49.

Fuhrmann, Helmut. *Der androgyne Mensch: "Bild" und "Gestalt" der Frau und des Mannes im Werk Goethes*. Würzburg: Königshausen & Neumann, 1995.

Gaiffe, Félix Alexandre. *Le Drame en France au XVIIIe siècle*. Paris: Librairie Armand Colin, 1910.

Gallagher, James. "Gene therapy: Glybera approved by European Commission." November 2, 2012. http://www.bbc.co.uk/news/health-20179561 (accessed June 24, 2013).

Galsworthy, John. *The Forsyte-Saga*. London, W. Heinemann, 1922.

Gaudillière, Jean-Paul, and Ilana Löwy. *Heredity and Infection: The History of Disease and Transmission*. London and New York: Routledge, 2001.

Gayon, Jean. "Entre force et structure: Genèse du concept naturaliste de l'hérédité." In *Le paradigme de la filiation*, edited by Jean Gayon and Jean-Jacques Wunenburger, 61–75. Paris: l'Harmattan, 1996.

Geertz, Clifford. "Thick Description: Toward an Interpretive Theory of Culture." In *The Interpretations of Cultures: Selected Essays*, 3–30. New York: Basic Books, 1973.

Gehlen, Arnold. *Man. His Nature and Place in the World*. New York: Columbia University Press, 1988.

Gericke, Friedrich Karl Gustav. *Praktische Anleitung zur Führung der Wirthschafts-Geschäfte für angehende Landwirthe. Erster Theil. Von der Viehzucht. Herausgegeben und mit einer Vorrede begleitet von Albrecht Thaer*. Edited by Albrecht D. Thaer. Berlin: Realschulbuchhandlung, 1804.

Germline Gene Therapy. [U.S. NIH]. March 2006. http://www.genome.gov/10004764 (accessed June 24, 2013).

Germline Therapy. Human Genetic Modification: Embryo Protection Law [Germany]. n.d. http://www.dnapolicy.org/policy.international.php?action=detail&laws_id=38 (accessed June 24, 2013).

Gigante, Denise. *Life: Organic Form and Romanticism*. London and New Haven, CT: Yale University Press, 2009.

Gilbert, Ruth. *Early Modern Hermaphrodites: Sex and Other Stories*. New York: Palgrave, 2002.

Girard, René. *Violence and the Sacred*. London and New York: Athlone Press, 2005.

Gliboff, Sander. *H. G. Bronn, Ernst Haeckel, and the Origins of German Darwinism: A Study in Translation and Transformation*. Cambridge, MA: MIT Press, 2008.

Goethe, Johann Wolfgang. *Conversations of Goethe with Eckermann*. Translated by John Oxenford. Edited by J. K. Moorhead, pseud. London: Dent, 1930.

———. *Goethe's Botanical Writings*. Translated by Bertha Mueller. With an Introduction by Charles J. Engard. Woodbridge, Connecticut: Ox Bow Press, 1989.

———. *From My Life. Poetry and Truth*. Part Four. Edited by Thomas P. Saine and Jeffrey L. Sammons, translated by Robert R. Heitner. New York: Suhrkamp, 1987.

———. *Sämtliche Werke. Briefe, Tagebücher und Gespräche*. Edited by Dieter Borchmeyer et al., 45 vols. Series 1, vols. 1–27; Series 2, vols. 1–18. Frankfurt: Deutscher Klassiker Verlag, 1985–1999.

———. *Sämtliche Werke nach Epochen seines Schaffens*. Edited by Karl Richtert et al., 20 vols. München: Carl Hanser, 1985–1998.

———. *Werke*. Edited by Erich Trunz, 14 vols. Hamburg: C. Wegner, 1948–1960.

———. *Wilhelm Meister's Journeyman Years or The Renunciants*.Translated by Krishna Winston. Edited by Jane K. Brown. New York: Suhrkamp, 1989.

———. *Wilhelm Meister's Apprenticeship*. Edited and translated by Eric A. Blackall in cooperation with Victor Lange. New York: Suhrkamp, 1989.

Goethe, Johann Wolfgang and Friedrich Schiller. *Correspondence between Goethe and Schiller 1794–1805*. Translated by Liselotte Dieckmann. New York: Peter Lang, 1994.

Goody, Jack. "Inheritance, Property and Women: some Comparative Considerations." In *Family and Inheritance: Rural Society in Western Europe, 1200–1800*, edited by Jack Goody, Joan Thirsk, and Edward P. Thompson, 10–36. Cambridge: Cambridge University Press, 1976.

Graf von Pfeil. "Ueber Voll-Veredlung der Schafe durch spanische Race (mit Tabellen)." In *Annalen des Ackerbaus 1807/6*, edited by Albrecht Daniel Thaer, 222–59. Berlin: Realschulbuchhandlung, 1807.

Gray, Richard T. *Money Matters: Economics and the German Cultural Imagination, 1770–1850*. Seattle: University of Washington Press, 2008.

Gray, Ronald Douglas. *Goethe the Alchemist: A Study of Alchemical Symbolism in Goethe's Literary and Scientific Works*. Cambridge: Cambridge University Press, 1952. Reprint 2010.

Greenblatt, Stephen. "Fiction and Friction." In *Reconstructing Individualism: Autonomy, Individuality, and the Self in Western Thought*, edited by Thomas C. Heller, Christine Brooke-Rose et al., 30–52. Stanford, CA: Stanford University Press, 1986.

Grene, Marjorie, and David J. Depew. *The Philosophy of Biology: An Episodic History*. New York and Cambridge: Cambridge University Press, 2004.

Grimm, Jacob, and Wilhelm Grimm. *Deutsches Wörterbuch*. 15 vols. Leipzig: S. Hirzel, 1854–1913.

Groß, Sabine. "Diskursregelung und Weiblichkeit: Mignon und ihre Schwestern." In *Goethes Mignon und ihre Schwestern*, edited by Gerhart Hoffmeister, 83–99. New York: Peter Lang, 1993.

Grolman, Karl Ludwig. *Begründung des Criminalrechts*. Gießen: Georg Friedrich Heyer, 1799.

Gustafson, Susan E. "The Cadaverous Bodies of Vampiric Mothers and the Genealogy of Pathology in E. T. A. Hoffmann's Tales." *German Life and Letters* 52, no. 2 (1999): 238–54.

Hallgrímsson, Benedikt, and Brian K. Hall, eds. *Epigenetics: Linking Genotype and Phenotype in Development and Evolution*. Berkeley: University of California Press, 2011.

Hankins, Thomas L. *Science and the Enlightenment*. Cambridge: Cambridge University Press, 1985.

Hansen, Julie V. "Resurrecting Death: Anatomical Art in the Cabinet of Dr. Frederik Ruysch." *Art Bulletin* 78, no. 4 (December 1996): 663–79.

Hederich, Benjamin. *Gründliches mythologisches Lexicon. Durchgesehen, ansehnlich vermehret und verbessert von Johann Joachim Schwaben*. Leipzig: In Gleditschens Handlung, 1770.

Helfer, Martha. "The Male Muse of Romanticism: The Poetics of Gender in Novalis, E. T. A. Hoffmann, and Eichendorff." *German Quarterly* 78, no. 3 (July 2005): 299–319.

Henkel, Arthur. *Entsagung; Eine Studie zu Goethes Altersroman*. Tübingen: M. Niemeyer, 1954.

Herder, Johann Gottfried. *Auch eine Philosophie der Geschichte zur Bildung der Menschheit.* Edited by Hans Dietrich Irmscher. Stuttgart: Philipp Reclam, 1990.

———. *Ideen zur Philosophie der Geschichte der Menschheit.* Berlin: Deutsche Bibliothek, 1914.

Hesse-Fink, Evelyne. *Études sur le thème de l'inceste dans la littérature française.* Bern and Frankfurt: Peter Lang, 1971.

Heuermann, D. Georg. "Bemerkungen von einigen in Seeland befindlichen Missgeburten, welche man gemeiniglich Zwitter (Hermaphroditi und Androgynos) benennet," In *Vermischte Bemerkungen und Untersuchungen der ausübenden Arzeneiwissenschaft*, edited by D. G. Heuermann. vol. 2. Koppenhagen and Leipzig: Gabriel Christian Rothens Wittwe, 1765.

Hipp, Dietmar. "Dangerous Love: German High Court Takes a look at Incest." *DER SPIEGEL*, March 11, 2008. http://www.spiegel.de/international/germany/dangerous-love-german-high-court-takes-a-look-at-incest-a-540831.html (accessed May 29, 2013).

Hippel, Theodor Gottlieb von. *Über die Ehe.* In *G. v. Hippel's Sämmtliche Werke.* vol. 5. Berlin: G. Reimer, 1828–1839.

———. *On Marriage.* Translated and with an Introduction by Timothy F. Sellner. Detroit: Wayne State UP, 1994.

Hoffmann, E. T. A. *Devil's Elixirs.* Translated by Ronald Taylor. New York: John Calder, 1963.

———. *Die Elixiere des Teufels. Werke 1814–1816.* In *Sämtliche Werke in sechs Bänden*, edited by Hartmut Steinecke and Wulf Segebrecht. vol. 2. Frankfurt am Main: Deutscher Klassiker Verlag, 2003.

———. *Frühe Prosa. Briefe. Tagebücher. Libretti. Juristische Schrift. Werke 1794–1813.* In *Sämtliche Werke in sechs Bänden*, edited by Hartmut Steinecke and Wulf Segebrecht. vol. 1. Frankfurt am Main: Deutscher Klassiker Verlag, 2003.

———. *Juristische Arbeiten.* München: Winkler Verlag, 1973.

———. *Nachtstücke. Klein Zaches. Prinzessin Brambilla. Werke 1816–1820.* In *Sämtliche Werke in sechs Bänden*, edited by Hartmut Steinecke and Wulf Segebrecht. vol. 3. Frankfurt am Main: Deutscher Klassiker Verlag, 2003.

———. *Sämtliche Werke in sechs Bänden*, edited by Hartmut Steinecke and Wulf Segebrecht. 6 vols. Frankfurt am Main: Deutscher Klassiker Verlag, 2003.

———. *Selected Letters.* Edited and translated by Johanna C. Sahlin, with an introduction by Leonard J. Kent and Sahlin. Chicago: University of Chicago Press, 1977.

———. *Späte Prosa. Briefe. Tagebücher und Aufzeichnungen. Juristische Schriften. Werke 1814–1822.* In *Sämtliche Werke in sechs Bänden*, edited by Hartmut Steinecke and Wulf Segebrecht. vol. 6. Frankfurt am Main: Deutscher Klassiker Verlag, 2004.

———. *Tales of Hoffmann.* Translated and edited by Christopher Lazare. New York: Grover Press, 1946.

Hohendahl, Peter Uwe. "Die Krise der Männlichkeit im späten 18. Jahrhundert. Eine Problemskizze." *Zeitschrift für Germanistik* 12, no. 2 (2002): 275–86.

Hölderlin, Friedrich. *Sämtliche Werke.* Edited by Friedrich Beißner. 8 vols. Stuttgart: W. Kohlhammer, 1946–1977.

Holland, Jocelyn. *German Romanticism and Science: The Procreative Poetics of Goethe, Novalis, and Ritter.* New York and London: Routledge, 2009.

Hommel, Karl Ferdinand. *Über die Belohnung und Strafe nach türkischen* Gesetzen, Bayreuth, Leipzig, Lübeck, 1772.

Horace. *Epistles, Book 1.* Edited by Oswald Ashton Wentworth Dilke. London: Methuen & Co. Ltd., 1954.

Hörisch, Jochen. *Die andere Goethezeit: Poetische Mobilmachung des Subjekts um 1800.* München: Fink, 1992.

———. *Gott, Geld und Glück: Zur Logik der Liebe in den Bildungsromanen Goethes, Kellers und Thomas Manns.* Frankfurt am Main: Suhrkamp, 1983.

Huet, Marie Hélène. *Monstrous Imagination.* Cambridge, MA: Harvard University Press, 1993.

Hufeland, Christoph Wilhelm. "Beschreibung und Abbildung eines zu Berlin beobachteten weiblichen Hermaphroditen." In *Journal der practischen Arzneykunde und Wundarzneykunst*, edited by Christoph Wilhelm Hufeland, 170–72. Berlin: Ungers Journalhandlung, 1803.

Hull, Gillian. "Caleb Hillier Parry 1755–1822: A notable provincial physician." *Journal of the Royal Society of Medicine* 91, no. 6 (June 1998): 335–38.

Humboldt, Wilhelm. "Über den Geschlechtsunterschied (1794)." In *Schriften zur Anthropologie und Geschichte*, edited by Andreas Flitner, 4th ed., vol. 1, 268–95. Darmstadt: Wissenschaftliche Buchgesellschaft, 2002.

Hunt, Lynn. *The Family Romance of the French Revolution.* Berkeley: University of California Press, 1992.

Hunter, John. "Account of the Free Martin." In *Observations on Certain Parts of the Animal Oeconomy*, 45–62. London: Sold at No. 13 Castle-Street, Leicester-Square, 1786.

Imhof, Arthur E. "Unsere Lebensuhr. Phasenverschiebungen im Verlaufe der Neuzeit." In *Liebe, Ehe, Tod. Zum Wandel der Familie, der Geschlechts- und Generationsbeziehungen in der Neuzeit*, edited by Peter Borscheid and Hans J. Teuteberg, 170–98. Münster: F. Copenrath, 1983.

Jacobs, Jürgen, and Markus Krause. *Der deutsche Bildungsroman: Gattungsgeschichte vom 18. bis zum 20. Jahrhundert.* München: C. H. Beck, 1989.

Jahn, Ilse. "Die Museologie als Lehr- und Forschungsdisziplin mit spezieller Berücksichtigung ihrer Funktion in naturhistorischen Museen, Geschichte, gegenwärtiger Stand und theoretische Grundlagen." *Neue Museumskunde*, 1979.

———. *Grundzüge der Biologiegeschichte.* Jena: G. Fischer, 1990.

Jaumann, Herbert, and Wilhelm Voßkamp. "Die Sozialutopie der Turmgesellschaft und ihre Ausgrenzungen." In *Johann Wolfgang Goethe. Sämtliche Werke. Briefe, Tagebücher und Gespräche*, edited by Herbert Jaumann and Wilhelm Voßkamp, vol. 9, 1374–78. Frankfurt am Main: Deutscher Klassiker Verlag, 1992.

———. "Stellenkommentar." In *Johann Wolfgang Goethe. Sämtliche Werke. Briefe, Tagebücher und Gespräche*, edited by Herbert Jaumann and Wilhelm Voßkamp, vol. 9, 1381–1502. Frankfurt am Main: Deutscher Klassiker Verlag, 1992.

Jones, Ann Rosalind, and Peter Stallybrass. "Fetishizing Gender: Constructing the Hermaphrodite in Renaissance Europe." In *Body Guards: The Cultural Politics of Gender Ambiguity*, edited by Julia Epstein and Kristina Straub, 80–111. New York and London: Routledge, 1991.

Jordanova, Ludmilla. "Interrogating the Concept of Reproduction in the Eighteenth Century." In *Conceiving the New World Order: The Global Politics of Reproduction*, edited by Faye D. Ginsburg and Rayna Rapp, 369–86. Berkeley: University of California Press, 1995.

Joris, Elisabeth. "Kinship and Gender: Property, Enterprise, and Politics." In *Kinship in Europe: Approaches to Long-Term Development (1300–1900)*, edited by David Warren Sabean, Simon Teuscher, and Jon Mathieu, 231–57. New York: Berghahn, 2007.

Jung, Carl Gustav. *Das Unbewußte im normalen und kranken Seelenleben: Ein Überblick über die moderne Theorie und Methode der analytischen Psychologie*. Zurich: Rascher, 1926.

Kagan, Jerome. *The Three Cultures: Natural Sciences, Social Sciences, and the Humanities in the 21st Century*. New York and Cambridge: Cambridge University Press, 2009.

Kames, Henry Home, Lord. *Sketches of History of Man*. 4 vols. Dublin: Printed by James Williams, 1775. Eighteenth Century Collections Online: http://find.galegroup.com.myaccess.library.utoronto.ca/ecco/infomark.do?&source=gale&prodId=ECCO&userGroupName=utoronto_main&tabID=T001&docId=CB128044381&type=multipage&contentSet=ECCOArticles&version=1.0&docLevel=FASCIMILE (accessed February 9, 2014 from University of Toronto Libraries).

Kaminski, Nicola. "Die 'Kirche in der Linde' oder Medialer Heiligen-Schein. Hoffmanns *Elixiere des Teufels*." In *Kreuz-Gänge. Romanexperimente der deutschen Romantik*, edited by Nicola Kaminski, 307–74. Paderborn: F. Schöningh, 2001.

Kant, Immanuel. "Beantwortung der Frage: Was ist Aufklärung?" Berliner Monatschrift. Dez. 1784. In *Schriften zur Anthropologie, Geschichtsphilosophie, Politik und Pädagogik*, edited by Wilhelm von Weischedel, vol. 11, 481–97. Frankfurt am Main: Suhrkamp, 1993.

———. "Bestimmung des Begriffs einer Menschenrasse. Berliner Monatsschrift, Nov. 1785." In *Pädagogik*, edited by Wilhelm von Weischedel, vol. 11, 65–82. Frankfurt am Main: Suhrkamp, 1993.

———. *Critique of Judgement*. Translated by Nicholas Walker. Oxford: Oxford University Press, 2007.

———. "Determination of the Concept of a Human Race (1785)." In *The Cambridge Edition of the Works of Immanuel Kant: Anthropology, History, and Education*, translated by Holly Wilson and Günter Zöller, edited by Günter Zöller and Robert B. Louden. Cambridge: Cambridge University Press, 2011. 143–59.

———. "Of the Different Races of Human Beings (1775)." In *The Cambridge Edition of the Works of Immanuel Kant: Anthropology, History, and Education*, translated by Holly Wilson and Günter Zöller, edited by Günter Zöller and Robert B. Louden. Cambridge: Cambridge University Press, 2011. 82–97.

———. *Kants Gesammelte Schriften*. Herausgegeben von der Königlich Preußischen Akademie der Wissenschaften. 29 vols. Berlin: Georg Reimer, Walter de Gruyter, 1912–1983.

———. *Kritik der Urteilskraft*. Edited by Wilhelm Weischedel. Frankfurt: Suhrkamp, 1974.

———. "Lectures on Pedagogy (1803)." In *The Cambridge Edition of the Works of Immanuel Kant: Anthropology, History, and Education*, translated by Holly Wilson and Günter Zöller, edited by Günter Zöller and Robert B. Louden. Cambridge: Cambridge University Press, 2011. 434–85.

———. *Natural Science*. Edited by Eric Watkins. Translated by Lewis Whitebeck, Jeffrey B. Edwards, Olaf Reinhardt, Martin Schönfeld, and Eric Watkins. Cambridge: Cambridge University Press, 2012.

———. "Von den verschiedenen Rassen der Menschen (1775)." In *Werkausgabe. Schriften zur Anthropologie, Geschichtsphilosophie, Politik und Pädagogik*, edited by Wilhelm von Weischedel, vol. 11, 11–30. Frankfurt am Main: Suhrkamp, 1993.

———. *Werke*. Edited by Wilhelm Weischedel. 12 vols. Frankfurt am Main: Suhrkamp, 1977.

Kay, Lily E. *Who Wrote the Book of Life? A History of the Genetic Code*. Stanford, CA: Stanford University Press, 2000.

Kielmeyer, Carl Friedrich. *Ueber die Verhältniße der organischen Kräfte unter einander in der Reihe der verschiedenen Organisationen, die Geseze und Folgen dieser Verhältniße. Eine Rede den 11ten Februar 1793 am Geburtstage des regierenden Herzogs Carl von Wirtemberg im großen akademischen Hörsale gehalten*. New unaltered reprint. Tübingen: Christian Friedrich Osiander, 1814.

Kittler, Friedrich A. *Aufschreibesysteme 1800/1900*. München: Fink, 1985.

———. "Das Phantom unseres Ichs und die Literaturpsychologie: E. T. A. Hoffmann, Freud, Lacan." In *Romantikforschung seit 1945*, edited by Klaus Peter, 335–56. Königstein/Ts.: Verlagsgruppe Athenäum, 1980.

———. "Der Dichter, die Mutter, das Kind: Zur romantischen Erfindung der Sexualität." In *Romantik in Deutschland: ein interdisziplinäres Symposion*, edited by Richard Brinkmann, 102–14. Stuttgart: Metzler, 1978.

———. "Die Laterna magica der Literatur: Schillers und Hoffmanns Medienstrategien." *Anthenäum. Jahrbuch für Romantik* 4 (1994): 219–37.

———. *Discourse Networks 1800/1900*. Stanford, CA: Stanford University Press, 1990.

———. "Über die Sozialisation Wilhelm Meisters." In *Dichtung als Sozialisationsspiel*, edited by Friedrich A. Kittler and Gerhard Kaiser, 13–124. Göttingen: Vandenhoeck & Ruprecht, 1978.

Klappert, Annina. *Die Perspektiven von Link und Lücke: Sichtweisen auf Jean Pauls Texte und Hypertexte*. Bielefeld: Aisthesis, 2006.

Kleist, Heinrich. *Die Marquise von O–: Die Verlobung in St.Domingo. Erzählungen*. Stuttgart: Reclam, 1961.

Klügel, Georg Simon. *Enzyklopädie oder zusammenhängender Vortrag der gemeinnützigsten Kenntnisse. Dritter Theil, Erste Abtheilung, welche das Naturrecht, die praktische Mechanik, die bürgerliche Baukunst, die Kriegsbaukunst und die Schiffsbaukunst enthält*. Berlin and Stetting: Friedrich Nicolai, 1784.

Knaupp, Michael. *Friedrich Hölderlin: Hyperion*. Stuttgart: P. Reclam, 1997.

Kniesche, Thomas W. "Die psychoanalytische Rezeption von Mignon." In *Goethes Mignon und ihre Schwestern*, edited by Gerhart Hoffmeister, 61–81. New York: Peter Lang, 1993.

Koelreuter, Joseph Gottlieb. *D.J.G.K.'s Vorläufige Nachricht von einigen das Geschlecht der Pflanzen betreffenden Versuch und Beobachtungen*. Leipzig: Gleditsch, 1761–1766.

Koepke, Wulf. "Bildung and the Transformation of Society: Jean Paul's *Titan* and *Flegeljahre*." In *Reflection and Action: Essays on the Bildungsroman*, edited by James N. Hardin, 228–53. Columbia: University of South Carolina Press, 1991.

Kofman, Sarah. "Vautour Rouge: Le double dans *Les Élixirs du diable* d'Hoffmann." In *Mimesis*, edited by Sylviane Agacinsky et al., 95–163. Paris: Flammarion, 1975.

Kolata, Gina. "After Setbacks, Small Successes for Gene Therapy." *New York Times*, November 5, 2009, A19. Also available online at http://www.nytimes.com/2009/11/06/health/06gene.html?_r=0 (accessed January 19, 2014).

Koschorke, Albrecht. *Körperströme und Schriftverkehr: Mediologie des 18. Jahrhunderts*. München: Fink, 1999.

———. "Inseminationen, Empfängnislehre, Rhetorik und christliche Verkündigung." In *Kunst—Zeugung—Geburt: Theorien und Metaphern ästhetischer Produktion in der Neuzeit*, edited by Christian Begemann and David E. Wellbery, 89–100. Freiburg im Breisgau: Rombach, 2002.

Koselleck, Reinhart. "Einleitung." In *Geschichtliche Grundbegriffe. Historisches Lexikon zur politisch-sozialen Sprache in Deutschland*, edited by Otto Brunner, Werner Conze, and Reinhart Koselleck. vol. 1, xiii-xxvii. Stuttgart: Klett, 1972.

———. *Preußen zwischen Reform und Revolution: Allgemeines Landrecht, Verwaltung und soziale Bewegung von 1791 bis 1848.* Stuttgart: Klett, 1967.

———. *Vergangene Zukunft. Zur Semantik geschichtlicher Zeit.* Frankfurt am Main: Suhrkamp, 1979.

Kraus, Johanne Ernesto Francisco. *Der Edle Gestütt-Garten, oder aufrichtige Anleitung, zur Gestütt- und Pferd-Zieglung, auch Maulthier-Zucht.* Nürnberg: Joh. Fried. Rüdiger, 1724.

Kremer, Detlef. "Flüchtige Familien-Phantome: 'Die Elixiere des Teufels.'" In *Romantische Metamorphosen: E. T. A. Hoffmanns Erzählungen*, 245–59. Stuttgart: Metzler, 1993.

Kremer, Elmar J. "Leibniz and the 'Disciples of Saint Augustine' on the Fate of Infants Who Die Unbaptized." In *The Problem of Evil in Early Modern Philosophy*, edited by Elmar J. Kremer and Michael J. Latzler, 119–37. Toronto: University of Toronto Press, 2001.

Krimmer, Elisabeth. "Mama's Baby, Papa's Maybe: Paternity and Bildung in Goethe's *Wilhelm Meisters Lehrjahre*." *German Quarterly* 77, no. 3 (Summer 2004): 257–77.

Krüger-Fürhoff, Irmela Marei. *Der versehrte Körper: Revisionen des klassistischen Schönheitsideals.* Göttingen: Wallstein, 2001.

Krupp, Anthony. *Reason's Children: Childhood in Early Modern Philosophy.* Lewisburg, PA: Bucknell University Press, 2009.

Kugler, Hans. "Kölreuter, Joseph Gottlieb." *Neue Deutsche Biographie* 12 (1979): 325–26.

Kuhn, Thomas S. "Second Thoughts on Paradigms." In *The Essential Tension. Selected Studies in Scientific Tradition and Change*, 293–319. Chicago: University of Chicago Press, 1977.

———. *The Structure of Scientific Revolutions.* Chicago: University of Chicago Press, 1962.

Kuper, Adam. *Incest & Influence: The Private Life of Bourgeois England.* Cambridge, MA: Harvard University Press, 2009.

Kutzer, Elisabeth. *Zum Stammbaumroman in der neueren Literatur.* PhD diss., Universität Leipzig, Leipzig: Gerhardt, 1929.

Lamarck, Jean-Baptiste Pierre Antoine de Monet. *Philosophie Zoologique ou exposition des considérations relatives à l'histoire naturelle des animaux.* J. B. Baillière, Libraire. Paris, 1809.

———. *Zoological Philosophy: An Exposition with Regard to the Natural History of Animals* (1809). Translated by Hugh Elliot. Chicago: University of Chicago Press, 1984.

Lange, Horst. "Goethe and Spinoza: A Reconsideration." *Goethe Yearbook* 18 (2001): 11–33.

Laqueur, Thomas Walter. *Making Sex: Body and Gender from the Greeks to Freud.* Cambridge, MA: Harvard University Press, 1990.

Larrett, William. "Wilhelm Meister and the Amazons: The Quest for Wholeness." *Puplications of the English Goethe Society*, n.s., 39 (1968–1969): 31–56.

Larson, James L. *Interpreting Nature: The Science of Living Forms from Linnaeus to Kant.* Baltimore: Johns Hopkins University Press, 1994.

Leisewitz, Carl. "Sturm, Karl Christ. Gottl." In *Allgemeine Deutsche Biographie*, vol. 37, 40–41. Leipzig: Duncker & Humblot, 1894.

Lepenies, Wolf. *Das Ende der Naturgeschichte: Wandel kultureller Selbstverständlichkeiten in den Wissenschaften des 18. und 19. Jahrhunderts*. München: C. Hanser, 1976.

Lessing, Gotthold Ephraim. *Emilia Galotti*. Edited by Jan-Dirk Müller. Stuttgart: P. Reclam, 1974.

———. *Laokoon Briefe, antiquarischen Inhalts: Text und Kommentar*. Edited by Wilfried Barner. Frankfurt am Main: Deutscher Klassiker Verlag, 2007.

Lévi-Strauss, Claude. *Structural Anthropology*. Translated by Claire Jacobson and Brooke Grundfest Schoepf. New York, Basic Books, 1963.

Lichtenberg, Georg Christoph. *Schriften und Briefe*. Edited by Wolfgang Promies. 4 vols. München: C. Hanser, 1968.

———. *Waste Books*. Translated and with an Introduction and Notes by R. J. Hollingdale. New York: New York Review Books, 1990.

Lima-de-Faria, A. *One Hundred Years of Chromosome Research and What Remains to be Learned*. Dordrecht and Boston: Kluwer Academic Publishers, 2003.

Lindemann, A., F. M. Rosenthal, S. Hase, P. Markmeyer, and R. Mertelsmann. "Guidelines for the Design and Implementation of Clinical Studies in Somatic Cell Therapy. German Working Group for Gene Therapy." *Journal for Molecular Medicine (Berlin, Germany)* 73, no. 4 (April 1995): 207–11.

Locke, John. *Two Treatises on Government*. Edited by Mark Goldie. London: J. M. Dent; Rutland, VT: C. E. Tuttle, 1993.

Long, Kathleen P. *Hermaphrodites in Renaissance Europe*. Aldershot, Hampshire and Burlington, VT: Ashgate, 2006.

López-Beltrán, Carlos. "Forging Heredity: From Metaphor to Cause, a Reification Story." *Studies in the History and Philosophy of Science* 25, no. 2 (April 1994): 211–35.

———. "Heredity Old and New. French Physicians and l'Hérédité Naturelle in Early 19th Century." In *A Cultural History of Heredity II: 18th and 19th Centuries*. Max Planck Institute for the History of Science. Preprint 247, 7–19. Berlin: Max-Planck-Inst. für Wiss.-Geschichte, 2003.

———. "The Medical Origins of Heredity." In *Heredity Produced: At the Crossroads of Biology, Politics, and Culture, 1500–1870*, edited by Staffan Müller-Wille and Hans-Jörg Rheinberger, 105–32. Cambridge, MA: MIT Press, 2007.

Lovejoy, Arthur O. *The Great Chain of Being: A Study of the History of and Idea*. Cambridge, MA: Harvard University Press, 1961.

Luhmann, Niklas. *The Differentiation of Society*. New York: Columbia University Press, 1982.

———. *Liebe als Passion: Zur Codierung von Intimität*. Frankfurt am Main: Suhrkamp, 1983.

———. *Love as Passion: The Codification of Intimacy*. Cambridge, MA: Harvard University Press, 1986.

MacLeod, Catriona. "Pedagogy and Androgyny in *Wilhelm Meisters Lehrjahre*." *Modern Languages Notes* 108, no. 3 (April 1993): 389–426.

Mahoney, Dennis F. "The Apprenticeship of the Reader: The Bildungsroman of the 'Age of Goethe.'" In *Reflection and Action: Essays on the Bildungsroman*, edited by James N. Hardin, 97–117. Columbia: University of South Carolina Press, 1991.

———. *Der Roman der Goethezeit, 1774–1829*. Stuttgart: Metzler, 1988.

Majer, Johann Christian. *Teutsche Staatskonstitution*, Hamburg: Carl Ernst Bohn, 1800.

Malaure, Marguerite. "Bittschrift der Margarethe Malaure an den König." In *Eine Frauensperson, die fälschlich für einen Zwitter gehalten wird. Erzählung sonderbarer Rechtshändel, sammt deren gerichtlichen Entscheidung*, edited by François Gayot von Pitaval, 404–15. Vierter Theil. Leipzig: Gottfried Kiesewetter, 1748.

Mangold, Hartmut. *Gerechtigkeit durch Poesie: Rechtliche Konfliktsituationen und ihre literarische Gestaltung bei E. T. A. Hoffmann*. Wiesbaden: Deutscher Universitäts Verlag, 1988.

Mann, Thomas. *Buddenbrooks; Verfall einer Familie. Roman*. Berlin: S. Fischer, 1901.

———. *Bekenntnisse des Hochstaplers Felix Krull*. Frankfurt a.M.: S. Fischer, 2012.

———. *Confessions of Felix Krull: Confidence Man*. Translated by Denver Lindley. New York: Knopf, 1955.

Margócsy, Dániel. "Advertising cadavers in the republic of letters: anatomical publications in the early modern Netherlands." *British Journal of The History of Science* 42, no. 2 (June 2009): 187–210.

Martin, Xavier. "The Paternal Role and the Napoleonic Code," *Paternity and Fatherhood: Myths and Realities*, edited by Lieve Spaas, 27–39. New York: St. Martin's Press, 1998.

Martini, Fritz. "Der Bildungsroman. Zur Gesichte des Wortes und der Theorie." In *Zur Geschichte des deutschen Bildungsromans*, edited by Rolf Selbmann, 239–64. Darmstadt: Wissenschaftliche Buchgesellschaft, 1988.

Mathieu, Jon. "Kin Marriages: Trends and Interpretations from the Swiss Example." In *Kinship in Europe: Approaches to Long-Term Developments (1300–1900)*, edited by David Warren Sabean, Simon Teuscher, and Jon Mathieu, 211–30. New York, Oxford: Berghahn Books, 2007.

Max Planck Institute for the History of Science. *A Cultural History of Heredity I: Seventeenth and Eighteenth Centuries*. Preprint 222. Berlin: Max-Planck-Inst. für Wiss.-Geschichte, 2002. http://www.mpiwg-berlin.mpg.de/Preprints/P222.PDF (accessed January 17, 2014).

Mayer, Gerhart. *Der deutsche Bildungsroman: Von der Aufklärung bis zur Gegenwart*. Stuttgart: Metzler, 1992.

Mayr, Ernst. *The Growth of Biological Thought: Diversity, Evolution, and Inheritance*. Cambridge, MA: Belknap of Harvard University Press, 1982.

Mazzolini, Renato. "Las Castas: Inter-Racial Crossing and Social Structure (1770–1835)." In *Heredity Produced: At the Crossroads of Biology, Politics, and Culture, 1500–1870, edited by Staffan Müller-Wille and Hans-Jörg Rheinberger*, 349–74. Cambridge, MA: MIT Press, 2007.

McGlathery, James M. *Mysticism and Sexuality: E. T. A. Hoffmann*. Las Vegas: Peter Lang, 1981.

Meyer-Krentler, Eckhard. "Erdichtete Verwandtschaft. Inzestmotiv, Aufklärungsmoral, Strafrecht in J. W. Goethes *Die Geschwister*." *Literatur für Leser* 4 (1982): 230–49.

Michaelis, Johann David. *Abhandlungen von den Ehegesetzen Mosis welche die Heyrathen in die nahe Freundschaft untersagen*. Diss., n.p. 1755. Reprint Zweite und vermehrte Auflage. Göttingen: Verlegt Abram Vandenhoecks seel. Witwe, 1768.

Michler, Markwart. "Lobstein, Johann Friedrich der Ältere." *Neue Deutsche Biographie* 14 (1985): 737–38.

Milburn, Colin. *Nanovision: Engineering the Future*, Durham, NC: Duke University Press, 2008.

Minden, Michael. "The Place of Inheritance in the Bildungsroman: *Agathon*, *Wilhelm Meisters Lehrjahre*, and *Der Nachsommer*." In *Reflection and Action: Essays on the Bildungsroman*, edited by James N. Hardin, 254–92. Columbia: University of South Carolina Press, 1991.

Montesquieu, Charles Louis de Secondat. *Lettres persanes* [1721]. Oxford: Voltaire Foundation, 2004.

Morelly Etienne-Gabriel. *Le naufrage des Isles flottantes ou Basiliade du célèbre Pilpai*. Messine. Paris: Société de libraires, 1753.

Morgan, Dianne. *Kant Trouble: the Obscurities of the Enlightened*. London: Routledge, 2000.

Moritz, Karl Philipp. *Anton Reiser. Ein Psychologischer Roman*. Stuttgart: Reclam, 1986.

———. "On the Artistic Imitation of the Beautiful" (1788). In *Classic and Romantic German Aesthetics*, edited by J. M. Bernstein, 131–44. Cambridge: Cambridge University Press, 2003.

———. *Schriften zur Ästhetik und Poetik*. Edited by Hans Joachim Schrimpf. Tübingen: Niemeyer, 1962.

Möser, Justus. *Sämtliche Werke*. Historisch-kritische Ausgabe in 14 Bänden. Edited by Ludwig Schirmeyer and Werner Kohlschmidt. vol. 7. Hamburg: Gerhard Stalling, 1943–1954.

Mücke, Dorothea. *Virtue and the Veil of Illusion: Generic Innovation and the Pedagogical Project in Eighteenth-Century Literature*. Stanford: Stanford University Press, 1991.

Müller, Adam. *Elemente der Staatskunst*. Jena: Fischer, 1922.

Müller, Gerhard H. "'Der Künstler setzt sich an die Stelle der Natur.' Goethes Alternativen in der Plastischen Anatomie (1832)." In *Anatomie: Sektionen einer medizinischen Wissenschaft im 18. Jahrhundert*, edited by Jürgen Helm, Karin Stukenbrock, 221–23. Wiesbaden: Franz Steiner Verlag, 2003.

Müller, Götz. "Die Einbildungskraft im Wechsel der Diskurse. Annotationen zu Adam Bernd, Karl Philipp Moritz und Jean Paul." In *Jean Paul im Kontext*, edited by Wolfgang Riedel, 140–64. Würzburg: Königshausen & Neumann, 1996.

———. *Jean Pauls Ästhetik und Naturphilosophie*. Tübingen: Niemeyer, 1983.

———. *Jean Pauls Exzerpte*. Würzburg: Königshausen & Neumann, 1988.

Müller, Johannes. *Bildungsgeschichte der Genitalien*. Düsseldorf: Arnz, 1830.

Müller-Seidel, Walter. "Naturforschung und Deutsche Klassik." In *Die Geschichtlichkeit der deutschen Klassik. Literatur und Denkformen um 1800*, 105–18. Stuttgart: Metzler, 1983.

Müller-Sievers, Helmut. *Self-Generation: Biology, Philosophy and Literature around 1800*. Stanford: Stanford University Press, 1997.

Müller-Wille, Staffan. "Genealogie, Naturgeschichte und Naturgesetz bei Linné und Buffon." In *Genealogie als Denkform im Mittelalter und Früher Neuzeit*, edited by Kilian Heck and Bernhard Jahn, 109–19. Tübingen: Niemeyer, 2000.

Müller-Wille, Staffan, and Vítezslav Orel. "From Linnaean Species to Mendelian Factors: Elements of Hybridism, 1751–1870." *Annals of Science* 64, no. 2 (April 2007): 171–215.

Müller-Wille, Staffan, and Hans-Jörg Rheinberger, eds. *Heredity Produced: At the Crossroads of Biology, Politics, and Culture, 1500–1870*. Cambridge, MA: MIT Press, 2007.

Münch, Ragnhild. "Theater des Todes—Museum des Lebens." In *Theater der Natur und Kunst: Theatrum naturae et artis; Wunderkammern des Wissens*, edited by Horst Bredekamp, Jochen Brüning, and Cornelia Weber, 135–42. Berlin: Henschel, 2000–2001.

Museum Waldenburg. Menschliche Präparate. n.d. http://www.museum-waldenburg.de/?id=af792&s=nature (accessed November 20, 2011).

Naumann-Beyer, Waltraud. "Ästhetik." In *Goethe Handbuch in vier Bänden*, edited by Hans-Dietrich Dahnke and Regine Otto, vol. 4, 8–13. Stuttgart and Weimar: Metzler, 1998.

Neickelio, C. F. [Caspar Friedrich Jenckel]. *Museographia oder Anleitung zum rechten Begriff und nützlicher Anregung der Museorum oder Raritätenkammern*. Leizpig and Breslau: Michael Hubert, 1727.

New American Standard Bible (NASB). http://www.biblegateway.com (accessed March 16, 2012).

Newman, William R. *Promethean Ambitions: Alchemy and the Quest to Perfect Nature*. Chicago: University of Chicago Press, 2004.

Nonnenmacher, Hartmut. *Natur und Fatum: Inzest als Motiv und Thema in der französischen und deutschen Literatur des 18. Jahrhunderts*. Frankfurt am Main and New York: Peter Lang, 2002.

Nora, Pierre. "Generation." In *Realms of Memory: Rethinking the French Past*, edited by Pierre Nora and Lawrence D. Kritzman. Translated by Aurthur Goldhammer, vol. 1, 499–531. New York: Columbia University Press, 1996–1998.

Oehler-Klein, Sigrid. *Die Schädellehre Franz Joseph Galls in Literatur und Kritik des 19. Jahrhunderts: zur Rezeptionsgeschichte einer medizinisch-biologisch begründeten Theorie der Physiognomik und Psychologie*. Stuttgart: Gustav Fischer Verlag, 1990.

Oken, Lorenz. *Die Zeugung*. Bamberg and Würzburg: Joseph Anton Goebhardt, 1805.

Olding, Paul. "The Genius Sperm Bank." *BBC News*.15 June 2006, http://news.bbc.co.uk/2/hi/5078800.stm (accessed March 17, 2014).

Olson, Susanne. "Das Wunderbare und seine psychologische Funktion in E. T. A. Hoffmanns *Die Elixiere des Teufels*." *Mitteilungen der E. T. A. Hoffmann-Gesellschaft* 24 (1978): 34–50.

Ovid. *Heroides; and Amores*. English translation by Grant Showerman. Cambridge, MA: Harvard University Press, 1986.

———. *Metamorphosen*. In deutsche Hexameter übertragen und mit dem Text herausgegeben von Erich Rösch. München: Ernst Heimeran, 1964.

Pabst, Rainer. *Schicksal bei E. T. A. Hoffmann: zur Erscheinungsform, Funktion und Entwicklung eines Interpretationsmusters*. Köln: Böhlau, 1989.

Paine, Thomas. *Rights of Man. Being an Answer to Mr. Burke's Attack on the French Revolution*. London: J. S. Jordan, 1791.

Park, Katherine. "The Rediscovery of the Clitoris." In *The Body in Parts: Fantasies of Corporeality in Early Modern Europe*, edited by David Hillman and Carla Mazzio, 171–93. London: Routledge, 1997.

Parnes, Ohad. "Generationen und die neue Epistemologie der Vererbung im 19. Jahrhundert." *Die Generation. Geneaologie des Konzepts—Konzepte von Genealogie*, edited by O. Parnes et al., 235–59. München: Fink, 2005.

Parry, Caleb Hillier. "Part II. History of the Merino-Ryeland Breed of the Author." In *Communications to the Board of Agriculture; on subjects relative to the husbandry, and internal improvement of the country.* vol. 5, pts. 1–2, 433–526. London: Bulmer and Co., 1806. Eighteenth Century Collections Online: http://quod.lib.umich.edu/e/ecco (accessed November 5, 2011).

Parsons, James. *A Mechanical and Critical Enquiry into the Nature of Hermaphrodites*. London: J. Walthoe, 1741.

Pearson, Sue, Hepeng Jia, and Keiko Kandachi. "China approves first gene therapy." *Nature Biotechnology* 22, no. 1 (2004): 3–4.

Pethes, Nicolas. *Zöglinge der Natur: Der literarische Menschenversuch des 18. Jahrhunderts*. Göttingen: Wallstein, 2007.

Pfotenhauer, Helmut. "Jean Pauls literarische Biologie. Zur Verschriftlichung von Zeugung und Tod, mit besonderer Berücksichtigung des Siebenkäs." In *Prägnanter Moment. Studien zur deutschen Literatur der Aufklärung und Klassik: Festschrift für Hans-Jürgen Schings*, edited by Peter-André Alt, Alexander Kosenina, Hartmut Reinhardt, and Wolfgang Riedel, 461–77. Würzburg: Königshausen & Neumann, 2002.

Pickstone, John V. *Ways of Knowing: A New History of Science, Technology and Medicine*. Manchester: Manchester University Press, 2000.

Pilkington, A. E. "'Nature' as Ethical Norm in the Enlightenment." In *Languages of Nature: Critical Essays on Science and Literature*, edited by L. J. Jordanova, 51–85. New Brunswick, NJ: Rutgers University Press, 1986.

Pinel, Philippe. *Traité médico-philosophique sur l'aliénation mentale*. Paris, 1801: Reprint 2nd ed. Paris: Brosson, 1809.

Piper, Andrew. *Dreaming in Books: The Making of the Bibliographic Imagination in the Romantic Age*. Chicago: University of Chicago Press, 2009.

Ploucquet, Wilhelm Gottfried. *Ueber die physische Erfordernisse der Erbfähigkeit der Kinder*. Tübingen: Heerbrandt, 1779.

Prandi, Julie D. *Dare to Be Happy!: A Study of Goethe's Ethics*. Lanham, MD: University Press of America, 1993.

Pross, Caroline. *Falschnamenmünzer: Zur Figuration von Autorschaft und Textualität im Bildfeld der Ökonomie bei Jean Paul*. Frankfurt am Main: Peter Lang, 1997.

Pufendorf, Samuel. *Gesammelt Werke*. Edited by Wilhelm Schmidt-Biggemann. Berlin: Akademie Verlag, 1996–.

Raijkov [Rajkov], Boris E. "Caspar Friedrich Wolff." *Zoologische Jahrbücher Systematik* 91 (1965): 51–65.

Rainer, Ulrike. "A Question of Silence: Goethe's Speechless Women." In *Goethes Mignon und ihre Schwestern*, edited by Gerhart Hoffmeister, 101–11. New York: Peter Lang, 1993.

Rank, Otto, ed. "Der Doppelgänger." In *Psychoanalytische Beiträge zur Mythenforschung: Gesammelte Studien aus den Jahren 1912 bis 1914*, 267–354. Leipzig and Vienna: Internationaler Psychoanalytischer Verlag, 1919.

Rankl, Maximilian. *Jean Paul und die Naturwissenschaft*. Frankfurt am Main: Peter Lang, 1987.

Rasch, William. *Niklas Luhmann's Modernity: The Paradoxes of Differentiation*. Stanford, CA: Stanford University Press, 2000.

Reber, Natalie. *Studien zum Motiv des Doppelgängers bei Dostojevskij und E. T. A. Hoffmann*. Gießen: Wilhelm Schmitz, 1964.

Reed, Terrence James. "Revolution and Rücknahme: *Wilhelm Meisters Lehrjahre* im Kontext der Französischen Revolution." *Goethe Jahrbuch* 107 (1990): 27–43.

Reil, Johann Christian. *Rhapsodien über die Anwendung der psychischen Curmethode auf Geisteszerrüttungen*. Halle: Curt, 1803. Reprint Amsterdam: Bonset, 1968.

Rheinberger, Hans-Jörg. "Naudin, Darwin, Nägeli: Bemerkungen zu den Vererbungsvorstellungen des 19. Jahrhunderts." *Medizinhistorisches Journal* 18, no. 3 (1983): 198–212.

Rheinberger, Hans-Jörg, Peter McLaughlin, and Staffan Müller-Wille. "Introduction." In: *A Cultural History of Heredity 1: 17th and 18th Centuries.* Max Planck Institute for the History of Science. Preprint 222, 3–5. Berlin, Max-Planck-Inst. für Wiss.-Geschichte, 2002.

Rheinberger, Hans-Jörg, and Staffan Müller-Wille. "Introduction." In *A Cultural History of Heredity II: 18th and 19th Centuries.* Max Planck Institutee for the History of Science. Preprint 247, 3–6. Berlin, Max-Planck-Inst. für Wiss.-Geschichte, 2003.

———. *Vererbung. Geschichte und Kultur eines biologischen Konzepts.* Frankfurt am Main: Fischer- Taschenbuch-Verl., 2009.

Richards, Robert J. *The Romantic Conception of Life: Science and Philosophy in the Age of Goethe.* Chicago: University of Chicago Press, 2002.

Richter, Jean Paul. *Werke.* Edited by Norbert Miller. München: Hanser, 1959–1985.

———. *Dr. Katzenbergers Badereise.* In *Jean Paul Richtler Werke*, edited by Norbert Miller, vol. 4, 77–363. München: Hanser, 1967.

———. *Flegeljahre.* In *Jean Paul Richtler Werke*, edited by Norbert Miller, vol. 2, 567–1065. München: Hanser, 1959.

———. *Flower, Fruit, and Thron Pieces; Or, the Wedded Life, Death, and Marriage of Firmian Stanislaus Siebenkaes.* Translated by Alexander Ewing. London: George Bell and Sons, 1897.

———. *Frage über das Entstehen der ersten Pflanzen, Thiere und Menschen* (1812). In *Jean Paul Richter* Sämtliche Werke. Historisch-kritische Ausgabe. Erste Abtheilung. *Sechzehnter Band*, edited by Preußische Akademie der Wissenschaften, series 1, 16: 51–75. Weimar: Böhlau Nachfolger, 1938.

———. *Kampaner Tal.* In *Jean Paul Richter Werke.*, edited by Norbert Miller, vol. 4, 561–716. München: Hanser, 1967.

———. *Komet.* In *Jean Paul Richter Werke.*, edited by Norbert Miller, vol. 4, 563–1036. München: Hanser, 1967.

———. "Rede des todten Christus vom Weltgebäude herab, daß kein Gott sey" [*Siebenkäs*]. In *Jean Paul Richter Sämtliche Werke. Historisch-kritische Ausgabe. Erste Abtheilung. Sechster Band*, edited by Kurt Schreinert, 247–57. Weimar: Böhlau Nachfolger, 1928. Reprint 1977.

———. *Sämtliche Werke.* Edited by Norbert Miller and Wilhelm Schmidt-Biggemann. 4 vols. München: Hanser, 1974–1985.

———. *Vorschule der Ästhetik.* In *Jean Paul Richter Werke.*, edited by Norbert Miller, vol. 5, 11–456. München: Hanser, 1963.

Richter, Melvin, and Michaela W. Richter. "Introduction. Translation of Reinhart Koselleck's 'Krise' in 'Geschichtliche Grundbegriffe.'" *Journal of the History of Ideas* 67, no. 2 (April 2006): 343–56.

Richter, Simon. *Missing the Breast: Gender, Fantasy, and the Body in the German Enlightenment.* Seattle: University of Washington Press, 2006.

Riedel, Wolfgang. "Zur Textgestalt." In *Johann Wolfgang Goethe Sämtliche Werke nach Epochen seines Schaffens*, ed. Karl Richter, 5: 699–708. München: Carl Hanser, 1988.

Riedl, Peter Philipp. "Die Zeichen der Krise. Erbe und Eigentum in Achim von Arnims 'Die Majoratsherren' und E. T. A. Hoffmanns 'Das Majorrat.'" *Aurora* 52 (1992): 17–50.

Roberts, Corey. "German Pietism and the Genesis of Literary Aesthetics: The Discourse of 'Erfahrung' in the 1700." *Deutsche Vierteljahrsschrift* 78, no. 2 (June 2004): 200–28.

Roe, Shirley A. *Matter, Life, and Generation: Eighteenth-Century Embryology and the Haller-Wolff Debate*. Cambridge: Cambridge University Press, 1981.

Rougemont, Joseph Claude. *Abhandlungen über die erblichen Krankheiten. Eine gekrönte Preißschrift*. Translated by Friedrich Georg Wegeler. Frankfurt am Main: Fleischer, 1794.

Rousseau, Jean Jacques. *Discourse on the Origin of Inequality*. Translated by Franklin Philip. Edited with an introduction and notes by Patrick Coleman.Oxford: Oxford University Press, 2009.

Rueff, Jacob. *Hebammen Buch*. Gedruckt zu Franckfort am Mayn: Sigmund Feyerabendt, 1580.

Ruisinger, Marion Maria. "Schön oder Hässlich? Die Ambivalenz medizinischer Museumsdinge." In *Schönheit: Traum–Kunst–Bildung*, edited by Eckart Liebau and Jörg Zirfas, 119–48. Bielefeld: transcript Verlag, 2007.

Sabean, David Warren, and Simon Teuscher. "Introduction." In *Kinship in Europe: Approaches to Long-Term Developments (1300–1900)*, edited by David Warren Sabean, Simon Teuscher, and Jon Mathieu, 1–32. New York, Oxford: Berghahn Books, 2007.

Saine, Thomas P. "Was Wilhelm Meisters Lehrjahre Really Supposed to Be a Bildungsroman?" In *Reflection and Action: Essays on the Bildungsroman*, edited by James N. Hardin, 118–41. Columbia: University of South Carolina Press, 1991.

Schaeder, Grete. *Gott und die Welt: Drei Kapitel Goethescher Weltanschauung*. Hameln: Fritz Seifert, 1947.

Schäffer, Jacob Christian. *Die Armpolypen in den süßen Wassern um Regensburg*. Ev. Prediger in Regensburg und Mitglied der Königl. Deutschen Gesellschaft in Göttingen. Nebst 3 Kupferplatten mit Figuren in Farben. Regensburg: E. A. Weiss, 1754.

Scheuchzer, Johann Jacob. *Kupfer-Bibel, in welcher die Physica Sacra oder geheiligte Natur-Wissenschafft derer in Heiliger Schrifft vorkommenden natürlichen Sachen deutlich erklärt und bewährt*. 4 vols. Augspurg [sic] und Ulm: Christian Ulrich Wagner, 1733–1735.

Schiebinger, Londa. *Nature's Body: Gender in the Making of Modern Science*. Boston: Beacon Press, 1993. Reprint New Brunswick, NJ: Rutgers University Press, 2004.

———. "Why Mammals Are Called Mammals." In *Nature's Body: Gender in the Making of Modern Science*, 40–74. Boston: Beacon Press, 1993.

Schiller, Friedrich. "Die Räuber." In *Werke und Briefe*, edited by Klaus Harro Hilzinger et al., vol. 2. Frankfurt am Main: Deutscher Klassiker Verlag, 1988.

———. "Kallias, oder über die Schönheit." In *Werke und Briefe*, edited by Rolf-Peter Janz, vol. 8, 276–329. Frankfurt am Main: Deutscher Klassiker, 1992.

———. "Kallias or Concerning Beauty: Letters to Gottfried Körner (1793)." In *Classic and Romantic German Aesthetics*. Edited by J.M. Bernstein. Cambridge: Cambridge University Press, 2003, 145–83.

———. "Verbrecher aus Infamie." In *Werke und Briefe*, edited by Otto Dann, vol. 7, 562–87. Frankfurt am Main: Deutscher Klassiker Verlag, 2002.

———. "The Criminal of Lost Honor." Translated by Jeffrey L. High. In *Schiller's Literary Prose Works. New Translations and Critical Essays*. Edited by Jeffrey L. High, Rochester: New York, 2008, 39–55.

Schiller, Friedrich, and Johann Wolfgang von Goethe. *Der Briefwechsel Schiller and Goethe*. Edited by Emil Staiger. Frankfurt am Main: Insel, 1977.

Schindler, Stephan K. *Das Subjekt als Kind: Die Erfindung der Kindheit im Roman des 18. Jahrhunderts*. Berlin: E. Schmidt, 1994.

Schings, Hans-Jürgen. "Einführung." In *Johann Wolfgang Goethe Sämtliche Werke nach Epochen seines Schaffens*, edited by Karl Richter, 5: 613–43. München: Carl Hanser, 1988.

———. "Kommentar." In *Johann Wolfgang Goethe Sämtliche Werke nach Epochen seines Schaffens*, edited by Karl Richter, 5: 711–856. München: Carl Hanser, 1988.

Schlaffer, Hannelore. *Wilhelm Meister: Das Ende der Kunst und die Wiederkehr des Mythos*. Stuttgart: Metzler, 1980.

Schlechta, Karl. *Goethes "Wilhelm Meister."* Mit einer Einleitung von Heinz Schlaffer. Frankfurt am Main: V. Klostermann, 1953. Reprint 1985.

Schlegel, August Wilhelm. "De l'origine des Hindous." In *Essais littereraires et historiques*, 441–518. Bonn: E. Weber, 1842.

Schlegel, Friedrich Wilhelm. *Lucinde. Ein Roman*. Jean: Diederichs, 1907.

———. *Lucinde and the Fragments*. Translated and with an introduction by Peter Firchow. Minneapolis: University of Minnesota Press, 1971.

Schlipphacke, Heidi M. "'Vaterschaft beruht nur überhaupt auf der Überzeugung': The Displaced Family in *Wilhelm Meisters Lehrjahre*." *Journal of English and Germanic Philology* (July 2003): 390–412.

Schmauß, Johann Jakob. *Neues Systema des Rechts der Natur*. Göttingen: Verlegts Abram Bandenhoecks seel. Witwe, 1754.

Schmid, Imtraut. "Amtliche Tätigkeit." In *Goethe Handbuch in vier Bänden*, edited by Hans-Dietrich Dahnke and Regine Otto, vol. 4, 33–46. Stuttgart and Weimar: Metzler, 1998.

Schmidt, Eberhard. *Einführung in die Geschichte der deutschen Strafrechtspflege*. 3rd ed. Göttingen: Vandenhoeck & Ruprecht, 1965.

Schmied-Kowarzik, Wolfdietrich. "Der Streit um die Einheit des Menschengeschlechts. Gedanken zu Forster, Herder und Kant." In *Georg Forster in interdisziplinärer Perspektive*, edited by Claus-Volker Klenke, 115–32. Berlin: Akademie Verlag, 1994.

Schnöpf, Markus. *Die Kunstkammer Peters des Großen. Geschichte der Sammlung*. 2000. http://www.schnoepf.de/m/kukammer/geschichte.html (November 20, 2011).

———. *Die Kunstkammer Peters des Großen. Präparate*. 2000. http://www.schnoepf.de/m/kukammer/praep.html (November 20, 2011).

Schochow, Maximilian. *Die Ordnung der Hermaphroditen-Geschlechter: Eine Genealogie des Geschlechtsbegriffs*. Berlin: Akademie Verlag, 2009.

Schor, Esther. *Bearing the Dead: The British Culture of Mourning from the Enlightenment to Victoria*. Princeton, NJ: Princeton University Press, 1994.

Schreiber, Elliott. *The Topography of Modernity: Karl Philipp Moritz and the Space of Autonomy*. Ithaca, NY: Cornell University Press, 2012.

Schreiber, Werner, and Friedrich Karl Mathys. *Infectio: Infectious Diseases in the History of Medicine*. 2nd ed. Translated by D. Q. Stephenson. Basle, Switzerland: Roche, 1987.

Schultka, Rüdiger, and Luminita Göbbel. "Präparationstechniken und Präparate im 18. und frühen 19. Jahrhundert, dargestellt an Beispielen aus den anatomischen Sammlungen zu Halle (Saale)." In *Anatomie: Sektionen einer medizinischen Wissenschaft im 18. Jahrhundert*, edited by Jürgen Helm and Karin Stukenbrock, 49–81. Stuttgart: Franz Steiner Verlag, 2003.

Schweickhard, C. L. "Geschichte eines lange Zeit hindurch für einen Hermaphroditen gehaltenen wahren Mannes." In *Journal der practischen Arzneykunde und Wundarzneykunst*, edited by Christoph Wilhelm Hufeland, 9–52. Berlin: Ungers Journalhandlung, 1803.

Schweikert, Uwe "Jean Paul und die Restauration." *Jahrbuch der Jean Paul Gesellschaft* 10 (1975): 123–49.

Sebright, John Saunders. *The Art of Improving the Breeds of Domestic Animals*. London: John Harding, 1809.

Segebrecht, Wulf. "Krankheit und Gesellschaft." In *Romantik in Deutschland: Ein interdisziplinäres Symposion*, edited by Rolf Dieter [Richard] Brinkmann, 267–90. Stuttgart: Metzler, 1978.

Selbmann, Rolf. *Der deutsche Bildungsroman*. Stuttgart: J. B. Metzler, 1984.

Shell, Marc. *Children of the Earth: Literature, Politics, and Nationhood*. Oxford and New York: Oxford University Press, 1993.

Shorter, Edward. *The Making of the Modern Family*. New York: Basic Books, 1975.

Sieder, Reinhard. *Sozialgeschichte der Familie*. Frankfurt am Main: Suhrkamp, 1987.

Simpson, James Y. "Hermaphroditism." In *The Cyclopaedia of Anatomy and Physiology*, edited by Robert B. Todd. vol. 2, 684–738. London: Sherwood, 1839.

Sinclair, John. *Grundsätze des Ackerbaues nebst Bemerkungen über Gartenbau, Obstbaumzucht, Forst-Cultur und Holzpflanzungen*. Vienna: Bei Heubner, 1823.

Smith, Justin E. H., ed. "Imagination and the Problem of Heredity in Mechanistic Embryology." In *The Problem of Animal Generation in Early Modern Philosophy*, 80–99. Cambridge: Cambridge University Press, 2006.

Snow, C. P. *The Two Cultures and the Scientific Revolution. The Rede Lecture 1959*. New York and London: Cambridge University Press, 1961.

Sölle, Dorothee. *Realisation: Studien zum Verhältnis von Theologie und Dichtung nach der Aufklärung*. Darmstadt: Luchterhand, 1973.

Sömmering, Samuel Thomas von. *Über die körperliche Verschiedenheit des Negers vom Europäer (1785)*. Edited by Sigrid Oehler-Klein. Stuttgart: G. Fischer, 1998.

Sprecher, Thomas and Monica Bussmann. *Bekenntnisse des Hochstaplers Felix Krull. Memoiren erster Teil. Kommentar*. Frankfurt a. M.: S. Fischer, 2012.

Spurzheim, Johann Gaspar. *The Physiognomical System of Drs. Gall and Spurzheim*. London: Baldwin, Cradock, and Joy, 1815.

———. *Observations on the Deranged Manifestations of the Mind; or, Insanity*. Boston: March, Capen & Lyon, 1833. Reprint Gainesville, FL: Scholars Facsimiles and Reprints, 1970.

Stafford, Barbara Maria. *Body Criticism: Imaging the Unseen in Enlightenment Art and Medicine*. Cambridge, MA: MIT Press, 1991.

Stark, D. Johann Christian, ed. "Nro. VI. Kurze Beschreibung eines sogenannten Hermaphroditen oder Zwitters, welcher aber mehr zum männlichen, als weiblichen Geschlechte zu rechnen ist, nebst einer Vorerinnerung von dem Herausgeber." In *Neues Archiv für die Geburtshülfe, Frauenzimmer- und Kinderkrankheiten mit Hinsicht auf die Physiologie, Diätektik und Chirurgie. Zweyten Bandes, drittes Stück*, 538–56. Jena: Wolfgang Strahl, 1801.

Steedman, Carolyn. *Strange Dislocations: Childhood and the Idea of Human Interiority, 1780–1930*. Cambridge, MA: Harvard University Press, 1995.

Steigerwald, Joan. "Figuring Nature, Figuring the (Fe)male: The Frontispiece to Humboldt's Ideas Towards a Geography of Plants." In *Figuring it Out: Science, Gender and*

Visual Culture, edited by Ann B. Shteir and Bernhard V. Lightman, 54–82. Hanover, NH: Dartmouth College Press, University Press of New England, 2006.

Steinecke, Hartmut. "Die Elixiere des Teufels" [Kommentar]. E. T. A. Hoffmann." In *Sämtliche Werke in sechs Bänden*, edited by Hartmut Steinecke and Wulf Segebrecht. vol. 2, 545–91. Frankfurt am Main: Deutscher Klassiker Verlag, 1988.

Stiegler, Bernd. "Die Spiegelreflexkamera-stammlinde. Bildsysteme in E. T. A. Hoffmanns *Die Elixiere des Teufels*." *Athenäum* 5 (1995): 235–52.

Strobel, Jochen. "Die Ahnenprobe des Medardus." *E. T. A. Hoffmann Jahrbuch* 13 (2005): 29–46.

Stumpf, Johann Georg. *Versuch einer pragmatischen Geschichte der Schäfereien in Spanien und der Spanischen in Sachsen, Anhalt-Dessau &c.* Leipzig: In der J.G. Müllerischen Buchhandlung, 1785.

Sturm, Karl Christian Gottlieb. *Beiträge zur teutschen Landwirtschaft und deren Hülfswissenschaften.* Bonn: Marcus, 1821–1826.

———. *Ueber Raçen, Kreuzungen und Veredlung der landwirthschaftlichen Hausthiere.* Eberfeld: Büschler, 1825.

Südeckum, Friederich. *Practische Bemerkungen über die Veredelung der Schafzucht auf Niedersächsischen Schäfereyen, sowohl über deren Hindernisse als Beförderungsmittel.* Braunschweig: Friedrich Vieweg, 1800.

Svarez, Gottlieb Carl. *Vorträge über Recht und Staat.* Edited by Hermann Conrad and Gerd Kleinheyer. Köln: Westdeutscher Verlag, 1960.

Tantillo, Astrida Orle. *The Will to Create: Goethe's Philosophy of Nature.* Pittsburgh, PA: University of Pittsburgh Press, 2002.

Taylor, Charles. *Sources of the Self: The Making of the Modern Identity.* Cambridge, MA: Harvard University Press, 1989.

Taylor, Thomas. *The Mystical Initiations; Or, Hymns of Orpheus.* Translated by Thomas Taylor. London: T. Payne and Son, 1787.

Thaer, Albrecht Daniel. *Einleitung zur Kenntnis der englischen Landwirthschaft und ihrer neueren practischen und theoretischen Fortschritte in Rücksicht auf Vervollkommnung deutscher Landwirthschaft für denkende Landwirthe und Cameralisten.* Hannover: Bey den Gebrüdern Hahn, 1798.

———. *Handbuch für die feinwollige Schafzucht.* Berlin: Ville and Krammer, 1811.

Thomasius, Christian. *Grundlehren des Natur- und Völckerrechts.* Hall im Magdeburgischen: Renger, 1709.

Tieck, Ludwig. *Der blonde Eckbert. Der Runenberg.* Edited by Hanne Castein. Stuttgart: P. Reclam, 1987.

Tiffany, Grace. *Erotic Beasts and Social Monsters: Shakespeare, Jonson, and Comic Androgyny.* London and Cranbury, NJ: Associated University Press, 1995.

Tisserand, Roger. *Au temps de l'Encyclopédie. L'Académie de Dijon de 1740 à 1793.* Paris: Boivin & Cie, 1939.

Tittmann, Carl August. *Handbuch der Strafrechtswissenschatft.* Halle: Bei Hemmerde und Schwetschke, 1806.

Tobien, Heinz. "Johann Heinrich Merck und Samuel Thomas Sömmering als Paläontologen im ausgehenden 18. Jahrhundert." In *Samuel Thomas Sömmering und die Gelehrten der Goethezeit: Beiträge eines Symposiums ein Mainz vom 19. bis 21. Mai 1983*, edited by. Gunter Mann and Franz Dumont, 119–48. Stuttgart: Fischer, 1985.

Tobin, Robert. "The Medicinalization of Mignon." In *Goethes Mignon und ihre Schwestern*, edited by Gerhart Hoffmeister, 43–60. New York: Peter Lang, 1993.

———. *Warm Brothers: Queer Theory and the Age of Goethe*. Philadelphia: University of Pennsylvania Press, 2000.

Trembley, Abraham. *Mémoires pour servir à l'histoire d'un genre de polypes d'eau douce: à bras en forme de cornes*. Leide: Jean & Herman Verbeek, 1744.

Treviranus, Gottfried Reinhold. *Biologie, oder Philosophie der lebenden Natur für Naturforscher und Aerzte*. 6 vols. Göttingen: Johann Friedrich Röwer, 1802–1822.

Trunz, Erich. "Anmerkungen des Herausgebers. *Wilhelm Meisters Wanderjahre oder Die Entsagenden*." In *Johann Wolfgang Goethe Werke*, edited by Erich Trunz, vol. 8, 579–732. Hamburg: Wegner, 1950.

Türk, Johannes. *Die Immunität der Literatur*. Frankfurt am Main: M. S. Fischer, 2011.

Valentini, M. B. [Michael Bernhard]. *Musaeum Musaeorum, oder vollständige Schaubühne aller Materialien und Specereyen, nebst deren natürlichen Beschreibung . . . Aus andern Material-Kunst- und Naturalien-Kammern, Oost- und West-Indischen Reiss-Beschreibungen*. Frankfurt: Heirs of Johann David Zunner and Johann Adam Jungen, 1714.

Vedder, Ulrike. "Literature and the Law of Succession." In *A Cultural History of Heredity II: 18th and 19th Centuries*. Max Planck Institute for the History of Science. Preprint 247, 175–85. Berlin, 2003.

Veltheim, Röttger von. *Abhandlungen über die Pferdezucht Englands, noch einiger Europäischer Länder, des Orients, u.s.w., in Beziehung auf Deutschland, nebst einer Revision der seit Mitte des 18. Jahrhunderts aufgestellten Systeme über die Pferdezucht*. Braunschweig: Druck und Verlag von Friedrich Vieweg, 1833.

Vielhauer, Inge. *Bruder und Schwester. Untersuchungen und Betrachtungen zu einem Urmotiv zwischenmenschlicher Beziehung*. Bonn: Bouvier, 1979.

Vodosek, Peter. "Eine Leihbibliothek der Goethe-Zeit. Das 'Königl Privilegierte neue Leseinstitute' des Carl Friedrich Kunz zu Bamberg." *Jahrbuch des Wiener Goethe-Vereins* 77 (1973): 110–33.

Wachsmuth, Andreas B. *Geeinte Zwienatur: Aufsätze zu Goethes naturwissenschaflichem Denken*. Berlin: Aufbau-Verlag, 1966.

Wachsmuth, Wilhelm. *Entwurf einer Theorie der Geschichte*. Halle: Hemmerde & Schwetschke, 1820.

Walker, Alexander. *Intermarriage: Or, the Mode in Which, and the Causes Why, Beauty, Health, and Intellect Result from Certain Unions, and Deformity, Disease and Insanity from Others*. London: John Churchill, Princes Street, Soho, 1838.

Walther, Karl Kraus. *Carl Friedrich Kunz: Ein literarischer Unternehmer aus Bamberg*. Bamberg: Universität Bamberg, 1994.

Watson, James D., and Francis H. C. Crick. "Molecular Structure of Nucleic Acids: A Structure of Deoxyribose Nucleic Acid." *Nature* 171 (April 25, 1953): 737–38.

Weber, Max. *Wirtschaft und Gesellschaft*. Edited by Wolfgang Schluchter. Tübingen: Mohr, Siebeck, 2009.

Weigel, Sigrid. "Generation, Genealogie, Geschlecht. Zur Geschichte des Generationenkonzepts und seiner wissenschaftlichen Konzeptionalisierung seit dem Ende des 18. Jahrhunderts." In *Kulturwissenschaften: Forschung, Praxis, Positionen*, edited by Lutz Musner and Gotthart Wunberg, 161–90.Wien: WUV Universitätsverlag, 2002.

———. *Genea-Logik: Generation, Tradition und Evolution zwischen Kultur- und Naturwissenschaften.* München: W. Fink, 2006.

Weinrich, Harald, ed. "Literaturgeschichte eines Weltereignisses: Das Erdbeben von Lissabon." In *Literatur für Leser. Essays und Aufsätze zur Literaturwissenschaft*, 74–90. Stuttgart: W. Kohlhammer, 1971.

Weissmann-Barrier. http://en.wikipedia.org/wiki/Weismann_barrier (accessed May 15, 2013).

Wichmann, August Christian, and Louis-Jean-Marie Daubenton. *Katechismus der Schaafzucht: Zum Unterrichte fuer Schaefer und Schaeferey-Herren.* Leipzig; Dessau: self-published, 1784.

Wieland, Christoph Martin. "Geschichte des weisen Danischmend und der drey Kalender." In *C. M. Wielands Sämtliche Werke,* edited by Hans Radspieler. Hamburg: Hamburger Stiftung zur Förderung von Wissenschaft und Kultur, 1984.

Wieden, Lotta. "Vater Unbekannt." *Die Zeit*, April 18, 2002.

Wilkinson, Elizabeth. "The Poet as Thinker: On the Varying Modes of Goethe's Thought." In *Goethe: Poet and Thinker*, by Elizabeth M. Wilkinson and Leonard A. Willoughby, 167–84. London: Edward Arnold, 1962.

Wilson, Daniel W. "Science, Natural Law, and Unwitting Incest in Eighteenth-Century Literature." In *Studies in Eighteenth-Century Culture*, edited by O. M. Brack, vol. 13, 249–70. Madison, WI, University of Wisconsin Press, 1984.

Wilson, Philip K. "Erasmus Darwin and the 'Noble' Disease (Gout): Conceptualizing Heredity and Disease in Enlightenment England." In *Heredity Produced: At the Crossroads of Biology, Politics, and Culture, 1500–1870*, edited by Staffan Müller-Wille and Hans-Jörg Rheinberger, 133–54. Cambridge, MA: MIT Press, 2007.

Winter, Georg Simon. *Gründliche und kurtz durchgehende Beschreibung, wie eine Stutherey anzurichten, dieselbe in gutem Esse zu erhalten, und wie man sich deroselben mit gutem Nutzen gebrauchen könne.* Onoltzbach: In Verlegung deß Authoris, 1670.

Wolff, Caspar Friedrich. *Theoria Generationis* (*1759*). Edited and translated by Paul Samassa. 2 vols. Leipzig: W. Engelmann, 1896.

———. *Theorie von der Generation: in zwo Abhandlungen erklärt und bewiesen.* Berlin: Friedr. W. Birnstiel, 1764.

Wolff, Eugen. *Mignon. Ein Beitrag zur Geschichte des Wilhelm Meister.* München: Beck, 1909.

Wolff, Larry. "When I Imagine a Child: The Idea of Childhood and the Philosophy of Memory in the Enlightenment." *Eighteenth-Century Studies* 31, no. 4 (Summer 1998): 377–401.

Wood, Roger J., and Vítězslav Orel. *Genetic Prehistory in Selective Breeding: A prelude to Mendel.* New York and Oxford: Oxford University Press, 2001.

Woodmansee, Martha. *The Author, Art, and the Market: Rereading the History of Aesthetics.* New York: Columbia University Press, 1994.

Zedler, Johann Heinrich. *Großes vollständiges Universal Lexicon aller Wissenschafften und Künste.* 68 vols. Leipzig and Halle: Zedler, 1731–1754.

Ziegenspeck, Hermann. "Camerarius, Rudolph Jakob." *Neue Deutsche Biographie* 3 (1957): 107–8.

Zola, Émile. *Les Rougon-Macquart. Histoire naturelle et sociale d'une famille sous le Second Empire* (*1871–1893*). Paris: Éditions du Seuil, 1969–1970.

Zucker, A. E. "The Genealogical Novel, a New Genre." *PMLA* 43, no. 2 (June 1928): 551–60.

———. "The Genealogical Novel Again." *PMLA* 44, no. 3 (September 1929): 925–27.

INDEX